Genetic Influences on Addiction

Genetic Influences on Addiction

An Intermediate Phenotype Approach

edited by James MacKillop and Marcus R. Munafò

The MIT Press
Cambridge, Massachusetts
London, England

MIT Press books may be purchased at special quantity discounts for business or sales promotional use. For information, please email special_sales@mitpress.mit.edu.

This book was set in Stone Sans and Stone Serif by Toppan Best-set Premedia Limited, Hong Kong. Printed and bound in the United States of America.

Library of Congress Cataloging-in-Publication Data

Genetic influences on addiction : an intermediate phenotype approach / edited by James MacKillop and Marcus R. Munafò.
  p. ; cm.
Includes bibliographical references and index.
ISBN 978-0-262-01969-9 (hardcover : alk. paper)
I. MacKillop, James, 1975– editor of compilation.   II. Munafò, Marcus, 1972– editor of compilation.
[DNLM: 1. Substance-Related Disorders—genetics.   2. Phenotype.   WM 270]
RM301.3.G45
616.86′0642—dc23
2013011177

10 9 8 7 6 5 4 3 2 1

To Emily and Annabelle—JM

To Eric and Carl—MRM

Contents

Contributors

John Acker   University of Georgia

Steven R. H. Beach   University of Georgia

Gene H. Brody   University of Georgia

Angela D. Bryan   University of Colorado, Boulder

Meghan J. Chenoweth   University of Toronto

Eske M. Derks   University of Amsterdam

Danielle M. Dick   Virginia Commonwealth University

Mary-Anne Enoch   National Institute on Alcohol Abuse and Alcoholism

Meg Gerrard   Dartmouth College

Frederick X. Gibbons   Dartmouth College

Thomas E. Gladwin   University of Amsterdam

Mark S. Goldman   University of South Florida

Markus Heilig   National Institute on Alcohol Abuse and Alcoholism

Kent E. Hutchison   University of Colorado, Boulder

Hollis C. Karoly   University of Colorado, Boulder

Steven M. Kogan   University of Georgia

Man Kit Lei   University of Georgia

Susan E. Luczak   University of California, San Diego

James MacKillop   University of Georgia

Renee E. Magnan   University of New Mexico

Leah M. Mayo   University of Chicago

Marcus R. Munafò   University of Bristol

Daria Orlowska   University of California, San Diego

Abraham A. Palmer   University of Chicago

Danielle Pandika   University of California, San Diego

Clarissa C. Parker   University of Chicago

Robert A. Philibert   University of Iowa

Lara A. Ray   University of California, Los Angeles

Richard R. Reich   University of South Florida, Sarasota–Manatee

Ronald L. Simons   University of Georgia

Courtney J. Stevens   University of Colorado, Boulder

Rachel E. Thayer   University of Colorado, Boulder

Rachel F. Tyndale   University of Toronto

Tamara L. Wall   University of California, San Diego

Reinout W. Wiers   University of Amsterdam

Michael Windle   Emory Unversity

Harriet de Wit   University of Chicago

1 An Intermediate Phenotype Approach to Addiction Genetics

James MacKillop and Marcus R. Munafò

It is further remarkable that drunkenness resembles certain hereditary, family, and contagious diseases.
—Benjamin Rush (1790)

Statistics show that over 50% of all inebriates and alcoholics come from stocks with inherited degenerations, and the inebriety is literally the expression of transmitted defects.
—T. D. Crothers (1912)

Although perspectives on drug addiction have historically emphasized moral failing, the accepted contemporary scientific perspective is that its causes are rooted in a matrix of biological, pharmacological, psychological, and sociocultural factors. Within the domain of biological influences, there is further consensus that genetic factors play a substantial role in an individual's vulnerability to developing drug addiction. Indeed, the apparent involvement of heredity has been noted from the start of the scientific study of addiction, as illustrated by the quotes above,[1] and there has been a steady accumulation of empirical evidence over the last several decades that has unequivocally demonstrated the importance of genetic influences. However, the strength of the evidence for the general importance of genetic influences belies a persistent paradox in the field, that the specific genetic variants that confer vulnerability or resilience have remained elusive. In other words, although the question *Does genetics matter in risk for addiction?* can be answered with a resounding *Yes*; the questions of *Which genes? And, why?* remain largely unanswered.

All the more troubling, a long-held presumption was that the specific genetic risk variables remained unknown in part because the sequence of the genome itself remained incompletely characterized and the prospect of examining the vast amount of possible variation was a technical impossibility. This is no longer the case. The full human genome has been sequenced for over a decade [1, 2], and the tools for genotyping a million or more polymorphisms across the genome are widely available. Moreover, data harmonization from large-scale projects and collaborations across laboratories has resulted in aggregated samples of thousands of research participants, creating exceptionally high levels of statistical power. Unfortunately, however, these revolutions in bioinformatics have not led to revolutions in understanding genetic influences on addiction. What has become clear is that there is no single "addiction gene," or even only a small number of variants that are substantially responsible for a person's level of risk. Indeed, among the polymorphisms that are best supported, the most reliable genetic loci contribute only a tiny fraction to the overall level of risk. Furthermore, the field has not only become replete with failures to replicate promising findings from earlier smaller studies, but meta-analyses reveal evidence of publication bias (i.e., the probability that positive findings in studies with small samples will be published but not the other way around; e.g., [3]). Thus, despite many substantial technological advances, the understanding of how genetic variation specifically contributes to the etiology of addiction remains largely obscure.

The goal of this volume is to contribute to the understanding of addiction genetics by reviewing the insights that can be gleaned using an intermediate phenotype approach. This strategy has its roots in the endophenotype approach, which originated from investigations into the genetics of schizophrenia [4], a domain which has had remarkably parallel challenges to those of addiction genetics [5, 6]. An endophenotype approach seeks to move the focus away from diagnosis as a phenotype and toward more focal, reliable, and mechanistically informative phenotypes that are established risk factors for the disorder in question. These intermediary processes are theorized to be more closely related to specific forms of genetic variation and also to be the inherited mechanisms by which those variants confer risk. The endophenotype approach has been increasingly elaborated into what is referred to as an intermediate phenotype approach, reflecting broadening of focus to all genetically informative phenotypes. There are potentially important semantic distinctions between these terms, discussed later in this chapter, but the fundamental premise is that distal genetic influences on addiction may be best understood by clarifying their influence on more discrete mechanistic phenotypes.

In this first chapter, we will review the fundamental concepts of an intermediate phenotype approach. We will provide a précis of the general evidence for substantial genetic influences on drug addiction (a full review could itself take up the entire volume), review the definitions of endophenotypes and intermediate phenotypes, discuss what these approaches can offer, and, in a concluding section, review reasons for caution that remain important. The following chapters then provide critical reviews or applications in a single domain of phenotypes. These range from the most established relationships in the field to domains that are promising but more exploratory in nature. The goal of the volume is to provide breadth of scope across chapters but depth of focus within chapters.

1.1 Evidence of Genetic Influences on Addiction

Evidence for a substantial role of genetic factors broadly comes from two domains, animal models using recombinant inbred strains and quantitative genetic studies using twin and adoption methodologies. In the first case, the earliest empirical evidence comes from clear variation in alcohol preference that was observed as early as the 1940s [7]. Strain-based differences are important because inbred strains are largely isogenetic within strains [8], the equivalent of identical twins, but differ substantially across strains. Thus, among rodents with no previous drug exposure and identical rearing environments, differences in drug consumption (or other phenotypes) observed across strains can be attributed to those genetic differences, and is present [9]. Indeed, strains of animals that exhibit very high or low drug preferences (e.g., alcohol preferring [P] and nonpreferring [NP] rat strains) [9] have been developed using selective breeding based on alcohol preference over generations, further demonstrating the direct influence of genetic factors over these phenotypes. Since the early work in this area, animal models have substantially advanced the field not only by demonstrating proof-of-concept in the preceding examples but increasingly by identifying the specific genetic variants responsible for strain-based differences and clarifying general versus drug-specific genetic contributions [9, 10].

In humans, there is long-standing and extensive evidence from family studies that addictive disorders "run in families" [11–14]. In other words, having a biological relative with an addictive disorder confers risk for having an addictive disorder, and the closer the relative, the higher the risk. However, family risk studies confound genetic factors and environmental factors. For example, an individual who has two biological parents with nicotine dependence probably experiences greater modeling of smoking by important role models and, at a practical level, has greater access to cigarettes, both of which could explain the association. To address this, quantitative genetics studies using twin and adoption methodologies can be used to infer the relative level of genetic influences on addiction vulnerability in humans with greater specificity. Twin studies are the most common methodology in this area and examine differences in the expression of a phenotype (e.g., presence or absence of a disorder) between monozygotic (identical) twins, who share almost 100% of genetic variation, and dizygotic (fraternal) twins, who share, on average, approximately 50% of variation. Twin studies have consistently revealed significantly higher diagnostic concordance for addictive disorders in monozygotic twins [15], indicating that as shared genetic variability increases, so too does the probability of both twins having the diagnosis. In a relatively recent aggregation of twin studies, the estimates across studies for addictive disorders ranged from 39% to 72%, with a median of slightly greater than half, indicating moderate to high levels of heritability [15]. The other common research strategy in quantitative genetics is an adoption approach, which examines diagnostic status (or other phe-notypes) in adoptees whose biological parents have been characterized diagnostically (positive or negative), but who are raised by adoptive parents who are diagnostically negative. Thus, adoption studies permit examination of the influences of inheriting approximately 50% genetic variation from a diagnostically positive parent in the absence of influences from a drug-related home environment. Like twin studies, adop-tion studies have consistently indicated significantly greater probabilities of developing addictive disorders among adoptees with a positive biological parent (e.g., [16–18]).

Animal models and quantitative genetic studies clearly indicate the substantial heritability of addiction but are opaque in terms of identifying the specific genetic variants that are responsible for this heritability at the molecular level. For animal studies, although there is considerable homology between rodent and human genomes, the extent to which the genetic risk variables in one species truly parallel those in the other is unclear. For example, the genetic factors responsible for differences in alcohol consumption in P and NP rats may not have clear counterparts in humans, or, if they do, the loci in humans may not be relevant to vulnerability to alcoholism. In other words, evidence that specific genes are relevant in a model system does not necessarily mean the findings will fully translate to humans. For quantitative genetic studies, although the issue of translation is not an issue, the methodologies reflect parsing of patterns of inheritance that cannot be further decomposed into the individual sources of genetic variation that are responsible for those patterns. Molecular genetic tech-niques are necessary to determine individual genotypes and to examine differences in frequencies that can implicate individual variants with a given diagnosis.

This is where the disconnect in addiction genetics exists, because studies using a diversity of molecular genetic approaches have struggled to identify the underlying individual loci that are responsible for the substantial heritability observed in quan-titative genetic studies. Even as sizes of samples have become larger and larger, and the bandwidth of bioinformatics has become wider and wider, the results have been modest, with numerous conflicting findings and failures to replicate (e.g., [3, 19]) and only a small number of consistent findings (e.g., [20–22]). Notably, this issue is not restricted to addiction genetics; the same paradox applies across mental disorders [6], and it has been archly dubbed "the case of the missing heritability" [23]. Bridging this gap between aggregate heritability and the particulate sources of genetic risk is argu-ably the largest current challenge to addiction genetics and psychiatric genetics more generally.

1.2   Diagnosis as a Suboptimal Phenotype

One potential explanation for the paradox we describe is that the major focus to date has been on genetic influences on the presence or absence of a substance use disorder diagnosis (e.g., nicotine dependence, alcohol dependence), which may be suboptimal. To clarify why this is the case, it is instructive to first consider the general objectives of a diagnostic system. Diagnosis is a critical clinical tool for providing accurate descriptions of the presentation of a condition in order to facilitate communication among clinicians and aid in the care of a patient. Further, diagnoses have other important practical applications in epidemiology (e.g., ascertaining the prevalence of a condition) and health care systems (e.g., quantifying aspects of treatment delivery). Ideally, diagnosis is intended to be informative about the course and trajectory of a condition [24]; however, even in the absence of it doing so, it is important to recall that the essential role of diagnosis is denominational—the descriptive naming of a cluster of symptoms for clinical and practical purposes.

From this perspective, it is clear why clinical diagnoses may not be not optimal phenotypes. To start, psychiatric diagnoses commonly comprise syndromes of symptoms, with a categorical diagnosis being given based on the presence of a minimum number. For example, table 1.1 presents the *Diagnostic and Statistical Manual of Mental Disorders* (5th ed.; *DSM–V*) alcohol use disorder diagnostic criteria, which parallel the criteria for other substance use disorders and for which at least two of eleven are needed for a diagnosis to be given. What is immediately apparent is that many different permutations of symptoms could be present, all leading to the same categorical diagnosis. In fact, there are 2,036 combinations of two or symptoms, ranging from the lowest severity profile combinations of exactly two symptoms all the way to the presence of all eleven symptoms. Five different patients could be given the same diagnosis, none having overlapping symptoms. Clearly, the manifestations of the conditions are highly heterogeneous.

Moreover, addictive disorders are well established to be dimensional conditions [25–28], reflecting a continuum of severity, not a true dichotomy between affected and nonaffected individuals. In addition, there is considerable evidence that within individuals who meet criteria, meaningfully discrete subgroups are present [29–31] and these may further differ by sex [32]. Thus, the commonly used categorical diagnostic phenotype eliminates a highly meaningful variation among individuals. A consequence of these issues is that studies using categorical phenotypes are implicitly searching to identify genetic loci that are specific to the presence of the condition but do not differentiate by level of severity or subgroups among individuals with the condition (e.g., are not specific only to very severe individuals). Finally, it is worth considering that diagnoses are given according to clinical judgment, which is by no means perfectly reliable [33], introducing further variability. For example, a recent study on substance use disorder diagnoses given using the Semi-Structured Assessment for Drug Dependence and Alcoholism found a median κ coefficient of .59 (range = .31–.91), indicating highly variable interrater reliability ranging from weak to strong [34]. Taken together, categorical diagnosis as a phenotype is problematic from the start

Table 1.1

Diagnostic and Statistical Manual of Mental Disorders (fifth ed.) criteria for alcohol use disorder syndrome (two or more symptoms are required for an individual to receive a diagnosis of an alcohol use disorder)

Diagnostic criteria

A. A problematic pattern of alcohol use leading to clinically significant impairment or distress.

B. Two (or more) of the following occurring within a 12-month period:

1. Alcohol is often taken in larger amounts or over a longer period than was intended
2. There is a persistent desire or unsuccessful effort to cut down or control alcohol use
3. A great deal of time is spent in activities necessary to obtain alcohol, use alcohol, or recover from its effects
4. Recurrent alcohol use resulting in a failure to fulfill major role obligations at work, school, or home (e.g., repeated absences or poor work performance related to alcohol use; substance-related absences, suspensions, or expulsions from school; neglect of children or household)
5. Continued alcohol use despite having persistent or recurrent social or interpersonal problems caused or exacerbated by the effects of the substance
6. Important social, occupational, or recreational activities are given up or reduced because of alcohol use
7. Recurrent alcohol use in situations in which it is physically hazardous (e.g., driving an automobile or operating a machine when impaired by substance use)
8. Alcohol use is continued despite knowledge of having a persistent or recurrent physical or psychological problem that is likely to have been caused or exacerbated by the substance
9. Tolerance, as defined by either or both of the following:
   a. A need for markedly increased amounts of alcohol to achieve intoxication or desired effect
   b. Markedly diminished effect with continued use of the same amount of the substance
10. Withdrawal, as manifested by either of the following:
    a. The characteristic withdrawal syndrome for alcohol
    b. The same (or a closely related) substance is taken to relieve or avoid withdrawal symptoms
11. Craving, or a strong desire or urge, to use alcohol

because it necessarily includes a great deal of heterogeneity and suboptimal reliability across clinicians.

The preceding issues essentially reflect the limitations of diagnostic syndromes as phenotypes when considered at a single time point, but multiple significant challenges apply to diagnosis as a phenotype when considered across time. At the broadest level, it is worth noting that psychiatric diagnosis in both the *DSM* and International Classification of Diseases (ICD) are atheoretical descriptive systems that do not purport to be immutable and valid in an ultimate sense. Rather, psychiatric nosology is generally accepted to be "work-in-progress," as reflected in the current transition from *DSM–IV* to *DSM–V*. For example, substance use disorders were largely classified as a form of personality pathology in earlier versions of the *DSM*, and, in *DSM–V*, the two previous diagnoses of substance abuse and dependence are merged together, with one symptom being dropped (legal problems) and another being added (craving). Thus, the definitions comprising diagnosis itself are not immutable over time, making them "moving targets." At the other end of the spectrum, at the narrowest level, diagnostic test–retest reliability is both variable and far from perfect across short periods of time that should not affect diagnostic status [34–36], introducing further ambiguity and imprecision.

Perhaps more importantly, though, is that a cross-sectional diagnosis leaves out a great deal of information about the chronology and course of a condition. There is a great deal of variability in developmental and clinical trajectories among individuals and, as addiction reflects recursive etiological processes in which the condition, once established, become self-perpetuating, meaningful differences early in the progression may be obscured by common clinical end states. For example, the typical progression in nicotine dependence is from experimentation with cigarettes to occasional smoking, leading to physiological tolerance and withdrawal symptoms, which further escalates cigarette consumption, and the establishment of the vicious cycle of the nicotine dependence syndrome. However, two different individuals with nicotine dependence may have differed substantially in terms of initial reactions to tobacco and trajectory but, ultimately, look diagnostically equivalent. Similarly, although some individuals have persistent problems across the life span, many either reduce problematic levels of use on their own or are successful via formal treatment [37, 38]. Individual differences in successfully (or unsuccessfully) changing behavior may also be influenced by genetic factors, which will be undetectable in designs that simply focus on categorical diagnostic status. Traditional studies focusing on the presence or absence of a diagnosis are implicitly investigating genetic factors that pertain to "ever affected" clinical status but may not capture not meaningful variation in addictive behavior across the life span.

One final issue is what could be referred to as the problem of probablism [39]. That is, even though addiction may be heritable, the relationship is highly probabilistic. Many people who have high genetic risk in terms of a dense family history do not develop addictive disorders, and many others who have virtually no familial risk nonetheless develop the condition. In the first case, this has been called "quiet transmission" or "unexpressed liability" [5], meaning that genetic risk may be present in a given individual and passed to the next generation but does not result in the disorder [5], either because of protective factors or the absence of a developmental or environmental eliciting factor. Similarly, genetic vulnerability appears not to be a necessary precondition of addiction, meaning that an individual may meet criteria of environmentally mediated factors, not genetically influenced mechanisms. In both these cases, the relationship between specific genetic factors and a clinical diagnosis phenotype is further obscured by the presence of false negatives (the high-risk unaffected individuals) and false positives (the low-risk affected individuals).

In summary, although diagnosis serves a critical role clinically, there are manifold limitations to it as a phenotype in addiction genetics. The common theme across the preceding issues is that an individual's categorical diagnostic status at a given time point largely oversimplifies highly heterogeneous and complex disorders. As a result, diagnosis may simply be too distal from the individual's genetic risk factors, which may partially explain the disconnect between aggregate heritability estimates and the individual genetic variants conferring risk and protection.

1.3 The Endophenotype Approach

The endophenotype approach is a strategy for understanding genetic vulnerability for a clinical condition by   identifying genetically influenced risk mechanisms—endophenotypes—that are responsible for conferring risk for the condition. These mechanisms are intended to be both comparatively closelyrelated to genetic variation but also significantly associated with the condition of interest. This is illustrated in figure 1.1. The essence of the endophenotype strategy is to aid in mapping the complex pathways from individual forms of genetic variation to overall risk for a clinical syndrome.

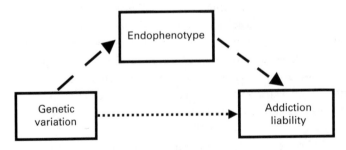

Figure 1.1

An endophenotype approach to addiction genetics. Distal genetic risk for addiction liability is mediated by more proximal influences on pleiotropic intervening processes.

Endophenotypes putatively "connect the dots" from the genome to the syndrome. The goal of the approach is to focus less on the direct relationship between specific variants and a clinical diagnosis, and more on triangulating interrelationships between specific genetic factors, intervening mechanistic phenotypes that confer liability, and disorder status.

The term *endophenotype* was first introduced in the context of psychiatric genetics approximately 40 years ago by Gottesman and Shields [4] with the intention of improving the understanding of the genetic basis of schizophrenia. It was adapted from a review on the geographic distribution of insects that made a distinction between exophenotypes, the clearly observable external phenotypes, and endophenotypes, which were "only knowable after aid to the naked eye, e.g., a biochemical test" [40, p. 19]. The term initially received relatively little discussion but was increasingly adapted to refer to genetically influenced components of disorder liability [39]. Its adaptation into psychiatric genetics has surged dramatically recently, in part because of the common challenges in identifying the specific genes underlying risk for heritable disorders. This is illustrated in figure 1.2, which depicts the number of articles

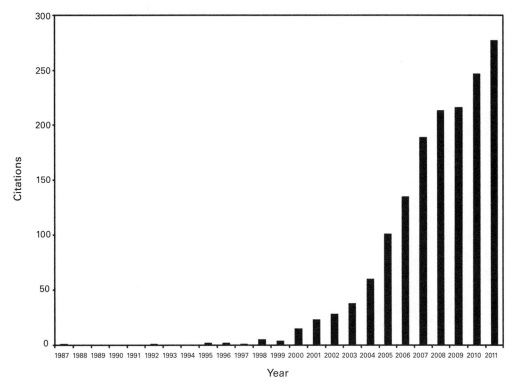

Figure 1.2
Search results for "endophenotype" from the PubMed database by year, 1987–2011.

resulting from a PubMed search for "endophenotype" for each year from 1987 (the first hit) through the end of 2011. Despite very little early activity, over 1,500 articles were identified to date and over 95% were in the last decade.

The defining characteristics of endophenotypes have also become increasingly formalized in the last ten years [39–42]. To start, the characteristic must itself be significantly associated with the condition of interest, ideally longitudinally, indicating that it is a risk factor for the condition, and it must be demonstrated to be heritable, indicating that it is at least partially determined by genetic factors. Also related to association with the disorder and heritability, endophenotypes are expected to exhibit cosegregation and familial transmission, meaning that the characteristic and the disorder will co-occur within affected families, and that biological relatives of an affected individual will exhibit the endophenotype at higher rates than those in the general population because of greater shared genetic variation. Importantly, endophenotypes are expressly not symptoms, subtypes, or other manifestations of the disorder but are separate long-standing risk factors. In addition, endophenotypes are putatively stable over time and state independent, meaning that they are trait-like and present in an individual regardless of the disorder. State independence is qualified, however, insofar as an endophenotype may be age normed, meaning it may only be detectable above a certain level of maturation, during a specific developmental period, or in the context of a test or biological challenge. For example, individual variations in reactions to different drugs are candidate endophenotypes (see chapters 5 and 6) but can only be observed via a laboratory drug challenge. In addition to these core criteria, it has been suggested that endophenotypes preferably have foundations in neuroscience [43] and, for psychological endophenotypes, have good psychometric properties [44].

Although these are high evidentiary standards, the successful identification of endophenotypes offers substantial benefits. As endophenotypes are putatively relatively closely related to underlying genetic variation, they are predicted to be more reliable and show larger effect sizes than clinical phenotypes. As described by Gottesman and Gould [39], "endophenotypes represent more defined and quantifiable measures that are envisioned to involve fewer genes, fewer interacting levels, and ultimately activation of a single set of neuronal circuits" (p. 115). At the first level, the proximal phenotype (e.g., effects on protein transcription) will exhibit the largest effect size differences, reflecting the comparatively close relationship between a given gene and the products for which it is responsible. At the level of the relevant system, relatively large differences are again expected to be present because of the close relationship between the system and its underlying genes. To the extent that the endophenotype reflects the functional differences in that system, similarly large effects are expected to be present. Finally, the relationship between the endophenotype and disorder liability is predicted to be comparatively large relative to the association between the distal polymorphism and the disorder. In addition, in clarifying the genetic bases of vulnerability, one of the promises of endophenotypes is to illuminate the etiological processes underlying psychiatric disorders to inform clinical advances. For example, established endophenotypes could aid diagnosis and classification, effectively serving as risk biomarkers. Moreover, understanding genetically mediated risk mechanisms could be leveraged in both targeted prevention and clinical intervention.

1.4   The Intermediate Phenotype Approach

Based on the preceding criteria, endophenotypes necessarily reflect a relatively narrow class of phenotypic characteristics. However, there has also been burgeoning interest in studying other novel phenotypes that are not necessarily endophenotypes, referred to as *intermediate phenotypes* and can be broadly defined as "mechanism-related manifestations of complex phenotypes" [45, p. 125]. This approach reflects a broadening of focus that subsumes endophenotypes but includes other phenotypic characteristics that are relevant to addictive disorders. Importantly, endophenotypes represent one class of intermediate phenotypes, but not the other way around. Intermediate phenotypes may be processes that are candidate endophenotypes but have not yet been validated on some of the criteria. Alternatively, some intermediate phenotypes may definitively *not* be endophenotypes but nonetheless may be informative about the genetic basis of a disorder. For example, endophenotypes are defined as being independent of a disorder and present in unaffected relatives, so manifestation-related phenotypes, such as disorder subtypes, motivational profiles, or behavioral features of the disorder, cannot be endophenotypes. Nonetheless, these characteristics may be associated with specific genetic factors and may illuminate genetic contributions to processes within the condition or potentially identify comparatively homogeneous subgroups, thus clarifying disorder liability. Whether as endophenotypes or manifestation-specific phenotypes, the defining characteristic of the intermediate phenotype approach is a focus on phenotypes that are *maximally mechanistically informative*.

In addition to addiction, an intermediate phenotype approach has been applied to an array of conditions, and a line of research in hypertension research provides an instructive parallel. Like addiction, hypertension is now recognized as being a condition in which diverse mechanisms are responsible for a common clinical end point, in this case, high blood pressure. Moreover, like addiction, there has been relatively little success in identifying individual genetic factors that contribute to the general clinical phenotype of hypertension [46]. However, among individuals with hypertension, a subgroup of patients has been identified who do not exhibit typical adrenal and vascular responses to the hormone angiotensin [47]. These individuals are referred to as nonmodulators and have been determined to be a discrete subgroup [48]. As such, nonmodulation has been hypothesized to be an intermediate phenotype for hypertension, and polymorphisms in the angiotensinogen, angiotensin-converting enzyme, and aldosterone synthase genes have indeed been found to significantly predict nonmodulation status [49]. Individuals with all three alleles implicated had a more than fivefold higher probability of being nonmodulators [50]. Thus, by identifying a clinical profile that is elicited by a biological challenge as a candidate intermediate phenotype, progress has been made in identifying the genetic vulnerability for this subgroup of hypertensive individuals.

In light of the evolving terminology, an obvious question is whether the distinction between intermediate phenotype and endophenotype is an important or meaningful one. On one hand, because the term endophenotype now has specific connotations and associated criteria are relatively established, a reasonable argument is that the distinction is useful because it accurately communicates this information. A characteristic that is described as an endophenotype would be assumed to be supported by evidence for at least most of the criteria detailed. On the other hand, it may be that these distinctions prove to be more semantic than informative. The position we take is that neither term is better or worse, but, ultimately, that what one calls these characteristics is less important than successfully identifying the phenotypes that are most useful for understanding the role of genetic factors in addiction.

1.5   Reasons for Caution and Critical Thinking

The work presented in the following chapters provides reason for optimism that intermediate phenotypes may be informative with respect to genetic influences on addiction liablity. Some chapters recapitulate the strongest lines of evidence for the utility of an intermediate phenotype approach, while other chapters, such as those on expectancies, behavioral economic variables, and differential susceptibility to drugs of abuse, suggest highly promising future directions. A theme which emerges is that there are three broad domains of interest: general susceptibility (irrespective of specific drug class), pharmacokinetics (peripheral and metabolic processing of the drug), and pharmacodynamics (actions of the drug within the central nervous system). In other words, there is evidence for genetic influences on vulnerability in general and vulnerability for developing specific forms of addiction.

However, there are some important considerations to be borne in mind when evaluating these findings. There is growing converging evidence that complex behavioral phenotypes share a common basic genetic architecture, comprising a very large number of common variants of very small effect [50–52], most likely together with a smaller number of rare variants of larger effect. Detecting both is problematic and requires very large sample sizes. This is notably lacking in the majority of studies on intermediate phenotypes, because of the hope that proximity to genetic antecedents will afford larger effect sizes. However, this assumption has been questioned [42] and is ultimately an empirical question for which data are largely lacking. Nevertheless, where they exist, these data suggest caution—a recent genome-wide association study of brain volume phenotypes indicated genetic effect sizes that were not qualitatively different from those reported for the more traditional phenotypes of clinical diagnosis [53]. This suggests, at the very least, that the assumption of larger effect sizes for intermediate phenotypes may not hold universally. Therefore, although these phenotypes may help to elucidate mechanism, this will only be the case if large samples are ascertained, in order to arrive at robust and replicable associations. In other words, the possibility of a "simpler" genetic architecture (and it is only a possibility) should not be taken to imply a "simple" genetic architecture.

Finally, there are three other issues to consider when evaluating the intermediate phenotype approach. First, there is certainly considerable complexity due to possible gene × gene and gene × environment interactions. While these are likely to exist, detecting reliable interaction effects is another matter and has been the subject of controversy [54–56], not least because the search for interaction effects dramatically increases the likelihood of false positives [57], in particular in small samples. Manifest phenotypes are presumably emergent representations of diverse underlying biological systems that are affected by many genes, and many combinations could lead to the system-level properties that give rise to the phenotype. It remains to be seen whether the intermediate phenotype approach will provide a partial solution to this problem, and some of the evidence we have to date suggests no difference in the effect size for behavioral and physiological phenotypes [58]. It is worth bearing in mind that even such proximal phenotypes as gene expression are polygenic and complex.

Second, most intermediate phenotype studies continue to rely on candidate genes which have been the focus of molecular genetic study for several years. However, ongoing genome-wide association studies are consistently revealing that these candidates are not emerging as major contributors with respect to diagnostic phenotypes. Information about genetic effects from whole genome approaches should (but does not always) inform the single locus analyses [59].

Third, many of the phenotypes used in intermediate phenotype studies may also have suboptimal psychometric properties. For example, behavioral tasks frequently used to assess attentional biases toward drug cues have been reported to show internal reliability coefficients well below accepted standards [60]. Unfortunately many of the measures used in intermediate phenotype studies have unknown (or unreported) psychometric properties. Greater measurement precision is critical if genetic effects are to be detected reliably [61]. Intermediate phenotypes can in principle address some of these issues, but this should not be taken for granted.

1.6 Conclusion

To recapitulate, the goal of this volume is to provide a critical review of the application of an intermediate phenotype approach to addiction genetics. The scope is maximally broad, hence the intentional use of the term intermediate phenotype rather than endophenotype. We begin with chapters reviewing the most established findings in the field, including brain electrophysiological profiles, variability in drug metabolism, and subjective reactions to direct drug effects. We then proceed to less well established but highly promising areas, including expectancies, attentional processing, and behavioral economic variables. Last, we move on to more exploratory territory, including an empirical report on an investigation of the differential susceptibility hypothesis, epigenetic modifications as potential intermediate phenotypes, and efforts to close the gap between mouse and human genetics.

Across the chapters, it is worthwhile to consider the assembled chapters as macro-level hypothesis testing. In the wake of small effect size and mixed findings in studies using clinical phenotypes, the intermediate phenotype approach reflects the hypothesis that these alternative phenotypes will advance the addiction genetics research enterprise, but the utility of shifting to more focal, mechanistic phenotypes, and thereby generating more reliable and larger magnitude effect sizes, is not assured at this point. As we have noted, there are reasons for caution, and the goal of the volume is not to prov ide definitive support for a foregone conclusion. Rather, we have assembled these critical works together for readers to judge the findings and prospects of the approach for themselves.

Note

1. Rush, B. (1790). *An inquiry into the effects of ardent spirits upon the human body and mind: With an account of the means of preventing, and of the remedies for curing them* (8th ed.). Boston: J. Loring [reprint 1823, p. 80]; and Crothers T. D. 1912. Neuro-psychoses in inebriety. *British Journal of Inebriety*, 9, 189–1897.

References

1. Venter, J. C., Adams, M. D., Myers, E. W., et al. (2001). The sequence of the human genome. *Science*, *291*, 1304–1351.

2. Lander, E. S., Linton, L. M., Birren, B., et al. (2001). Initial sequencing and analysis of the human genome. *Nature*, *409*, 860–921.

3. Munafò, M. R., Matheson, I. J., & Flint, J. (2007). Association of the DRD2 gene Taq1A polymorphism and alcoholism: a meta-analysis of case-control studies and evidence of publication bias. *Molecular Psychiatry*, *12*, 454–461.

4. Gottesman, I. I., & Shields, J. (1973). Genetic theorizing and schizophrenia. *British Journal of Psychiatry, 122*, 15–30.

5. Lenzenweger, M. F. (2010). *Schizotypy and Schizophrenia: The View from Experimental Psychopathology.* New York: Guilford.

6. Turkheimer, E. (2011). Still missing. *Research in Human Development, 8*, 227–241.

7. Mardones, J., & Segovia-Riquelme, N. (1983). Thirty-two years of selection of rats by ethanol preference: UChA and UChB strains. *Neurobehavioral Toxicology and Teratology, 5*, 171–178.

8. Beck, J. A., Lloyd, S., Hafezparast, M., et al. (2000). Genealogies of mouse inbred strains. *Nature Genetics, 24*, 23–25.

9. Crabbe, J. C., Phillips, T. J., & Belknap, J. K. (2010). The complexity of alcohol drinking: studies in rodent genetic models. *Behavior Genetics, 40*, 737–750.

10. Belknap, J. K., Metten, P., Beckley, E. H., & Crabbe, J. C. (2008). Multivariate analyses reveal common and drug-specific genetic influences on responses to four drugs of abuse. *Trends in Pharmacological Sciences, 29*, 537–543.

11. Cotton, N. S. (1979). The familial incidence of alcoholism: a review. *Journal of Studies on Alcohol, 40*, 89–116.

12. Merikangas, K. R., Stolar, M., Stevens, D. E., et al. (1998). Familial transmission of substance use disorders. *Archives of General Psychiatry, 55*, 973–979.

13. Johnson, J. L., & Leff, M. (1999). Children of substance abusers: overview of research findings. *Pediatrics, 103*, 1085–1099.

14. Tyrfingsson, T., Thorgeirsson, T. E., Geller, F., et al. (2010). Addictions and their familiality in Iceland. *Annals of the New York Academy of Sciences, 1187*, 208–217.

15. Goldman, D., Oroszi, G., & Ducci, F. (2005). The genetics of addictions: uncovering the genes. *Nature Reviews. Genetics, 6*, 521–532.

16. Agrawal, A., & Lynskey, M. T. (2006). The genetic epidemiology of cannabis use, abuse and dependence. *Addiction, 101*, 801–812.

17. McGue, M. (1997). A behavioral-genetic perspective on children of alcoholics. *Alcohol Health and Research World, 21*, 210–217.

18. Munafò, M. R., & Johnstone, E. C. (2008). Genes and cigarette smoking. *Addiction, 103*, 893–904.

19. Arias, A., Feinn, R., & Kranzler, H. R. (2006). Association of an Asn40Asp (A118G) polymorphism in the mu-opioid receptor gene with substance dependence: a meta-analysis. *Drug and Alcohol Dependence, 83*, 262–268.

20. Bierut, L. J. (2010). Convergence of genetic findings for nicotine dependence and smoking related diseases with chromosome 15q24–25. *Trends in Pharmacological Sciences, 31*, 46–51.

21. Ware, J. J., van den Bree, M. B., & Munafò, M. R. (2011). Association of the CHRNA5–A3-B4 gene cluster with heaviness of smoking: a meta-analysis. *Nicotine & Tobacco Research, 13*, 1167–1175.

22. Luczak, S. E., Glatt, S. J., & Wall, T. L. (2006). Meta-analyses of ALDH2 and ADH1B with alcohol dependence in Asians. *Psychological Bulletin, 132*, 607–621.

23. Maher, B. (2008). Personal genomes: the case of the missing heritability. *Nature, 456*, 18–21.

24. Robins, E., & Guze, S. B. (1970). Establishment of diagnostic validity in psychiatric illness: its application to schizophrenia. *American Journal of Psychiatry, 126*, 983–987.

25. Kahler, C. W., & Strong, D. R. (2006). A Rasch model analysis of DSM–IV alcohol abuse and dependence items in the National Epidemiological Survey on Alcohol and Related Conditions. *Alcoholism, Clinical and Experimental Research, 30*, 1165–1175.

26. Borges, G., Ye, Y., Bond, J., et al. (2010). The dimensionality of alcohol use disorders and alcohol consumption in a cross-national perspective. *Addiction, 105*, 240–254.

27. Strong, D. R., & Kahler, C. W. (2007). Evaluation of the continuum of gambling problems using the DSM–IV. *Addiction, 102*, 713–721.

28. Strong, D. R., Kahler, C. W., Ramsey, S. E., & Brown, R. A. (2003). Finding order in the DSM–IV nicotine dependence syndrome: a Rasch analysis. *Drug and Alcohol Dependence, 72*, 151–162.

29. Leggio, L., Kenna, G. A., Fenton, M., Bonenfant, E., & Swift, R. M. (2009). Typologies of alcohol dependence: from Jellinek to genetics and beyond. *Neuropsychology Review, 19*, 115–129.

30. Moss, H. B., Chen, C. M., & Yi, H. Y. (2007). Subtypes of alcohol dependence in a nationally representative sample. *Drug and Alcohol Dependence, 91*, 149–158.

31. Ball, S. A., Carroll, K. M., Babor, T. F., & Rounsaville, B. J. (1995). Subtypes of cocaine abusers: support for a type A–type B distinction. *Journal of Consulting and Clinical Psychology, 63*, 115–124.

32. Nixon, S. J. (1993). Recent developments in alcoholism: typologies in women. *Recent Developments in Alcoholism, 11*, 305–323.

33. Ustun, B., Compton, W., Mager, D., et al. (1997). WHO study on the reliability and validity of the alcohol and drug use disorder instruments: overview of methods and results. *Drug and Alcohol Dependence, 47*, 161–169.

34. Pierucci-Lagha, A., Gelernter, J., Feinn, R., et al. (2005). Diagnostic reliability of the Semi-structured Assessment for Drug Dependence and Alcoholism (SSADDA). *Drug and Alcohol Dependence, 80*, 303–312.

35. Feinn, R., Gelernter, J., Cubells, J. F., Farrer, L., & Kranzler, H. R. (2009). Sources of unreliability in the diagnosis of substance dependence. *Journal of Studies on Alcohol and Drugs, 70*, 475–481.

36. Pierucci-Lagha, A., Gelernter, J., Chan, G., et al. (2007). Reliability of DSM–IV diagnostic criteria using the semi-structured assessment for drug dependence and alcoholism (SSADDA). *Drug and Alcohol Dependence*, *91*, 85–90.

37. Brandon, T. H., Vidrine, J. I., & Litvin, E. B. (2007). Relapse and relapse prevention. *Annual Review of Clinical Psychology*, *3*, 257–284.

38. Klingemann, H., Sobell, M. B., & Sobell, L. C. (2010). Continuities and changes in self-change research. *Addiction*, *105*, 1510–1518.

39. Gottesman, I. I., & Gould, T. D. (2003). The endophenotype concept in psychiatry: etymology and strategic intentions. *American Journal of Psychiatry*, *160*, 636–645.

40. John, B., & Lewis, K. R. (1966). Chromosome variability and geographic distribution in insects. *Science*, *152*, 711–721.

41. Hines, L. M., Ray, L., Hutchison, K., & Tabakoff, B. (2005). Alcoholism: the dissection for endophenotypes. *Dialogues in Clinical Neuroscience*, *7*, 153–163.

42. Flint, J., & Munafò, M. R. (2007). The endophenotype concept in psychiatric genetics. *Psychological Medicine*, *37*, 163–180.

43. Doyle, A. E., Faraone, S. V., Seidman, L. J., et al. (2005). Are endophenotypes based on measures of executive functions useful for molecular genetic studies of ADHD? *Journal of Child Psychology and Psychiatry, and Allied Disciplines*, *46*, 774–803.

44. Waldman, I. D. (2005). Statistical approaches to complex phenotypes: evaluating neuropsychological endophenotypes for attention-deficit/hyperactivity disorder. *Biological Psychiatry*, *57*, 1347–1356.

45. Goldman, D., and Ducci, F. (2007). Deconstruction of vulnerability to complex diseases: enhanced effect sizes and power of intermediate phenotypes. Scientific World Journal, 7, 124–130.

46. Agarwal, A., Williams, G. H., & Fisher, N. D. (2005). Genetics of human hypertension. *Trends in Endocrinology and Metabolism*, *16*, 127–133.

47. Shoback, D. M., Williams, G. H., Moore, T. J., et al. (1983). Defect in the sodium-modulated tissue responsiveness to angiotensin II in essential hypertension. *Journal of Clinical Investigation*, *72*, 2115–2124.

48. Williams, G. H., Dluhy, R. G., Lifton, R. P., et al. (1992). Non-modulation as an intermediate phenotype in essential hypertension. *Hypertension*, *20*, 788–796.

49. Kosachunhanun, N., Hunt, S. C., Hopkins, P. N., et al. (2003). Genetic determinants of non-modulating hypertension. *Hypertension*, *42*, 901–908.

50. Munafò, M. R., & Flint, J. (2011). Dissecting the genetic architecture of human personality. *Trends in Cognitive Sciences*, *15*, 395–400.

51. Benjamin, D. J., Cesarini, D., van der Loos, M. J., et al. (2012). The genetic architecture of economic and political preferences. *Proceedings of the National Academy of Sciences of the United States of America*, *109*, 8026–8031.

52. Wray, N. R., & Visscher, P. M. (2010). Narrowing the boundaries of the genetic architecture of schizophrenia. *Schizophrenia Bulletin*, *36*, 14–23.

53. Stein, J. L., Medland, S. E., Vasquez, A. A., et al. (2012). Identification of common variants associated with human hippocampal and intracranial volumes. *Nature Genetics*, *44*, 552–561.

54. Munafò, M. R., Durrant, C., Lewis, G., & Flint, J. (2009). Gene × environment interactions at the serotonin transporter locus. *Biological Psychiatry*, *65*, 211–219.

55. Munafò, M. R., Durrant, C., Lewis, G., & Flint, J. (2010). Defining replication: a response to Kaufman and colleagues. *Biological Psychiatry*, *67*, 21–23.

56. Kaufman, J., Gelernter, J., Kaffman, A., Caspi, A., & Moffitt, T. (2010). Arguable assumptions, debatable conclusions. *Biological Psychiatry*, *67*, e19–20; author reply e21–3.

57. Ioannidis, J. P. (2005). Why most published research findings are false. *PLoS Medicine*, *2*, e124.

58. Veyrieras, J. B., Kudaravalli, S., Kim, S. Y., et al. (2008). High-resolution mapping of expression-QTLs yields insight into human gene regulation. *PLOS Genetics*, *4*, e1000214.

59. Flint, J., & Munafò, M. R. (2013). Candidate and non-candidate genes in behavior genetics. *Current Opinion in Neurobiology*, *23*, 57–61.

60. Ataya, A. F., Adams, S., Mullings, E., et al. (2012). Internal reliability of measures of substance-related cognitive bias. *Drug and Alcohol Dependence*, *121*, 148–151.

61. Munafò, M. R., Timofeeva, M. N., Morris, R. W., et al. (2012). Association between genetic variants on chromosome 15q25 locus and objective measures of tobacco exposure. *Journal of the National Cancer Institute*, *104*, 740–748.

2 Electrophysiological Intermediate Phenotypes for the Detection of Genetic Influences on Alcoholism

Mary-Anne Enoch

Like most psychiatric diseases, alcoholism is a complex multifactorial disorder with both genetic and environmental risk factors. The heritability is approximately 50% [1]. The effect of etiological heterogeneity is to dilute power to detect genetic influences on diagnostic phenotypes. This has lead to an interest in the use of dimensionally or qualitatively measured intermediate phenotypes (or endophenotypes) that are heritable, stable traits that are relevant to disease and likely to be influenced by variation at fewer genes [2]. The electroencephalogram (EEG) accesses brain electrical activity and may represent quantitative correlates of disease liability that are more amenable to genetic analysis than the psychiatric diagnosis itself. Electrophysiological phenotypes, including resting EEG power, event-related potentials, event-related oscillations, EEG coherence and more recently functional connectivity, are proving to be successful in detecting genetic variation underlying vulnerability to alcoholism and comorbid disorders.

Hans Berger succeeded in obtaining the first scalp recording of the human EEG in 1924, but the importance of his discovery was not recognized by the scientific community and adopted into practice until 1938. The EEG is a recording of the rhythmical, electrical activity of the brain, and approximately 75 years ago this was the first insight into brain activity. The scalp-recorded EEG is thought to derive from extracellular current flow associated with summated postsynaptic potentials in synchronously activated, radially oriented pyramidal cells in the cerebral cortex [3, 4]. The EEG records millisecond changes in the balance between excitation and inhibition in neuronal circuitry. The patterns of EEG oscillations are constantly changing depending on mental activity, relaxation, drowsiness, and sleep, and this dynamic process is therefore an index of cortical activation, cognitive function, and consciousness. The EEG may thus be considered an intermediate phenotype for complex behaviors and psychopathology in which arousal is implicated, such as anxiety, depression, and alcoholism. However, it should be noted that the EEG is not clinically useful for the diagnosis of any one specific psychiatric disorder.

2.1 The Resting EEG

In the healthy awake adult two sinusoidal wave patterns, alpha (8–13 Hz) and beta (13–30 Hz), dominate the resting EEG. Lower frequency rhythms theta (4–8 Hz) and delta (0.1–4 Hz) are less prominent [4, 5]. Alpha is the predominant resting (sometimes described as spontaneous or background) EEG waveform. The amplitude of the alpha rhythm is maximal posteriorly during alert relaxation and mental inactivity, when eyes are closed. Higher alertness attenuates or suppresses the alpha rhythm, which is then supplanted by desynchronized low-voltage fast activity. The alpha rhythm has also been implicated in information processing in a variety of cognitive tasks [6]. Low-voltage alpha (LVA), characterized by the virtual absence of the alpha rhythm, is a distinctive EEG trait found in 7% to 14% of individuals that may be transmitted in an autosomal dominant fashion with high penetrance [7] and resembles a high-arousal EEG [8]. It is therefore likely that brain electrical activity in LVA individuals cannot be synchronized as effectively. Beta, a lower voltage, irregular waveform has a diffuse distribution but is most evident frontally. The beta rhythm predominates in mentally active individuals. The theta frequency is important in infancy, childhood, drowsiness, and light sleep and is the dominant rhythm in nonprimates. The generator of theta oscillations is in the hippocampus, a brain region that is a constituent of the memory/conditioning neuronal circuitry of addiction [9, 10]. Theta rhythm is thought to be implicated in memory and learning [11]. The delta frequency is commonly seen in infants and during deep sleep in adults. Delta waves are almost always pathological in the waking state in adults.

The resting EEG is recorded from electrodes applied to scalp sites (either directly or embedded in a cap) according to the International 10–20 system of placement. Scalp recorded electrical activity is quite low, measured in microvolts. Changes can be detected over milliseconds. Typically, the recording at each electrode is referenced to a single reference electrode (monopolar or referential montage) located in an electrically quiet region. However some studies, for example the Collaborative Study on the Genetics of Alcoholism (COGA), have used a bipolar montage where the recordings at adjacent electrodes of the 10–20 system are compared. The bipolar montage is good for analyzing low- to medium-amplitude waveforms that are highly localized. The disadvantage is that cancellation can occur when two inputs are equipotential, thereby producing a flat EEG.

The resting EEG waveform can be characterized by amplitude and frequency, but for research purposes the recording is usually fast Fourier transformed to yield spectral power per frequency band. Some studies have shown that decreased alpha power and increased theta and beta power is associated with psychopathology, including addiction [12]. A family history of alcoholism has been shown to predict reduced alpha power [13]. LVA EEG has been shown to be more common in alcoholics with anxiety disorders [8]. Increased beta activity has been associated with alcoholism in men [14], a family history of alcoholism [15], and relapse in abstinent alcoholics [16, 17]. Alcoholics have been shown to have increased theta power relative to controls [18].

Although there is considerable interindividual variability, the distinctive resting EEG pattern of an individual tends to be stable in healthy adults [7]. Test–retest reliability for resting EEG power is high, even after two years (r = 0.5–0.9) [19–22]. The heritability of eyes closed, resting EEG power using a monopolar montage tends to be the same at all scalp locations [22–24] and has been shown in twin studies to be approximately 0.80–0.85 for theta, alpha, and beta power and 0.40 for delta [23–26]. The heritability for theta, alpha, and beta frequency bands derived from studies with nontwin sibling pairs is approximately the same whether using a bipolar electrode montage as in the COGA study [27] or using a monopolar montage as in a Plains Indian study [19]. Genetic correlations between alpha, beta, theta, and delta power bands (0.55 to 0.75) indicate that a substantial portion of the genetic variance can be attributed to a common source [19, 22]; however, frequency band–specific and region-specific genetic factors are also influential [24].

2.1.1 Genetic Effects on the Resting EEG

The evidence for both shared and specific genetic influences on the resting EEG is compelling. The stability and heritability of the resting EEG makes it highly suitable for genetic analysis. Because the resting EEG is an indicator of cortical activation, it may be regarded as an intermediate phenotype for anxiety, stress, and addiction. Therefore candidate genes for the resting EEG are likely to encompass genes influencing arousal-related behavior—for example, stress-related genes such as those encoding the corticotropin releasing hormone (CRH), its binding protein (CRHBP) and receptors (CRHR1, CRHR2), and anxiety-related genes such as GABAergic genes, serotonergic genes, and catechol-O-methyltransferase (*COMT*).

Candidate genes for the resting EEG will also include genes that are directly implicated in current flow, synchronization, and rhythm generation. The resting EEG derives from transmembrane currents originating in parietal and occipital regions for alpha and the hippocampus and limbic system for theta. The rhythm generator for alpha originates in the thalamus [28–30] and in the hippocampus for theta [3] although the thalamus can act as an independent pacemaker for both alpha and theta rhythms [28]. Cholinergic and GABAergic afferents contribute to hippocampal theta activity [3]. The beta rhythm is thought to be mediated by GABAA interneurons [31, 32]. Benzodiazepine-induced enhanced GABA transmission results in increased beta power [33]. Clearly, variation in GABAergic and cholinergic pathway genes may well influence patterns of electrical activity in the brain.

Results for Individual Genes

GABRA2

One of the first studies to identify a resting EEG genetic association came from COGA. This is a large, multisite U.S. study dedicated to the identification of genetic vulnerability loci for the development of alcoholism and comorbid disorders. COGA systematically ascertained and studied a large collection of predominantly Caucasian families starting with treatment-seeking alcoholic probands with at least two additional alcohol-dependent first-degree relatives [34]. The COGA study found significant linkage of bipolar derivations of beta EEG power to a region of chromosome (chr) 4 that includes the GABAA receptor subunit gene complex: *GABRG1, GABRA2, GABRA4, GABRB1* [35]. Using the same data set, another study found significant linkage peaks for beta power on chromosomes 1, 4, 5, and 15 [36]. This chr 4 linkage finding was subsequently tracked to the *GABRA2* gene. COGA demonstrated an association between *GABRA2* single nucleotide polymorphisms (SNPs) and haplotypes located within a haplotype block distal to intron 3 and both beta EEG power and alcohol dependence (AD) [37]. However, it is not clear from this paper whether the *GABRA2* SNP variants that associated with AD also associated with *higher* beta power, as would be expected from earlier studies [14]. There was no association with the other three GABAA receptor subunit genes in the chr 4 complex. Numerous studies have subsequently shown an association between *GABRA2* variation and AD (reviewed in Enoch, 2008 [38]).

A recent study in British Caucasian alcoholics, abstinent for at least two years (with cirrhosis but no encephalopathy), found a significant relationship between *GABRA2* SNPs and increased beta power but in this case no relationship between *GABRA2* variation and the heterogeneous AD phenotype [39].

This resting EEG association with *GABRA2* is logical since the generation of beta rhythm results from the balance between networks of excitatory pyramidal cells and networks of inhibitory interneurons as pacemakers, gated by GABA [32]. Alcoholics and their alcohol-naive offspring have been shown to have elevated beta power, and this could indicate an imbalance between neuronal excitation and inhibition that might be a risk factor for alcoholism [40, 41].

CRHBP

We undertook a dense whole genome linkage scan using nearly 4,000 unlinked SNPs in a large pedigree derived from a population sample of Plains American Indians [19]. Probands were initially ascertained at random from the tribal register, and the families of alcoholic probands were extended. This was a community sample rather than a treatment-seeking sample of alcoholics. There were no linkage signals for any alcoholism-related phenotypes in this data set. However, linkage peaks for EEG power in all three frequency bands converged on chr 5q13–14 with genome-wide significant logarithm (base 10) of odds (LOD) scores (3.5) for alpha and beta power and suggestive LOD scores (2.2) for theta power. These convergent findings on chr 5 suggest a common genetic origin for alpha, beta, and theta EEG power. The *CRHBP* gene was located at the apex of these convergent linkage peaks. *CRHBP* was significantly associated with alpha power in the Plains Indians and also in a replication community sample of U.S. Caucasians, together with a trend in the same direction for an association with beta and theta power. Moreover, the same *CRHBP* SNPs and haplotypes, located within the distal *CRHBP* haplotype block, were also associated with anxiety disorders in the Plains Indians and alcohol use disorders in the Caucasians. In summary, the haplotype that was associated with lower alpha power in both populations was more common in U.S. Caucasian alcoholics but less common in Plains Indians with anxiety disorders. *CRHBP* has subsequently been associated with alcoholism and other stress-related disorders in several independent data sets (reviewed in Roy et al., 2012 [42]).

Prior to the identification of *CRHBP* as a candidate gene in this study, publications had largely focused on the important role that CRHBP plays in modulating CRH and the stress response in pregnancy and childbirth. However, in thinking about the wider role of the CRH binding protein, it becomes clear that it is indeed logical that the *CRHBP* gene might influence stress-related disorders, including alcoholism and the resting EEG. CRHBP modulates the activity of CRH, the primary mediator of the neuroendocrine stress response. In the ventral tegmental area (VTA) CRHBP has been shown to modulate the effects of CRH on stress-induced relapse to drug abuse [43]. CRHBP may influence the EEG via modulation of CRH activity in the locus coeruleus (LC) and the VTA. CRH is known to increase the firing rate of LC neurons and to influence global forebrain EEG activity [44, 45]. LC activation induces frontocortical and hippocampal EEG changes [46] and influences the firing properties of thalamic neurons that are implicated in the generation of the alpha rhythm [28, 47]. It is possible that in the VTA *CRHBP* variation might influence not only addiction phenotypes [43] but also the resting EEG via the VTA dopamine–LC–thalamic neuronal oscillator pathway [19].

This study in Plains Indians also identified suggestive linkage peaks for theta and alpha EEG power (but not beta power) on chr 4. However, these linkage peaks were nowhere near the GABAA receptor gene complex. Moreover, there was no overlap with COGA's chr 5 linkage peak [36], nor were linkage peaks detected on chromosomes 1 and 15 as in the COGA study [36]. Also, COGA found no evidence for linkage for alpha or theta EEG power. Reasons for a lack of comparability between the Plains Indian linkage study and the COGA study may include differences in montage (monopolar vs. bipolar montage), derivation of EEG spectral power, and sample characteristics (community vs. clinical samples of alcoholics).

HTR3B

In the Plains Indian whole genome linkage scan, a suggestive peak (LOD score = 2.2) for alpha power was found on chr 11. The *HTR3A* and *HTR3B* genes that encode the 5-HT3 serotonin receptor were located at the apex of the peak [19]. The same *HTR3B* SNP and haplotype was associated with antisocial alcoholism in a sample of incarcerated Finnish Caucasian men and lower alpha power in the community sample of U.S. Caucasians. Associations between alpha power and two other *HTR3B* SNPs and another haplotype were also observed among the Plains Indians [48]. Activation of the 5-HT3 receptor, the only serotonin receptor that is a ligand-gated ion channel, results in rapid neuronal depolarization. One possible mechanism for the influence of the 5-HT3 receptor on resting EEG is through the modulation of GABAergic neurons. Activation of GABAergic interneurons via 5-HT3 receptors has been shown to result in a GABA-mediated inhibition of pyramidal cells in the hippocampus [49], and similar mechanisms are likely to exist in other brain regions [49, 50].

COMT

The *COMT* gene encodes the enzyme COMT that is largely responsible for the metabolism of dopamine and norepinephrine in the prefrontal cortex. A functional polymorphism, *COMT* Val158Met, is responsible for a fourfold difference in COMT enzyme activity in humans [51]. In a study of the *COMT* Val158Met polymorphism in the aforementioned samples of Plains Indians and U.S. Caucasians, female Met/Met homozygotes had higher dimensional anxiety (harm avoidance) and were more likely to have the LVA EEG trait [52]. Previously, the LVA trait had been associated with alcoholism comorbid with anxiety disorders in a predominantly Caucasian sample [8]. In a study in healthy men, it was shown that the Val158Met polymorphism predicted stable interindividual variation in alpha-band oscillations: the peak frequency was 1.4 Hz lower in Val/Val homozygotes than in Met/Met homozygotes. Moreover, Val/Val homozygotes exhibited lower EEG fast alpha power (11–13 Hz) than Met/Met homozygotes [53]. In a large EEG data set of healthy individuals derived from the Brain Resource International Database, researchers have shown that the *COMT* Val allele was associated with increased posterior versus frontal delta/theta power together with increased extraversion [54].

Genome-wide Association Study

The first published genome-wide association study (GWAS) on resting EEG power was performed in the same sample of 322 Plains American Indians in order to take advantage of the genetic and environmental homogeneity of this population isolate [55]. The diagnostic phenotype of alcoholism did not generate statistical signals approaching genome-wide significance. However, the GWAS identified three genes (*SGIP1*, *ST6GALNAC3*, and *UGDH*) that showed association with alpha or theta power. *SGIP1* accounted for 8.8% of variance in theta power in the Plains Indians and 3.5% of the theta power variance in the replication U.S. Caucasian data set. In addition, *SGIP1* was associated with alcoholism, an effect that may be mediated via the same brain mechanisms accessed by theta EEG power. Additionally the *BICD1* gene, identified at subthreshold significance level in the GWAS, showed significant association with alpha power [55]. The fact that these genes act within particular pathways [55] suggests that neuronal excitability in the brain is determined in part by the ability to recycle neuronal membrane components. Although the genome-wide significant findings of the GWAS were for genes that solely influence EEG power, convergence of findings at the subthreshold level with previous findings in genetic studies of addictions suggest that the intermediate phenotype approach can potentially identify genes that have a general effect on addiction even in data sets of modest size, notably of population isolates.

Results from the Plains Indian GWAS showed some overlap with findings from both the Plains Indian and the COGA linkage scans for resting EEG power. The GWAS identified two chromosomal regions, each containing multiple candidate genes for variability in resting EEG power. *SGIP1* and *ST6GALNAC3* lie within the same chr 1p peak that was linked to beta power in the COGA study [36]. The *UGDH, LIAS, RPL9,* and *KLB* genes lie in the chr 4 region where the Plains Indian scan identified suggestive linkage peaks for theta and alpha power [19]. However, the GWAS did not replicate the key chr 5q13–14 finding from the Plains Indian linkage scan, perhaps because of the relatively small effect of this locus on EEG power. A subsequent COGA study of a large sample of Caucasian and African American alcoholics did not find an association between *SGIP1* and AD or theta power [56]. There may be several reasons for this lack of replication; for example, the two studies differed in the samples of alcoholics (community vs. treatment seeking), the EEG montages (monopolar vs. bipolar), and ethnicities (Native American vs. Caucasian/African American).

2.2 Event-Related Potentials

The P300 event-related potential (ERP) is a scalp-recorded voltage change that is elicited in response to infrequent target stimuli occurring among frequent nontargets. The P300 ERP amplitude is measured as the voltage difference between the prestimulus baseline and the largest positive peak in the latency window of 300–600 ms after stimulus onset. It is thought to reflect the degree to which attention is allocated and incoming information is processed and incorporated into working memory [57]. The peak amplitude of the P300 ERP, elicited in response to auditory and visual stimuli, is moderately heritable (0.3 to 0.8) with the greatest amplitude and heritability at posterior leads [58–61].

One of the first scientists to detect that abstinent male alcoholics, together with their alcohol-naive young biological sons, tended to have low P300 ERPs was Henri

Begleiter [62]. Since then, numerous studies have shown that, compared with controls, P300 amplitude is lower in abstinent alcoholics [63, 64], children of alcoholics [62, 63, 65–69], family history positive alcoholics [70, 71], and nonalcoholic relatives of alcoholics [63].

The majority of these studies have focused on men. Female alcoholics have been found to have lower P300 amplitudes in some studies [63, 64, 72, 73] but not others [65, 74–76]. However, not all studies have replicated the association with alcoholism [67], particularly for auditory paradigms [77]. Possible reasons for this include differences in stimulus modality and task difficulty [78, 79] or subject populations [80–83].

Anxious individuals tend to experience increased arousal, heightened awareness of their surroundings, and behavioral inhibition and therefore might be expected to respond more strongly to rare target stimuli. Only a few studies have looked at the relationship between anxiety and P300. Elevated auditory P300 amplitudes have been found in normal volunteers exposed to anxiety-provoking situations [84], in unprovoked anxious community-ascertained individuals [85, 86], and in acutely ill unmedicated patients with obsessive–compulsive disorder [87], indicating a hyperactivated cortical state. In contrast, a community study found that women with anxious mood, the majority of whom were alcoholic, had diminished visual P300 amplitude [74]. Indeed, we have shown, also in a community sample, that auditory P300 amplitude was highest in individuals with pure anxiety disorders, lower in individuals with alcohol use disorders only or major depression only, and lowest in alcoholics with anxiety disorders [88, 89].

2.2.1 Genetic Effects on ERPs

In one of the first studies of genetic influences on ERPs, COGA found evidence for linkage of both alcoholism as a discrete phenotype and P300 amplitude to a chr 4 region near the class I alcohol dehydrogenase locus ADH3. In other words, the same quantitative-trait locus influenced both risk of alcoholism and the amplitude of the P300 ERP [90]. Other genome-wide linkage scans have shown evidence of linkage to chr 5 [91] and chr 2 [92].

Again in the COGA data set, analysis of variation in the *CRHR1* gene that encodes the CRHR1 receptor showed a haplotype and SNP association with AD and also an association with low-amplitude P300 ERP in nonalcoholics and alcoholics alike [93]. Variation in the cannabinoid receptor 1 gene, *CNR1*, has been associated with low-amplitude P300 in a group of individuals with comorbid alcohol and drug dependence; this genetic variation contributed to 20% of the variance of the frontal lobe P300 ERP amplitude [94]. Moreover, results indicated that *CNR1* variation might influence the acute effects of cannabinoids on P300 generation in healthy individuals [95].

2.3 Event-Related Oscillations

For many years it was not known whether ERPs resulted from stimulus-evoked brain events or stimulus-induced changes in ongoing brain electrical activity. Studies eventually emerged showing that the background or resting EEG is actively involved in ERP signal production as a feature of attention allocation and memory processing [96–98]. Time-frequency decomposition of ERPs has shown that multiple consecutive and overlapping components (evoked delta, theta, alpha) are induced in response to targets. Theta and delta oscillations are maximally induced in posterior regions and largely account for P300 amplitude [99–100]. Wavelet entropy analysis has shown that there is a prominent dominance of theta ERP components over other frequency bands [101]. Theta oscillations have been associated with attention and memory; delta oscillations have been related to signal detection and decision making [41]. It has been shown that non-phase-locked event-related oscillation (ERO) power measures are an alternative and comparable representation of reduced P300 amplitude in alcoholics [102]. There is evidence to suggest that induced theta ERO power measures may be more powerful than P300 amplitude in discriminating between alcoholics and controls [103]. Alcoholics have been shown to have reduced theta and delta ERO power [12, 104], including during reward processing, compared with controls [105]. Moreover, compared with controls, adolescent offspring of alcoholic fathers have shown decreased total theta and delta ERO power in response to visual targets, and this appears to be more robust in differentiating between the two groups than P300 amplitude [106].

2.3.1 Genetic Effects on EROs

P300 generation and the underlying EROs are though to be influenced in part by interactions between GABAergic, glutamatergic, and cholinergic systems [107]. In a genome-wide linkage scan of theta ERO in response to target visual stimuli, COGA found a significant linkage peak (LOD = 3.5) on chr 7 [108]. Significant association was found between SNPs in *CHRM2*, the gene encoding the muscarinic acetylcholine receptor gene that is located under the chr 7 linkage peak, and theta EROs [109], together with the diagnosis of AD and depression [110]. *GRM8*, a gene encoding the group III metabotropic glutamate receptor subunit mGluR8, also located in the chr 7 linkage peak region, was associated with theta EROs and AD [111].

Genome-wide Association Scan

COGA recently reported the results of the first GWAS for theta EROs evoked by visual targets in a sample of approximately 1,000 individuals [112]. After reviewing the results from the primary genome-wide association analysis that revealed no findings of genome-wide significance, a total of 42 SNPs were genotyped in a family-based sample of approximately 1,000 individuals: thirty SNPs were selected from the top 250 ranking SNPs located within or near genes deemed to be of interest, and 12 SNPs, outside the top 250 but at p < 0.05, were identified as potentially relevant to neuro-electrical activity together with SNPs in the serotonin receptor genes *HTR2A* and *HTR7*. Four SNPs showed nominally significant associations with theta ERO power in both data sets. The strongest result was in *ARID5A*; however, a SNP within *HTR7* was associated both with theta ERO power and with AD. Significant effects were detected for AD in both samples, with the *HTR7* risk allele corresponding to reduced theta ERO power among homozygotes. These results suggest that the serotonergic system, in particular the 5-HTR7 receptor, may play a role in the neurophysiological underpinnings of theta EROs and AD. These results reinforce the utility of brain oscillations in understanding the genetic complexity underlying this heterogeneous psychiatric disorder.

2.4   EEG Coherence

EEG coherence measures the synchronization in electrical activity between two scalp locations and is a means of assessing connectivity between cortical regions. Significant pathways of coherence are primarily anterior–posterior (A-P) interhemispheric or perpendicular to the A-P axis [113]. Theta-band coherence involves distinct midline and temporal sources, the latter showing A-P differentiation; alpha-band coherence has a distinct posterior focus while beta activity shows no clear global structure [113]. One twin study has shown that the heritability of coherence along an A-P axis within each hemisphere is 0.60 for theta, alpha, and beta bands but lower in delta with little evidence of contribution from shared environment [114]. The COGA study used nontwin sibships to show that the heritability for 15 coherence pairs across all four frequency bands ranged from 0.22 to 0.63. The heritability was greatest for the alpha band; the heritability in beta and theta bands was comparable. The alpha band also had the highest coherence followed by theta and beta [115]. Thalamocortical circuitry has been implicated in theta-band coherence [116].

It is likely that there may be differences in cortical connectivity between alcoholics and controls. Indeed, one study has shown that bilateral, intrahemispheric, posterior coherences were significantly increased in the alpha and beta bands both in long-term abstinent and currently drinking alcoholics relative to controls, particularly in the depressiveness subtype [117]. So far, only one published study has looked at genetic effects on coherence: variation in the *GABBR1* gene that encodes the GABAB receptor was associated with variation in EEG posterior coherence in healthy volunteers [118].

2.5   Functional Connectivity

EEG coherence is a measure of linear correlations between pairs of signals within frequency bands. However a better understanding of fundamental neurophysiology may be derived from analyzing dynamic interactions or functional connectivity (FC) between different brain regions. Disease states may be linked to dysfunctional connectivity. There are an increasing number of studies that are using functional magnetic resonance imaging (fMRI) measures to investigate resting state FC in behavioral disorders. For example, it has recently been shown that the "default mode network" that is maximally active when the mind is at rest and is thought to characterize basal neural activity is disrupted in alcoholics compared with controls [119]. In resting-state fMRI, spontaneous low-frequency (<0.1 Hz) blood-oxygen-level-dependent (BOLD) signal fluctuations in functionally related gray matter regions show strong correlations. However, the BOLD signal lags the neuronal events triggering it by one to two seconds. Due to high temporal resolution (milliseconds), the resting EEG can be used for measures of FC within all frequency bands. FC can be mathematically reduced to nodes and their connections [120]. FC in the EEG frequency bands has been described in terms of "small world" networks that combine high clustering with low average path length [121]. Heritability of components of these networks across various frequency bands is moderate to high, ranging from 0.37 to 0.89; heritability appears to be highest in the alpha frequency band [122, 123]. A study of resting EEG measured in nearly 1,500 participants ages 16–50 years showed that FC is more random in adolescence and older ages and more structured in middle-age; heritability of FC parameters ranged from 0.3 to 0.8. In the alpha band, FC parameters reflected stable and moderately to highly heritable traits, and therefore FC in the alpha band may represent a good endophenotype for behavioral disorders [124].

Looking to the future, it is possible that resting EEG FC and resting state fMRI FC as separate analyses may be superseded by "multimodal functional network connectivity" (mFNC)—that is, the fusion of EEG and fMRI in network space. Investigation of both synthetic and real data has demonstrated that mFNC has the potential to reveal the underlying neural networks of each modality separately and in their combination [125]. With mFNC, exploration of neural activities and metabolic responses in a specific task or neurological state might reveal comprehensive relationships between functional networks [125].

2.6 Conclusion

Electrophysiological phenotypes, including resting EEG power, ERP amplitude, ERO power, EEG coherence, and functional connectivity within EEG frequency bands, are stable traits that are moderately to highly heritable. These phenotypes are relevant to psychiatric diseases, including alcoholism, a heterogeneous disorder. Moreover, these phenotypes appear to be predictors for disease vulnerability since they are also found in alcohol-naive biological children of alcoholics. Therefore, these electrophysiological phenotypes have the required characteristics of intermediate phenotypes.

Having established this fact, how well have these intermediate phenotypes lived up to the expectations for their utility in gene discovery? There have been several important findings for the resting EEG intermediate phenotype, notably as predicted in arousal/stress-related genes including *GABRA2*, *CRHBP*, and *COMT*. Pathway analysis of genes identified by a GWAS suggests that efficient recycling of neuronal membrane components may be important for brain electrical activity. This may also have relevance to alcoholism. Genetic studies of the P300 ERP have not been as fruitful. Instead, studies have moved on to time-frequency decomposition of the ERP and to analyses of the underlying theta and delta EROs. This approach has led to the detection of cholinergic, glutamatergic, and serotonergic genes that are associated both with electrophysiological phenotype and disease. Finally, we look to the future for studies such as GWAS in EEG functional connectivity or even multimodal functional network connectivity for gene detection. In conclusion, studies to date have provided validation for the use of the EEG as an intermediate phenotype for the detection of genetic influences on arousal-related behaviors, including alcoholism.

References

1. Goldman, D., Oroszi, G., & Ducci, F. (2005). The genetics of addictions: uncovering the genes. *Nature Reviews. Genetics, 6*, 521–532.

2. Gottesman, I. I., & Gould, T. D. (2003). The endophenotype concept in psychiatry: etymology and strategic intentions. *American Journal of Psychiatry, 160*, 636–645.

3. Chapman, C. A., & Lacaille, J. C. (1999). Cholinergic induction of theta-frequency oscillations in hippocampal inhibitory interneurons and pacing of pyramidal cell firing. *Journal of Neuroscience, 19*, 8637–8645.

4. Miller, R. (2007). Theory of the normal waking EEG: from single neurones to waveforms in the alpha, beta and gamma frequency ranges. *International Journal of Psychophysiology, 64*, 18–23.

5. Niedermeyer, E. (1993). The normal EEG of the waking adult. In F. Lopes da Silva (Ed.), *Electroencephalography: Basic Principles, Clinical Applications, and Related Fields* (pp. 97–117). Baltimore, MD: Williams and Wilkins.

6. Basar, E., Basar-Eroglu, C., Karakas, S., & Schurmann, M. (2001). Gamma, alpha, delta, and theta oscillations govern cognitive processes. *International Journal of Psychophysiology, 39*, 241–248.

7. Vogel, F. (1970). The genetic basis of the normal human electroencephalogram. *Humangenetik, 10*, 91–114.

8. Enoch, M.-A., White, K. V., Harris, C. R., Robin, R., Ross, J., Rohrbaugh, J. W., et al. (1999). Association of low voltage alpha EEG with a subtype of alcohol use disorders. *Alcoholism, Clinical and Experimental Research, 23*, 1312–1319.

9. Koob, G. F., & Volkow, N. D. (2010). Neurocircuitry of addiction. *Neuropsychopharmacology*, *35*, 217–238.

10. Volkow, N. D., Fowler, J. S., & Wang, G. J. (2004). The addicted human brain viewed in the light of imaging studies: brain circuits and treatment strategies. *Neuropharmacology*, 47 Suppl 1, 3–13.

11. Buzsáki, G. (2002). Theta oscillations in the hippocampus. *Neuron*, *33*, 325–340.

12. Campanella, S., Petit, G., Maurage, P., Kornreich, C., Verbanck, P., & Noël, X. (2009). Chronic alcoholism: insights from neurophysiology. *Neurophysiologie Clinique*, *39*, 191–207.

13. Finn, P. R., & Justus, A. (1999). Reduced EEG alpha power in the male and female offspring of alcoholics. *Alcoholism, Clinical and Experimental Research*, *23*, 256–262.

14. Rangaswamy, M., Porjesz, B., Chorlian, D. B., Wang, K., Jones, K. A., Bauer, L. O., et al. (2002). Beta power in the EEG of alcoholics. *Biological Psychiatry*, *52*, 831–842.

15. Ehlers, C. L., & Schuckit, M. A. (1990). EEG fast frequency activity in the sons of alcoholics. *Biological Psychiatry*, *27*, 631–641.

16. Bauer, L. O. (2001). Predicting relapse to alcohol and drug abuse via quantitative electroencephalography. *Neuropsychopharmacology*, *25*, 332–340.

17. Saletu-Zyhlarz, G. M., Arnold, O., Anderer, P., Oberndorfer, S., Walter, H., Lesch, O. M., et al. (2004). Differences in brain function between relapsing and abstaining alcohol-dependent patients, evaluated by EEG mapping. *Alcohol and Alcoholism*, *39*, 233–240.

18. Rangaswamy, M., Porjesz, B., Chorlian, D. B., Choi, K., Jones, K. A., Wang, K., et al. (2003). Theta power in the EEG of alcoholics. *Alcoholism, Clinical and Experimental Research*, *27*, 607–615.

19. Enoch, M.-A., Shen, P.-H., Ducci, F., Yuan, Q., Liu, J., Albaugh, B., et al. (2008). Common genetic origins for EEG, alcoholism and anxiety: the role of CRH-BP. *PLOS ONE*, *3*(10), e3620.

20. Pollock, V. E., Schneider, L. S., & Lyness, S. A. (1991). Reliability of topographic quantitative EEG amplitude in healthy late-middle-aged and elderly subjects. *Electroencephalography and Clinical Neurophysiology*, *79*, 20–26.

21. Salinsky, M. C., Oken, B. S., & Morehead, L. (1991). Test–retest reliability in EEG frequency analysis. *Electroencephalography and Clinical Neurophysiology*, *79*, 382–392.

22. Smit, D. J., Posthuma, D., Boomsma, D. I., & Geus, E. J. (2005). Heritability of background EEG across the power spectrum. *Psychophysiology*, *42*, 691–697.

23. van Beijsterveldt, C. E., Molenaar, P. C., de Geus, E. J., & Boomsma, D. (1996). Heritability of human brain functioning as assessed by electroencephalography. *American Journal of Human Genetics*, *58*, 562–573.

24. Zietsch, B. P. Hansen, J. L., Hansell, N. K., Geffen, G. M., & Martin, N. G. (2007). Common and specific genetic influences on EEG power bands delta, theta, alpha, and beta. *Biological Psychology*, *75*, 154–164.

25. Smit, C. M., Wright, M. J., Hansell, N. K., Geffen, G. M., & Martin, N. G. (2006). Genetic variation of individual alpha frequency (IAF) and alpha power in a large adolescent twin sample. *International Journal of Psychophysiology*, *61*, 235–243.

26. van Beijsterveldt, C. E., & van Baal, G. C. (2002). Twin and family studies of the human electroencephalogram: a review and a meta-analysis. *Biological Psychology*, *61*, 111–138.

27. Tang, Y., Chorlian, D. B., Rangaswamy, M., O'Connor, S., Taylor, R., Rohrbaugh, J., et al. (2007). Heritability of bipolar EEG spectra in a large sib-pair population. *Behavior Genetics*, *37*, 302–313.

28. Hughes, S. W., & Crunelli, V. (2007). Just a phase they're going through: the complex interaction of intrinsic high-threshold bursting and gap junctions in the generation of thalamic alpha and theta rhythms. *International Journal of Psychophysiology*, *64*, 3–17.

29. Goldman, R. I., Stern, J. M., Engel, J., Jr., & Cohen, M. S. (2002). Simultaneous EEG and fMRI of the alpha rhythm. *Neuroreport*, *13*, 2487–2492.

30. Feige, B., Scheffler, K., Esposito, F., Di Salle, F., Hennig, J., & Seifritz, E. (2005). Cortical and subcortical correlates of electroencephalographic alpha rhythm modulation. *Journal of Neurophysiology*, *93*, 2864–2872.

31. McCarthy, M. M., Moore-Kochlacs, C., Gu, X., Boyden, E. S., Han, X., & Kopell, N. (2011). Striatal origin of the pathologic beta oscillations in Parkinson's disease. *Proceedings of the National Academy of Sciences of the United States of America*, *108*, 11620–11625.

32. Whittington, M. A., Traub, R. D., Kopell, N., Ermentrout, B., & Buhl, E. H. (2002). Inhibition-based rhythms: experimental and mathematical observations on network dynamics. *International Journal of Psychophysiology*, *38*, 315–336.

33. Jensen, O., Goel, P., Kopell, N., Pohja, M., Hari, R., & Ermentrout, B. (2005). On the human sensorimotor-cortex beta rhythm: sources and modeling. *NeuroImage*, *26*, 347–355.

34. Foroud, T., Edenberg, H. J., Goate, A., Rice, J., Flury, L., Koller, D. L., et al. (2000). Alcoholism susceptibility loci: confirmation studies in a replicate sample and further mapping. *Alcoholism, Clinical and Experimental Research*, *24*, 933–945.

35. Porjesz, B., Almasy, L., Edenberg, H. J., Wang, K., Chorlian, D. B., Foroud, T., et al. (2002). Linkage disequilibrium between the beta frequency of the human EEG and a GABAA receptor gene locus. *Proceedings of the National Academy of Sciences of the United States of America*, *99*, 3729–3733.

36. Ghosh, S., Begleiter, H., Porjesz, B., Chorlian, D. B., Edenberg, H. J., Foroud, T., et al. (2003). Linkage mapping of beta 2 EEG waves via non-parametric regression. *American Journal of Medical Genetics. Part B, Neuropsychiatric Genetics*, *118*, 66–71.

37. Edenberg, H. J., Dick, D. M., Xuei, X., Tian, H., Almasy, L., Bauer, L. O., et al. (2004). Variations in GABRA2, encoding the alpha 2 subunit of the GABA(A) receptor, are associated with alcohol dependence and with brain oscillations. *American Journal of Human Genetics*, *74*, 705–714.

38. Enoch, M.-A. (2008). The role of GABAA receptors in the development of alcoholism. *Pharmacology, Biochemistry, and Behavior, 90*, 95–104.

39. Lydall, G. J., Saini, J., Ruparelia, K., Montagnese, S., McQuillin, A., Guerrini, I., et al. (2011). Genetic association study of GABRA2 single nucleotide polymorphisms and electroencephalography in alcohol dependence. *Neuroscience Letters, 500*, 162–166.

40. Begleiter, H., & Porjesz, B. (1999). What is inherited in the predisposition towards alcoholism? A proposed model. *Alcoholism, Clinical and Experimental Research, 23*, 1125–1135.

41. Porjesz, B., & Rangaswamy, M. (2007). Neurophysiological endophenotypes, CNS disinhibition, and risk for alcohol dependence and related disorders. *TheScientificWorldJournal, 7*, 131–141.

42. Roy, A., Hodgkinson, C. A., Deluca, V., Goldman, D., & Enoch, M. A. (2012). Two HPA axis genes, CRHBP and FKBP5, interact with childhood trauma to increase the risk for suicidal behavior. *Journal of Psychiatric Research, 46*, 72–79.

43. Wang, B., You, Z. B., Rice, K. C., & Wise, R. A. (2007). Stress-induced relapse to cocaine seeking: roles for the CRF(2) receptor and CRF-binding protein in the ventral tegmental area of the rat. *Psychopharmacology (Berl), 193*, 283–294.

44. Page, M. E., Berridge, C. W., Foote, S. L., & Valentino, R. J. (1993). Corticotropin-releasing factor in the locus coeruleus mediates EEG activation associated with hypotensive stress. *Neuroscience Letters, 164*, 81–84.

45. Jedema, H. P., & Grace, A. A. (2004). Corticotropin-releasing hormone directly activates noradrenergic neurons of the locus ceruleus recorded in vitro. *Journal of Neuroscience, 24*, 9703–9713.

46. Berridge, C. W., & Foote, S. L. (1994). Locus coeruleus-induced modulation of forebrain electroencephalographic (EEG) state in halothane-anesthetized rat. *Brain Research Bulletin, 35*, 597–605.

47. McCormick, D. A. (1992). Neurotransmitter actions in the thalamus and cerebral cortex and their role in neuromodulation of thalamocortical activity. *Progress in Neurobiology, 39*, 337–388.

48. Ducci, F., Enoch, M. A., Yuan, Q., Shen, P. H., White, K., Hodgkinson, C., et al. (2009). Genetic variation within HTR3B predicts alcoholism with comorbid antisocial personality disorder and resting EEG power. *Alcohol, 43*, 73–84.

49. Ropert, N., & Guy, N. (1991). Serotonin facilitates GABAergic transmission in the CA1 region of rat hippocampus in vitro. *Journal of Physiology, 441*, 121–136.

50. Lee, S., Hjerling-Leffler, J., Zagha, E., Fishell, G., & Rudy, B. (2010). The largest group of superficial neocortical GABAergic interneurons expresses ionotropic serotonin receptors. *Journal of Neuroscience, 30*, 16796–16808.

51. Weinshilboum, R. M., Otterness, D. M., & Szumlanski, C. L. (1999). Methylation pharmacogenetics: catechol-O-methyltransferase, thiopurine methyltransferase, and histamine N methyltransferase. *Annual Review of Pharmacology and Toxicology, 39*, 19–52.

52. Enoch, M.-A., Xu, K., Ferro, E., Harris, C. R., & Goldman, D. (2003). Genetic origins of anxiety in women; a role for a functional COMT polymorphism. *Psychiatric Genetics, 13*, 33–41.

53. Bodenmann, S., Rusterholz, T., Dürr, R., Stoll, C., Bachmann, V., Geissler, E., et al. (2009). The functional Val158Met polymorphism of COMT predicts interindividual differences in brain alpha oscillations in young men. *Journal of Neuroscience, 29*, 10855–10862.

54. Wacker, J., & Gatt, J. M. (2010). Resting posterior versus frontal delta/theta EEG activity is associated with extraversion and the COMT VAL(158)MET polymorphism. *Neuroscience Letters, 478*, 88–92.

55. Hodgkinson, C. A., Enoch, M.-A., Srivastava, V., Cummins-Oman, J. S., Ferrier, C., Iarikova, P., et al. (2010). Genome-wide association identifies candidate genes that influence the human electroencephalogram. *Proceedings of the National Academy of Sciences of the United States of America, 107*, 8695–8700.

56. Derringer, J., Krueger, R. F., Manz, N., Porjesz, B., Almasy, L., Bookman, E., et al. (2011). Nonreplication of an association of SGIP1 SNPs with alcohol dependence and resting theta EEG power. *Psychiatric Genetics, 21*, 265–266.

57. Polich, J., & Herbst, K. L. (2000). P300 as a clinical assay: rationale, evaluation, and findings. *International Journal of Psychophysiology, 38*, 3–19.

58. Almasy, L., Porjesz, B., Blangero, J., Chorlian, D. B., O'Connor, S. J., Kuperman, S., et al. (1999). Heritability of event-related brain potentials in families with a history of alcoholism. *American Journal of Medical Genetics. Part B, Neuropsychiatric Genetics, 88*, 383–390.

59. Hada, M., Porjesz, B., Begleiter, H., & Polich, J. (2000). Auditory P3a assessment of male alcoholics. *Biological Psychiatry, 48*, 276–286.

60. Katsanis, J., Iacono, W. G., Mcgue, M. K., & Carlson, S. R. (1997). P300 event-related potential heritability in monozygotic and dizygotic twins. *Psychophysiology, 34*, 47–58.

61. O'Connor, S. J., Morzorati, S., Christian, J. C., & Li, T. K. (1994). Heritable features of the auditory oddball event-related potential: peaks, latencies, morphology and topography. *Electroencephalography and Clinical Neurophysiology, 92*, 115–125.

62. Begleiter, H., Porjesz, B., Bihari, B., & Kissin, B. (1984). Event-related brain potentials in boys at risk for alcoholism. *Science, 225*, 1493–1495.

63. Porjesz, B., & Begleiter, H. (1998). Genetic basis of event-related potentials and their relationship to alcoholism and alcohol use. *Journal of Clinical Neurophysiology, 15*, 44–57.

64. Hill, S. Y., & Steinhauer, S. R. (1993). Event-related potentials in women at risk for alcoholism. *Alcohol, 10*, 349–354.

65. Hill, S. Y., & Steinhauer, S. R. (1993). Assessment of prepubertal and postpubertal boys and girls at risk for developing alcoholism with P300 from a visual discrimination task. *Journal of Studies on Alcohol, 54*, 350–358.

66. Iacono, W. G., Carlson, S. R., Malone, S. M., & McGue, M. (2002). P3 event-related potential amplitude and the risk for disinhibitory disorders in adolescent boys. *Archives of General Psychiatry*, *59*, 750–757.

67. Reese, C., & Polich, J. (2003). Alcoholism risk and the P300 event-related brain potential: modality, task, and gender effects. *Brain and Cognition*, *53*, 46–57.

68. van der Stelt, O., Gunning, W. B., Snel, J., & Kok, A. (1998). Event-related potentials during visual selective attention in children of alcoholics. *Alcoholism, Clinical and Experimental Research*, *22*, 1877–1889.

69. Hill, S. Y., Muka, D., Steinhauer, S., & Locke, J. (1995). P300 amplitude decrements in children from families of alcoholic female probands. *Biological Psychiatry*, *38*, 622–632.

70. Cohen, H. L., Wang, W., Porjesz, B., & Begleiter, H. (1995). Auditory P300 in young alcoholics: regional response characteristics. *Alcoholism, Clinical and Experimental Research*, *19*, 469–475.

71. Patterson, B. W., Williams, H. L., McLean, G. A., Smith, L. T., & Schaeffer, K. W. (1987). Alcoholism and family history of alcoholism: effects on visual and auditory event-related potentials. *Alcohol*, *4*, 265–274.

72. Prabhu, V. R., Porjesz, B., Chorlian, D. B., Wang, K., Stimus, A., & Begleiter, H. (2001). Visual P3 in female alcoholics. *Alcoholism, Clinical and Experimental Research*, *25*, 531–539.

73. Suresh, S., Porjesz, B., Chorlian, D. B., Choi, K., Jones, K. A., Wang, K., et al. (2003). Auditory P3 in female alcoholics. *Alcoholism, Clinical and Experimental Research*, *27*, 1064–1074.

74. Bauer, L. O., Costa, L., & Hesselbrock, V. M. (2001). Effects of alcoholism, anxiety and depression on P300 in women: a pilot study. *Journal of Studies on Alcohol*, *62*, 571–579.

75. Justus, A. N., Finn, P. R., & Steinmetz, J. E. (2001). P300, disinhibited personality, and early-onset alcohol problems. *Alcoholism, Clinical and Experimental Research*, *25*, 1457–1466.

76. Parsons, O. A., Sinha, R., & Williams, H. L. (1990). Relationships between neuropsychological test performance and event-related potentials in alcoholic and non-alcoholic samples. *Alcoholism, Clinical and Experimental Research*, *14*, 746–755.

77. Hill, S. Y., Steinhauer, S. R., & Locke, J. (1995). Event-related potentials in alcoholic men, their high-risk male relatives, and low-risk male controls. *Alcoholism, Clinical and Experimental Research*, *19*, 567–576.

78. Polich, J., Pollock, V. E., & Bloom, F. E. (1994). Meta-analysis of P300 amplitude from males at risk for alcoholism. *Psychological Bulletin*, *115*, 55–73.

79. Whipple, S. C., Berman, S. M., & Noble, E. P. (1991). Event-related potentials in alcoholic fathers and their sons. *Alcohol*, *8*, 321–327.

80. Ehlers, C. L., Garcia-Andrade, C., Wall, T. L., Sobel, D. F., & Phillips, E. (1998). Determinants of P3 amplitude and response to alcohol in Native American Mission Indians. *Neuropsychopharmacology*, *18*, 282–292.

81. Ehlers, C. L., Wall, T. L., Garcia-Andrade, C., & Phillips, E. (2001). Auditory P3 findings in Mission Indian youth. *Journal of Studies on Alcohol, 62*, 562–570.

82. Hill, S. Y., Locke, J., & Steinhauer, S. R. (1999). Absence of visual and auditory P300 reduction in nondepressed male and female alcoholics. *Biological Psychiatry, 46*, 982–989.

83. Holguin, S. R., Corral, M., & Cadaveira, F. (1998). Visual and auditory event-related potentials in young children of alcoholics from high- and low-density families. *Alcoholism, Clinical and Experimental Research, 22*, 87–96.

84. Grillon, C., & Ameli, R. (1994). P300 assessment of anxiety effects on processing novel stimuli. *International Journal of Psychophysiology, 17*, 205–217.

85. Boudarene, M., & Timsit-Berthier, M. (1997). Stress, anxiety and event related potentials. *L'Encéphale, 23*, 237–250.

86. Chattopadhyay, P., Cooke, E., Toone, B., & Lader, M. (1980). Habituation of physiological responses in anxiety. *Biological Psychiatry, 15*, 711–721.

87. Gohle, D., Juckel, G., Mavrogiorgou, P., Pogarell, O., Mulert, C., Rujescu, D., et al. (2008). Electrophysiological evidence for cortical abnormalities in obsessive–compulsive disorder—a replication study using auditory event-related P300 subcomponents. *Journal of Psychiatric Research, 42*, 297–303.

88. Enoch, M.-A., White, K. V., Waheed, J., & Goldman, D. (2008). Neurophysiological and genetic distinctions between pure and comorbid anxiety disorders. *Depression and Anxiety, 25*, 383–392.

89. Enoch, M.-A., White, K. V., Harris, C. R., Rohrbaugh, J. W., & Goldman, D. (2001). Alcohol use disorders and anxiety disorders: relation to the P300 event related potential. *Alcoholism, Clinical and Experimental Research, 25*, 1293–1300.

90. Williams, J. T., Begleiter, H., Porjesz, B., Edenberg, H. J., Foroud, T., Reich, T., et al. (1999). Joint multipoint linkage analysis of multivariate qualitative and quantitative traits: II. alcoholism and event-related potentials. *American Journal of Human Genetics, 65*, 1148–1160.

91. Almasy, L., Porjesz, B., Blangero, J., Goate, A., Edenberg, H. J., Chorlian, D. B., et al. (2001). Genetics of event-related brain potentials in response to a semantic priming paradigm in families with a history of alcoholism. *American Journal of Human Genetics, 68*, 128–135.

92. Porjesz, B., Begleiter, H., Wang, K., Almasy, L., Chorlian, D. B., Stimus, A. T., et al. (2002). Linkage and linkage disequilibrium mapping of ERP and EEG phenotypes. *Biological Psychology, 61*, 229–248.

93. Chen, A. C., Manz, N., Tang, Y., Rangaswamy, M., Almasy, L., Kuperman, S., et al. (2010). Single-nucleotide polymorphisms in corticotropin releasing hormone receptor 1 gene (CRHR1) are associated with quantitative trait of event-related potential and alcohol dependence. *Alcoholism, Clinical and Experimental Research, 34*, 988–996.

94. Johnson, J. P., Muhleman, D., MacMurray, J., Gade, R., Verde, R., Ask, M., et al. (1997). Association between the cannabinoid receptor gene (CNR1) and the P300 event-related potential. *Molecular Psychiatry*, *2*, 169–171.

95. Stadelmann, A. M., Juckel, G., Arning, L., Gallinat, J., Epplen, J. T., & Roser, P. (2011). Association between a cannabinoid receptor gene (CNR1) polymorphism and cannabinoid-induced alterations of the auditory event-related P300 potential. *Neuroscience Letters*, *496*, 60–64.

96. Intriligator, J., & Polich, J. (1994). On the relationship between background EEG and the P300 event-related potential. *Biological Psychology*, *37*, 207–218.

97. Başar, E., Yordanova, J., Kolev, V., & Başar-Eroglu, C. (1997). Is the alpha rhythm a control parameter for brain responses? *Biological Cybernetics*, *76*, 471–480.

98. Makeig, S., Westerfield, M., Jung, T. P., Enghoff, S., Townsend, J., Courchesne, E., et al. (2002). Dynamic brain sources of visual evoked responses. *Science*, *295*, 690–694.

99. Klimesch, W., Doppelmayr, M., Schwaiger, J., Winkler, T., & Gruber, W. (2000). Theta oscillations and the ERP old/new effect: independent phenomena? *Clinical Neurophysiology*, *111*, 781–793.

100. Karakaş, S., Erzengin, O. U., & Başar, E. (2000). The genesis of human event-related responses explained through the theory of oscillatory neural assemblies. *Neuroscience Letters*, *285*, 45–48.

101. Yordanova, J., Kolev, V., Rosso, O. A., Schürmann, M., Sakowitz, O. W., Ozgören, M., et al. (2002). Wavelet entropy analysis of event-related potentials indicates modality-independent theta dominance. *Journal of Neuroscience Methods*, *117*, 99–109.

102. Andrew, C., & Fein, G. (2010). Event-related oscillations versus event-related potentials in a P300 task as biomarkers for alcoholism. *Alcoholism, Clinical and Experimental Research*, *34*, 669–680.

103. Andrew, C., & Fein, G. (2010). Induced theta oscillations as biomarkers for alcoholism. *Clinical Neurophysiology*, *121*, 350–358.

104. Jones, K. A., Porjesz, B., Chorlian, D., Rangaswamy, M., Kamarajan, C., Padmanabhapillai, A., et al. (2006). S-transform time-frequency analysis of P300 reveals deficits in individuals diagnosed with alcoholism. *Clinical Neurophysiology*, *117*, 2128–2143.

105. Kamarajan, C., Rangaswamy, M., Manz, N., Chorlian, D. B., Pandey, A. K., Roopesh, B. N., et al. (2012). Topography, power, and current source density of theta oscillations during reward processing as markers for alcohol dependence. *Human Brain Mapping*, *33*, 1019–1039..

106. Rangaswamy, M., Jones, K. A., Porjesz, B., Chorlian, D. B., Padmanabhapillai, A., Kamarajan, C., et al. (2007). Delta and theta oscillations as risk markers in adolescent offspring of alcoholics. *International Journal of Psychophysiology*, *63*, 3–15.

107. Polich, J., & Criado, J. R. (2006). Neuropsychology and neuropharmacology of P3a and P3b. *International Journal of Psychophysiology*, *60*, 172–185.

108. Jones, K. A., Porjesz, B., Almasy, L., Bierut, L., Goate, A., Wang, J. C., et al. (2004). Linkage and linkage disequilibrium of evoked EEG oscillations with CHRM2 receptor gene polymorphisms: implications for human brain dynamics and cognition. *International Journal of Psychophysiology, 53*, 75–90.

109. Jones, K. A., Porjesz, B., Almasy, L., Bierut, L., Dick, D., Goate, A., et al. (2006). A cholinergic receptor gene (CHRM2) affects event-related oscillations. *Behavior Genetics, 36*, 627–639.

110. Wang, J. C., Hinrichs, A. L., Stock, H., Budde, J., Allen, R., Bertelsen, S., et al. (2004). Evidence of common and specific genetic effects: association of the muscarinic acetylcholine receptor M2 (CHRM2) gene with alcohol dependence and major depressive syndrome. *Human Molecular Genetics, 13*, 1903–1911.

111. Chen, A. C., Tang, Y., Rangaswamy, M., Wang, J. C., Almasy, L., Foroud, T., et al. (2009). Association of single nucleotide polymorphisms in a glutamate receptor gene (GRM8) with theta power of event-related oscillations and alcohol dependence. *American Journal of Medical Genetics. Part B, Neuropsychiatric Genetics, 150B*, 359–368.

112. Zlojutro, M., Manz, N., Rangaswamy, M., Xuei, X., Flury-Wetherill, L., Koller, D., et al. (2011). Genome-wide association study of theta band event-related oscillations identifies serotonin receptor gene HTR7 influencing risk of alcohol dependence. *American Journal of Medical Genetics. Part B, Neuropsychiatric Genetics, 156B*, 44–58.

113. Chorlian, D. B., Rangaswamy, M., & Porjesz, B. (2009). EEG coherence: topography and frequency structure. *Experimental Brain Research, 198*, 59–83.

114. van Beijsterveldt, C. E., Molenaar, P. C., de Geus, E. J., & Boomsma, D. I. (1998). Genetic and environmental influences on EEG coherence. *Behavior Genetics, 28*, 443–453.

115. Chorlian, D. B., Tang, Y., Rangaswamy, M., O'Connor, S., Rohrbaugh, J., Taylor, R., et al. (2007). Heritability of EEG coherence in a large sib-pair population. *Biological Psychology, 75*, 260–266.

116. Sarnthein, J., Morel, A., von Stein, A., & Jeanmonod, D. (2005). Thalamocortical theta coherence in neurological patients at rest and during a working memory task. *International Journal of Psychophysiology, 57*, 87–96.

117. Winterer, G., Enoch, M.-A., White, K. V., Saylan, M., Coppola, R., & Goldman, D. (2003). EEG phenotype in alcoholism: increased coherence in the depressive subtype. *Acta Psychiatrica Scandinavica, 108*, 51–60.

118. Winterer, G., Smolka, M., Samochowiec, J., Ziller, M., Mahlberg, R., Gallinat, J., et al. (2003). Association of EEG coherence and an exonic GABA(B)R1 gene polymorphism. *American Journal of Medical Genetics. Part B, Neuropsychiatric Genetics, 117B*, 51–56.

119. Chanraud, S., Pitel, A. L., Pfefferbaum, A., & Sullivan, E. V. (2011). Disruption of functional connectivity of the default-mode network in alcoholism. *Cerebral Cortex, 21*, 2272–2281.

120. Watts, D. J., & Strogatz, S. H. (1998). Collective dynamics of "small-world" networks. *Nature, 393*, 440–442.

121. Micheloyannis, S., Pachou, E., Stam, C. J., Vourkas, M., Erimaki, S., & Tsirka, V. (2006). Using graph theoretical analysis of multi channel EEG to evaluate the neural efficiency hypothesis. *Neuroscience Letters, 402*, 273–277.

122. Smit, D. J., Stam, C. J., Posthuma, D., Boomsma, D. I., & de Geus, E. J. (2008). Heritability of "small-world" networks in the brain: a graph theoretical analysis of resting-state EEG functional connectivity. *Human Brain Mapping, 29*, 1368–1378.

123. Posthuma, D., de Geus, E. J., Mulder, E. J., Smit, D. J., Boomsma, D. I., & Stam, C. J. (2005). Genetic components of functional connectivity in the brain: the heritability of synchronization likelihood. *Human Brain Mapping, 26*, 191–198.

124. Smit, D. J., Boersma, M., van Beijsterveldt, C. E., Posthuma, D., Boomsma, D. I., Stam, C. J., et al. (2010). Endophenotypes in a dynamically connected brain. *Behavior Genetics, 40*, 167–177.

125. Lei, X., Ostwald, D., Hu, J., Qiu, C., Porcaro, C., Bagshaw, A. P., et al. (2011). Multimodal functional network connectivity: an EEG–fMRI fusion in network space. *PLOS ONE, 6*(9), e24642.

3 Differential Metabolism of Alcohol as an Intermediate Phenotype of Risk for Alcohol Use Disorders: Alcohol and Aldehyde Dehydrogenase Variants

Tamara L. Wall, Susan E. Luczak, Daria Orlowska, and Danielle Pandika

Twin studies, comprised predominantly of participants of European Caucasian ancestry, have established major roles for both genetics and environment in the etiology of alcohol dependence [1, 2]. The familial pattern indicates it is a non-Mendelian disorder involving multiple genes, each accounting for only a small proportion of the variation in risk, as well as complex gene–gene and gene–environment interactions [3–5]. With the completion of the human genome sequence, there has been significant progress demonstrating associations between candidate genes and alcohol dependence, with a growing list of replicable results [6–10].

The *ADH* (alcohol dehydrogenase) and *ALDH* (aldehyde dehydrogenase) genes are the key genes of the primary alcohol metabolism pathway (see [11] for review). Alcohol is first oxidized into acetaldehyde by the ADH enzymes, and acetaldehyde is subsequently oxidized into acetate by the ALDH enzymes. There are multiple ADH and ALDH enzymes that are encoded by different genes. Based on gene variants, the enzymes encoded differ in the rate at which they metabolize alcohol or acetaldehyde or in the levels at which they are produced. Associations between ADH and ALDH gene variants are among the strongest and most consistent gene associations that have been found for alcohol dependence.

3.1 *ALDH2*

To date, the gene with the most widely reproduced association with alcohol dependence is *ALDH2* (rs671) located on chromosome 12 (12q24.2). *ALDH2* encodes mitochondrial ALDH2, the primary liver isoenzyme involved in the metabolism of acetaldehyde to acetate [12]. The common form of the allele *ALDH2*1* (*ALDH2*Glu487*) is functional, while the variant *ALDH2*2* allele (*ALDH2*Lys487*) is inactive. The *ALDH2*2* allele is almost exclusive to northeastern Asian populations (Chinese, Koreans, Japanese), with about 30% to 50% heterozygous (*ALDH2*1/*2* genotype) and a small percentage (about 5%) homozygous for this allele (*ALDH2*2/*2* genotype) [13]. A meta-analysis of the *ALDH2* gene with alcohol dependence indicates having one

ALDH2*2 allele was associated with a four- to five-fold reduction in alcohol dependence (odds ratio, OR = 0.22) and having two *ALDH2*2* alleles was associated with an eight- to nine-fold reduction (OR = 0.12) [14].

Phenotypic correlates of *ALDH2*2* have led to a proposed mechanism for how this genetic variant results in lower rates of alcohol use disorders (AUDs, i.e., alcohol abuse and dependence), and data supporting this mechanism are described next. Other *ADH* and *ALDH* genes that also may relate to AUDs via differences in alcohol metabolism and other endophenotypes are then considered. These findings highlight the value of trying to elucidate the mechanisms by which genes ultimately give rise to differences in AUDs through the examination of mediating variables.

In complex genetically influenced disorders, endophenotypes have been conceptualized as measurable components more proximal to the genes than the disorders themselves, which reflect the etiologic process by which genetic expression gives rise to differential vulnerability for the disorder [15]. Three major theoretical models have been proposed for how alcohol involvement develops—pharmacological vulnerability, deviance proneness, and affect regulation models [16]. Although neither comprehensive nor mutually exclusive, these models provide useful frameworks for studying the etiology of AUDs. The pharmacological vulnerability model posits that there are individual differences in sensitivity to the reinforcing and punishing effects of alcohol that lead to differences in AUD risk [16]. This model arose out of findings from twin studies that an individual's level of response to alcohol is genetically influenced (see [17]). The *ALDH2*2* allele, with its well-established effect on alcohol metabolism, provides a natural quasi-experimental design (i.e., Mendelian randomization; [18]) that can be used to test pharmacological vulnerability toward alcohol involvement and help to establish endophenotypes (mediators) associated with AUD risk.

A proposed mechanism for the effect of the *ALDH2*2* allele on AUDs is displayed in figure 3.1. It is hypothesized that the allele encodes a protein subunit that has deficient enzyme activity and results in increased acetaldehyde during alcohol metabolism [12, 19]. Acetaldehyde is then proposed to cause enhanced reactions to alcohol and to decrease positive alcohol expectancies and increase negative alcohol expectancies, which in turn reduces the likelihood of frequent and heavy drinking, alcohol-related consequences, and AUDs [20]. Specific *ALDH2* endophenotypes are shown in figure 3.2.

Figure 3.1
Hypothesized mechanism of influence for ALDH2*2 on the development of alcohol use disorders (AUDs).

↓ ALDH2 enzyme activity
- ↑ Acetaldehyde levels
- ↑ Response to alcohol
  - ➤ Flushing and other symptoms (e.g., headaches, nausea, sleepiness)
  - ➤ Ratings of intoxication (subjective)
  - ➤ Measures of intoxication (e.g., cardiovascular, psychomotor, neurophysiological)
  - ➤ Ratings of post-intoxication (hangover)
- ↓ Positive alcohol expectancies and ↑ negative expectancies
- ↓ Alcohol consumption (use and heavy use)
  - ➤ Quantity (drinks/occasion)
  - ➤ Frequency (drinking days)
  - ➤ Binge drinking
  - ➤ Peak BAC
  - ➤ Maximum drinks ever consumed in a 24-hour period
- ↓ Negative consequences
  - ➤ Subthreshold alcohol-related problems
  - ➤ Hangover
  - ➤ Blackouts
- ↓ AUDs
- ↑ Alcohol-related disease
  - ➤ Esophageal cancer, other cancers
  - ➤ Liver disease, pancreatitis
  - ➤ Alzheimer's disease

Figure 3.2
ALDH2 endophenotypes: From allele to complex behavior. BAC, blood alcohol concentration.

Based on their kinetic properties, *ALDH2*2* should lead to decreased ALDH2 enzyme activity and slower removal of acetaldehyde compared with *ALDH2*1* [21]. In vitro and in vivo studies find *ALDH2*1/*2* genotype results in 12% to 20% of the enzyme activity in the liver produced by *ALDH2*1/*1* genotype, and *ALDH2*2/*2* genotype results in no enzyme activity. Alcohol challenge studies have demonstrated one *ALDH2*2* allele is associated with elevated acetaldehyde and possession of two *ALDH2*2* alleles is associated with even higher levels, although only men have been included in these studies [22–29].

The next step of the proposed mechanism, that *ALDH2*2* leads to heightened responses to alcohol, is consistent with data from both self-report and alcohol challenge studies. *ALDH2*2* has been related to alcohol sensitivity as measured by retrospective self-reports of alcohol-induced flushing and other symptoms, such as nausea, headaches, and palpitations [30–36], and by a retrospective self-rating measure of level of response to alcohol [31, 37, 38]. These data are consistent with actual alcohol challenge studies, which have related *ALDH2*2* to heightened responses of flushing, pulse rate, hormonal changes, psychomotor performance, and neurophysiological reactivity, despite equivalent alcohol levels in those with *ALDH2*1/*1* genotype [25, 26, 36, 39–51]. Subjectively, individuals with one *ALDH2*2* allele report more intense positive, negative, and neutral reactions following an alcohol challenge compared to individuals matched on drinking history without this allele. The presence of two *ALDH2*2* alleles results in even more intense responses and appears to cause aversive reactions such as hypotension, tachycardia, nausea, and vomiting even at low doses. Most studies only evaluated men, but similar levels of response to alcohol associated with *ALDH2*2* have been found in both men and women when alcohol was dosed to reach equivalent levels across sex [42].

The third step of the pathway proposes *ALDH2*2* lowers positive alcohol expectancies and increases negative alcohol expectancies. Positive expectancies are consistently predictive of drinking in both cross-sectional and prospective studies, but associations of negative expectancies with drinking are less well established [52, 53]. Theories have conceptualized expectancies as mediators between biopsychosocial vulnerability factors and alcohol use, such that any individual difference factor that affects one's exposure and experience with alcohol can alter expectancies of its reinforcing and punishing effects and, in turn, alter alcohol use (i.e., reciprocal relationships; [54–57]). Four studies have examined *ALDH2*2* in relation to expectancies and tested their mediation with alcohol use. Three of the studies evaluated college student samples [58–60], and one evaluated a treatment-seeking sample [61]. In both studies by McCarthy and colleagues [58, 59], *ALDH2*2* was related to reduced positive expectancies and was unrelated to negative expectancies. In the first study, expectancies mediated the relationship of *ALDH2*2* with quantity of alcohol use in women, supporting the hypothesis that *ALDH2*2* may exert its influence on alcohol use by reducing the positive reinforcement anticipated from drinking [58]. This study, however, tested mediation using cross-sectional data, making it impossible to confirm that *ALDH2*2* is associated with the developmental trajectories of expectancies and alcohol use. The second study by McCarthy and colleagues [59] tested if it is the more intense response to alcohol experienced by those with *ALDH2*2* that affects the acquisition of expectancies. Using alcohol challenge data, it was found that the *ALDH2*–expectancy relationship was fully explained by level of response to alcohol in men, but not women. These findings suggest that expectancies associated with *ALDH2*2* appear to be caused by differences in response to alcohol but may vary by sex. Hendershot and colleagues [60] developed a measure of physiological expectancies to assess an expectancy phenotype specific to the mechanism by which *ALDH2* mediates differences in drinking behavior. In this study, individuals with *ALDH2*2* alleles reported greater negative expectancies and greater expectancies for the physiological effects of alcohol compared to those with the *ALDH2*1/*1* genotype. In addition, both *ALDH2* genotype and expectancy variables explained unique variance in drinking outcomes [60]. It is important to note, however, that the relationship between both positive and negative expectancies and alcohol involvement may differ in individuals who have developed alcohol dependence. In a cross-sectional study of treatment-seeking Chinese, Hahn and colleagues [61] found that alcohol-dependent individuals with *ALDH2*2* had lower negative expectancies and higher positive expectancies than those without this allele even though the groups did not differ on recent alcohol use.

The fourth step in the mechanistic pathway proposes that *ALDH2*2* leads to lower rates of drinking and heavy drinking. Studies have shown that one *ALDH2*2* allele relates to lower quantity and frequency of alcohol use, lower rates of binge drinking, lower estimated peak blood alcohol concentration, and a lower maximum number of drinks ever consumed in a 24-hour period; two *ALDH2*2* alleles relates to even lower levels of use on these variables [32, 37, 62–70]. These associations between *ALDH2*2* and drinking behavior, however, have been studied more consistently and are more pronounced in men than women. As a means of investigating the influence of *ALDH2* on the development of alcohol behavior, retrospective information about the onset and alcohol use was collected in a cross-sectional study of 21- to 26-year-old college students [70]. *ALDH2*2* was not related to the age at which alcohol use was initiated. This finding was replicated in a study of Korean American adolescents [63]. These results are consistent with studies of predominantly Caucasian twins that have found that alcohol initiation is primarily influenced by environmental and not genetic factors [71–73].

The fifth step in the pathway proposes that *ALDH2*2* leads to fewer alcohol-related adverse consequences. Individuals with *ALDH2*2* have been found to have lower scores on measures of hazardous alcohol use and of alcohol problems that are subthreshold for an AUD, such as the Rutgers Alcohol Problem Index, Alcohol Use Disorders Identification Test, and Young Adult Alcohol Problems Screening Test [37, 74]. Two specific consequences of heavy drinking, hangovers and blackouts, also have been inversely associated with ALDH2*2. In a study of anticipated hangovers, individuals with ALDH2*2 expected to have more severe hangover symptoms than those without ALDH2*2 if the same amount of alcohol was consumed [75]. Another study found ALDH2*2 was not associated with hangover frequency, but the amount of alcohol leading to hangover was significantly less for those with ALDH2*2 [76]. These studies highlight the importance of distinguishing between frequency and severity of consequences and taking into account differences in consumption (i.e., self-dosing). Another study found an inverse relationship between ALDH2*2 and alcohol-induced blackouts, even after controlling for lifetime maximum drinks in a 24-hour period [14] This suggests the possibility that individuals with ALDH2*2 not only drink less than those without this allele but also may have an altered consumption pattern that results in a less rapid intake and slower rising phase of the blood alcohol curve. Such an intake pattern should lead to decreased consequences from alcohol.

The last step of the mechanism proposes ALDH2*2 results in lowered rates of AUDs. A meta-analysis of the associations of ALDH2 with alcohol dependence used data from 1,980 cases and 2,550 controls (primarily men) and intentionally excluded samples with alcohol-related diseases, for example, liver disease, cancer, and pancreatitis [77]. One ALDH2*2 allele reduced risk for alcohol dependence to about one fourth (OR = 0.22), and two ALDH2*2 alleles reduced risk to about one eighth (OR = 0.12). The difference between one and two ALDH2*2 alleles approached, but did not reach, statistical significance, although this likely resulted from low power; only 5% of controls (n = 129) and 0.1% of cases (n = 3) were ALDH2*2/*2. Reports of the three cases with ALDH2*2/*2 suggest they each developed alcohol dependence through a low-quantity, high-frequency pattern of alcohol intake [78, 79]. These findings indicate that the protection against alcohol dependence afforded by the ALDH2*2/*2 genotype is powerful, but not complete as previously thought.

Consistent with the idea that individuals with ALDH2*2 might develop consequences at lower levels of drinking, there is evidence that alcoholics with ALDH2*2 develop alcohol dependence at lower alcohol intake [80] and their clinical course of alcohol-related events (e.g., habitual drinking, withdrawal) was delayed between one and five years [81]. Another study, however, found that the age of onset of alcoholism was lower for women with ALDH2*2 than for those without this allele, but the age of onset did not vary by ALDH2 genotype for men [82].

It is important to point out that the majority of studies evaluating associations of ALDH2*2 with alcohol consumption and AUDs have used young adult or treatment samples, and that two studies of adolescent samples did not find consistent associations. A study of Korean Americans with an average age of 18 years found *ALDH2*2* status related to quantity and frequency of alcohol use and an alcohol abuse diagnosis, but not related to alcohol dependence diagnosis or symptoms counts due to the low frequency of these behaviors [63]. In addition, a prospective study of 18- to 19-year-old Chinese and Korean American college students found *ALDH2*2* was not associated with recent alcohol use or past two-week binge drinking during the first year of college [83] but emerged as a significant protective factor for alcohol progression by the second year of college [84]. A similar finding was observed in a study of Korean young adults. *ALDH2* was not associated with drinking behaviors in university freshmen but was associated with drinking outcomes at a six-year follow-up [41]. Taken together, both retrospective and prospective data suggest that the protective effect of *ALDH2* only becomes relevant after initiation, with the transition to regular and heavy alcohol consumption inhibited in those with *ALDH2*2* [37, 70, 84]. These findings are consistent with twin studies that have found once alcohol use is initiated, differences in drinking patterns are strongly influenced by genetic factors [85], and that the influence of genetics increases with alcohol experience [86].

Finally, it also is important to acknowledge that in the presence of alcohol dependence or at lower levels of alcohol use, individuals with *ALDH2*2* alleles appear to be significantly more vulnerable to alcohol-related pathologies, particularly head and neck cancers but also liver disease, pancreatitis, and Alzheimer's disease, consistent with a role of acetaldehyde in the pathogenesis of organ damage [87–94]. Thus, the influence of *ALDH2*2* may change over the course of drinking such that *ALDH2*2* may be protective at one stage of alcohol use (e.g., progression to heavy drinking) but become a risk factor at another stage (e.g., progression to alcohol-related medical problems). Prospective studies with multiple phenotypes and endophenotypes are needed to determine how the effect of this gene may change over the life span.

There are also data to suggest that several *ADH* and other *ALDH* polymorphisms are associated with AUDs and other alcohol-related variables, but the effects of each appear to be small and may differ across ethnic subgroups and across phenotypes and endophenotypes. The human ADH gene cluster on chromosome 4 (4q21–25) is comprised of seven genes [11]. These genes, in order, are *ADH7*, *ADH1C*, *ADH1B*, *ADH1A*, *ADH6*, *ADH4*, and *ADH5*. The three class I genes (*ADH1A*, *ADH1B*, and *ADH1C*) are in a tighter cluster and are believed to code for the primary liver isoenzymes involved in the conversion of alcohol to acetaldehyde. They account for about 70% of the alcohol oxidizing capacity of the liver whereas the class II isoenzyme (encoded by *ADH4*) accounts for about 30% [95, 96]. The functionality and relationship to alcohol endophenotypes and phenotypes is less well-known for *ADH5*, *ADH6*, and *ADH7* [11]. It also should be noted that the nomenclature for these genes has changed several times and some studies use alternate nomenclature (e.g., [97]).

3.2 ADH1B

After *ALDH2*, the gene with the most widely reproduced association with alcohol dependence is *ADH1B* (rs1229984). The variant *ADH1B*2* allele (*ADH1B*47His*) has been associated with lower rates of alcohol dependence compared with *ADH1B*1* (*ADH1B*47Arg*). The prevalence of the *ADH1B*2* allele varies widely across populations [98]. General population samples indicate 80% or more of northeast Asians, 50% of Russians and Jews, and 10% or less of Caucasians of European ancestry possess an *ADH1B*2* allele [98, 99].

*ADH1B*2* consistently has been associated with lower rates of alcohol dependence. This relationship has been found after controlling for *ALDH2*2* in Asian subgroups (see [14]) and has been found in European Caucasian subgroups [20, 100–104]. Three meta-analyses have examined the relationship of *ADH1B*2* with alcohol dependence [77, 105, 106]. Whitfield [106] found Europeans with *ADH1B*1/*2* were about half as likely (OR = 0.47) to be alcohol dependent as *ADH1B*1/*1* individuals. In a meta-analysis of Asians, the effect of *ALDH2*2* and *ADH1B*2* alleles in combination on the risk for alcohol dependence was examined [77]. In *ALDH2*1/*1* individuals, the presence of one *ADH1B*2* allele was associated with about one fourth (OR = 0.26) and the presence of two *ADH1B*2* alleles was associated with about one fifth (OR = 0.20) the risk of alcohol dependence compared with individuals with no *ADH1B*2* alleles. In *ALDH2*1/*2* individuals, the presence of one *ADH1B*2* allele was associated with about one sixth (OR = 0.17) and the presence of two *ADH1B*2* alleles was associated with about one eleventh (OR = 0.09) the risk of alcohol dependence compared with individuals with no *ADH1B*2* alleles. These results suggest both *ALDH2* and *ADH1B* each contribute unique protective effects on alcohol dependence, and the level of protection may be even stronger in conjunction than alone (gene × gene interaction).

The mechanism by which *ADH1B* polymorphisms are thought to reduce AUD risk is via elevations in acetaldehyde [19]. Based on their kinetic properties in vitro, *ADH1B*2* should lead to faster production of acetaldehyde than *ADH1B*1* [21]. Similar to the effects of *ALDH2*2*, increased acetaldehyde is hypothesized to lead to heightened alcohol responses and lower alcohol use and problems, but the data supporting this mechanism for *ADH1B*2* has been less consistent than for *ALDH2*2* (see [20]). In vitro studies suggest rates of elimination produced by *ADH1B*2* should be approximately three times that of the *ADH1B*1*, but in vivo studies have found only twice the rate [107–109]. Despite this higher elimination rate, *ADH1B*2* has not been associated with acetaldehyde concentrations after controlling for *ALDH2*2* in Asians [24, 46, 110, 111]. Eriksson [19], however, has noted the difficulty of detecting changes in acetaldehyde at the low concentrations predicted by *ADH1B*2*.

Some studies have related *ADH1B*2* to increased sensitivity to alcohol such as self-reports of flushing and associated symptoms [20, 30, 32, 34, 101, 112, 113], but others have not [114, 115]. A study by Luczak and colleagues [31] found an interaction between *ALDH2* and *ADH1B* on self-reported sensitivity to alcohol in Asians, such that *ADH1B*2* only was associated with heightened sensitivity in those with *ALDH2*2* alleles. Alcohol challenge studies also have demonstrated mixed results. A study of Asians found an association between an increased level of response to alcohol and *ADH1B*2*, but only among individuals who were heterozygous for *ALDH2*2*, again consistent with a gene × gene interaction [39]. An alcohol challenge study of non-Asians (Caucasians and African Americans) found *ADH1B*2* was associated with a more intense response to alcohol [116], but other studies of Asians [46] and Caucasians [17] have not found significant effects of *ADH1B*2* on measures of intoxication. Finally, there have been inconsistent results relating *ADH1B*2* to alcohol intake. Some studies have related *ADH1B*2* to lower alcohol consumption particularly in samples of heavier drinkers, such as men, those with *ALDH2*1/*1* genotype, and treatment samples [20, 32, 101, 103, 104, 113–115, 117–122], but other studies have not [35, 62, 66].

3.3   Other ADH and ALDH Genes

Genome-wide studies have found evidence of linkage in the region of chromosome 4q that includes the ADH gene cluster using a variety of phenotypes (e.g., AUD diagnoses and symptoms) and endophenotypes (e.g., maximum drinks, craving) in a variety of ethnically diverse (although non-Asian) samples. These include Caucasian and African American participants from the Collaborative Study on the Genetics of Alcoholism [123–126], two Native American tribes [127–129], and an Irish sample [130]. Genome scans of Caucasian and African American participants also suggest protection against illicit substance use near the same area of chromosome 4q [131].

The *ADH1C*1* allele (*ADH1C*349Ile*) also has been well studied with respect to alcohol metabolism and alcohol dependence. Based on its kinetic properties in vitro, *ADH1C*1* should lead to faster production of acetaldehyde than *ADH1C*2* [21]. Studies of Asians [107, 132, 133], Mexicans [134], and Native Americans [135] found *ADH1C*1* was associated with lower rates of alcohol dependence, but studies of Europeans did not [136, 137]. More recent studies using multiple regression analyses showed *ADH1C*1* and *ADH1B*2* are in linkage disequilibrium, suggesting associations of *ADH1C*1* with alcohol dependence are likely due to correlation with *ADH1B*2* [20, 100, 138–140]. Choi et al. [141] genotyped 36 single nucleotide polymorphisms (SNPs) in the *ADH1B* and *ADH1C* genes in a Korean sample and found 14 SNPS were associated with alcohol dependence. The pattern of association, however, suggested *ADH1B*2* was the only locus responsible for susceptibility to alcohol dependence and that the other associations were due to linkage disequilibrium with *ADH1B*2*.

Birley et al. [142] found that the region of chromosome 4 containing the ADH genes was linked to the level of blood and breath alcohol at an early time point after drinking in Australian twins. They also determined that this quantitative trait loci accounted for 64% of the variation in in vivo alcohol metabolism but was not due to either *ADH1B*2* or *ADH1C*1* as would be predicted by in vitro enzyme kinetics [143]. These findings suggest polymorphisms other than *ADH1B*2* and *ADH1C*1* account for this variability.

Edenberg et al. [7] genotyped for 110 SNPs across the seven ADH genes and analyzed associations with alcohol behavior in a sample of Caucasian and African American participants from COGA. Twelve SNPs in and around *ADH4* were associated with alcohol dependence phenotypes. Analyses have revealed modest but not consistent associations of various phenotypes with SNPs in *ADH1A* and *ADH1B*, suggesting these genes also may contribute to susceptibility to alcohol dependence. *ADH1B*3* is found predominantly in individuals of African ancestry [7, 143] and also has been detected in low prevalence in Mission Indians, a Native American population with genetic admixture [28, 144]. In African Americans, *ADH1B*3* was associated with protection from alcohol dependence consistent with research in Native Americans [144]. The findings of a significant association of *ADH4* with alcohol dependence are consistent with other studies in Caucasian and African American participants [140, 145, 146]. Luo et al. [140, 145] found significant associations for seven *ADH4* SNPs with alcohol dependence in a sample of unrelated Caucasians and African Americans. In a subsequent study of this sample, Luo et al. [147] genotyped for four SNPs across the *ALDH2* gene and 16 SNPs across the remaining *ADH* genes and found *ADH5* genotypes and diplotypes of *ADH1A*, *ADH1B*, *ADH7*, and *ALDH2* were associated with alcohol dependence. Osier et al. [148] reported that an *ADH7* SNP was associated with alcohol dependence in Taiwanese and in a subsequent study determined that their previously reported association with alcohol dependence was not due to linkage disequilibrium with *ADH1B* [149]. Other studies have found significant associations between several ADH SNPs and alcohol dependence including *ADH5* and *ADH6* [101, 150–152].

The *ALDH1A1* gene on chromosome 9 (9q21.12) also codes for an important enzyme in alcohol metabolism [153]. *ALDH1A1* enzyme activity has been associated with alcohol phenotypes and endophenotypes including flushing and alcohol sensitivity in both Asians and Caucasians [19, 154–156]. Spence et al. [157] identified two *ALDH1A1* promotor polymorphisms, *ALDH1A1*2* and *ALDH1A1*3*, with varying allele frequencies across Asian American, African American, and Jewish and non-Jewish Caucasian samples. A study of Native Americans found *ALDH1A1*2* was associated with lower rates of alcohol dependence, a lower maximum number of drinks ever consumed, lower alcohol expectancies, and lower rates of tobacco use [158]. Other studies have found associations between other *ALDH1A1* SNPs [159] as well as other *ALDH2* SNPs [160] with alcohol dependence.

3.4   Summary

Taken together, these findings confirm the importance of the *ADH* and *ALDH* gene clusters in alcohol dependence. Several *ADH* and *ALDH* polymorphisms are associated with alcohol dependence and other alcohol-related behaviors, but other than *ALDH2*, the effects of each may be small and may not be replicable across ethnic subgroups and across phenotypes and endophenotypes. Mustavich et al. [161] have described a pharmacokinetic model of how *ADH1B*, *ADH1C*, *ADH7*, *ALDH2*, and *TAS2R38* affect alcohol consumption as well as alcohol and acetaldehyde levels over time in various tissues of individuals with particular genotypes to predict their susceptibility to alcohol dependence. Such a model exemplifies the complexity of gene × gene additive and interactive effects and their mechanistic pathways in the development of AUDs and can be used as a prototype for future analyses. Additionally, prospective studies with multiple phenotypes and endophenotypes will best elucidate how gene effects change over the life span.

References

1. Heath, A. C., Bucholz, K. K., Madden, P. A., et al. (1997). Genetic and environmental contributions to alcohol dependence risk in a national twin sample: consistency of findings in women and men. *Psychological Medicine, 27*, 1381–1396.

2. Prescott, C. A., Aggen, S. H., & Kendler, K. S. (1999). Sex differences in the sources of genetic liability to alcohol abuse and dependence in a population-based sample of U.S. twins. *Alcoholism, Clinical and Experimental Research, 23*, 1136–1144.

3. Schork, N. J., & Schork, C. M. (1998). Issues and strategies in the genetic analysis of alcoholism and related addictive behaviors. *Alcohol, 16*, 71–83.

4. Schuckit, M. A. (2000). Biological phenotypes associated with individuals at high risk for developing alcohol-related disorders. Part 2. *Addiction Biology, 5*, 23–36.

5. Young-Wolff, K. C., Enoch, M. A., & Prescott, C. A. (2011). The influence of gene–environment interactions on alcohol consumption and alcohol use disorders: a comprehensive review. *Clinical Psychology Review, 31*, 800–816.

6. Dick, D. M., & Foroud, T. (2003). Candidate genes for alcohol dependence: a review of genetic evidence from human studies. *Alcoholism, Clinical and Experimental Research, 27*, 868–879.

7. Edenberg, H. J., & Foroud, T. (2006). The genetics of alcoholism: identifying specific genes through family studies. *Addiction Biology, 11*, 386–396.

8. Higuchi, S., Matsushita, S., & Kashima, H. (2006). New findings on the genetic influences on alcohol use and dependence. *Current Opinion in Psychiatry, 19*, 253–265.

9. Kalsi, G., Prescott, C. A., Kendler, K. S., & Riley, B. P. (2009). Unraveling the molecular mechanisms of alcohol dependence. *Trends in Genetics, 25*, 49–55.

10. Kohnke, M. D. (2008). Approach to the genetics of alcoholism: a review based on pathophysiology. *Biochemical Pharmacology, 75*, 160–177.

11. Edenberg, H. J. (2007). The genetics of alcohol metabolism: role of alcohol dehydrogenase and aldehyde dehydrogenase variants. *Alcohol Research & Health, 30*, 5–13.

12. Li, T. K. (2000). Pharmacogenetics of responses to alcohol and genes that influence alcohol drinking. *Journal of Studies on Alcohol, 61*, 5–12.

13. Eng, M. Y., Luczak, S. E., & Wall, T. L. (2007). ALDH2, ADH1B, and ADH1C genotypes in Asians: a literature review. *Alcohol Research & Health, 30*, 22–27.

14. Luczak, S. E., Glatt, S. J., & Wall, T. L. (2006). Meta-analyses of ALDH2 and ADH1B with alcohol dependence in Asians. *Psychological Bulletin, 132*, 607–621.

15. Gottesman, I. I., & Gould, T. D. (2003). The endophenotype concept in psychiatry: etymology and strategic intentions. *American Journal of Psychiatry, 160*, 636–645.

16. Sher, K. J. (1991). *Children of Alcoholics: A Critical Appraisal of Theory and Research.* Chicago: University of Chicago Press.

17. Heath, A. C., Madden, P. A., Bucholz, K. K., et al. (1999). Genetic differences in alcohol sensitivity and the inheritance of alcoholism risk. *Psychological Medicine, 29*, 1069–1081.

18. Smith, G. D., & Ebrahim, S. (2003). "Mendelian randomization": can genetic epidemiology contribute to understanding environmental causes of disease? *Journal of Epidemiology, 32*, 1–22.

19. Eriksson, C. J. (2001). The role of acetaldehyde in the actions of alcohol (update 2000). *Alcoholism, Clinical and Experimental Research, 25*, 15S–32S.

20. Wall, T. L., Shea, S. H., Luczak, S. E., Cook, T. A., & Carr, L. G. (2005). Genetic associations of alcohol dehydrogenase with alcohol use disorders and endophenotypes in White college students. *Journal of Abnormal Psychology, 114*, 456–465.

21. Bosron, W. F., &. Li, T.-K. (1986). Genetic polymorphism of human alcohol and aldehyde dehydrogenases, and their relationship to alcohol metabolism. *Hepatology, 6*, 502–510.

22. Enomoto, N., Takase, S., Yasuhara, M., & Takada, A. (1991). Acetaldehyde metabolism in different aldehyde dehydrogenase-2 genotypes. *Alcoholism, Clinical and Experimental Research, 15*, 141–144.

23. Luu, S. U., Wang, M. F., Lin, D. L., et al. (1995). Ethanol and acetaldehyde metabolism in Chinese with different aldehyde dehydrogenase-2 genotypes. *Proceedings of the National Science Council, Republic of China. Part B, Life Sciences, 19*, 129–136.

24. Mizoi, Y., Yamamoto, K., Ueno, Y., Fukunaga, T., & Harada, S. (1994). Involvement of genetic polymorphism of alcohol and aldehyde dehydrogenases in individual variation of alcohol metabolism. *Alcohol and Alcoholism, 29*, 707–710.

25. Nishimura, F. T., Fukunaga, T., Nishijo, H., et al. (2001). Electroencephalogram spectral characteristics after alcohol ingestion in Japanese men with aldehyde dehydrogenase-2 genetic variations: comparison with peripheral changes. *Alcoholism, Clinical and Experimental Research, 25,* 1030–1036.

26. Peng, G. S., Wang, M. F., Chen, C. Y., et al. (1999). Involvement of acetaldehyde for full protection against alcoholism by homozygosity of the variant allele of mitochondrial aldehyde dehydrogenase gene in Asians. *Pharmacogenetics, 9,* 463–476.

27. Takeshita, T., & Morimoto, K. (2000). Accumulation of hemoglobin-associated acetaldehyde with habitual alcohol drinking in the atypical ALDH2 genotype. *Alcoholism, Clinical and Experimental Research, 24,* 1–7.

28. Wall, T. L., Garcia-Andrade, C., Thomasson, H. R., Carr, L. G., & Ehlers, C. L. (1997). Alcohol dehydrogenase polymorphisms in Native Americans: identification of the ADH2*3 allele. *Alcohol and Alcoholism, 32,* 129–132.

29. Yoshihara, H., Sato, N., Kamada, T., & Abe, H. (1983). Low Km isoenzyme ALDH isoenzyme and alcoholic liver injury. *Pharmacology, Biochemistry, and Behavior, 18*(Suppl. 1), 425–428.

30. Chen, W. J., Chen, C. C., Yu, J. M., & Cheng, A. T. (1998). Self-reported flushing and genotypes of ALDH2, ADH2, and ADH3 among Taiwanese Han. *Alcoholism, Clinical and Experimental Research, 22,* 1048–1052.

31. Luczak, S. E., Pandika, D., Shea, S. H., et al. (2011). ALDH2 and ADH1B interactions in retrospective reports of low-dose reactions and initial sensitivity to alcohol in Asian American college students. *Alcoholism, Clinical and Experimental Research, 35,* 1238–1245.

32. Matsuo, K., Wakai, K., Hirose, K., et al. (2006). Alcohol dehydrogenase 2 His47Arg polymorphism influences drinking habit independently of aldehyde dehydrogenase 2 Glu487Lys polymorphism: analysis of 2,299 Japanese subjects. *Cancer Epidemiology, Biomarkers & Prevention, 15,* 1009–1013.

33. Shibuya, A., Yasunami, M., & Yoshida, A. (1989). Genotype of alcohol dehydrogenase and aldehyde dehydrogenase loci in Japanese alcohol flushers and nonflushers. *Human Genetics, 82,* 14–16.

34. Takeshita, T., Mao, X. Q., & Morimoto, K. (1996). The contribution of polymorphism in the alcohol dehydrogenase beta subunit to alcohol sensitivity in a Japanese population. *Human Genetics, 97,* 409–413.

35. Takeshita, T., Yang, X., & Morimoto, K. (2001). Association of the ADH2 genotypes with skin responses after ethanol exposure in Japanese male university students. *Alcoholism, Clinical and Experimental Research, 25,* 1264–1269.

36. Wall, T. L., Thomasson, H. R., & Ehlers, C. L. (1996). Investigator-observed alcohol-induced flushing but not self-report of flushing is a valid predictor of ALDH2 genotype. *Journal of Studies on Alcohol, 57,* 267–272.

37. Hendershot, C. S., Collins, S. E., George, W. H., et al. (2009). Associations of ALDH2 and ADH1B genotypes with alcohol-related phenotypes in Asian young adults. *Alcoholism, Clinical and Experimental Research, 33*, 839–847.

38. Wall, T. L., Johnson, M. L., Horn, S. M., et al. (1999). Evaluation of the self-rating of the effects of alcohol form in Asian Americans with aldehyde dehydrogenase polymorphisms. *Journal of Studies on Alcohol, 60*, 784–789.

39. Cook, T. A., Luczak, S. E., Shea, S. H., et al. (2005). Associations of ALDH2 and ADH1B genotypes with response to alcohol in Asian Americans. *Journal of Studies on Alcohol, 66*, 196–204.

40. Hara, K., Terasaki, O., & Okubo, Y. (2000). Dipole estimation of alpha EEG during alcohol ingestion in males genotypes for ALDH2. *Life Sciences, 67*, 1163–1173.

41. Kim, S. K., Lee, S. I., Shin, C. J., Son, J. W., & Ju, G. (2010). The genetic factors affecting drinking behaviors of Korean young adults with variant aldehyde dehydrogenase 2 genotype. *Psychiatry Investigations, 7*, 270–277.

42. Luczak, S. E., Elvine-Kreis, B., Shea, S. H., Carr, L. G., & Wall, T. L. (2002). Genetic risk for alcoholism relates to level of response to alcohol in Asian-American men and women. *Journal of Studies on Alcohol, 63*, 74–82.

43. Minami, J., Todoroki, M., Ishimitsu, T., et al. (2002). Effects of alcohol intake on ambulatory blood pressure, heart rate, and heart rate variability in Japanese men with different ALDH2 genotypes. *Journal of Human Hypertension, 16*, 345–351.

44. Nishimura, F. T., Fukunaga, T., Kajiura, H., et al. (2002). Effects of aldehyde dehydrogenase-2 genotype on cardiovascular and endocrine responses to alcohol in young Japanese subjects. *Autonomic Neuroscience, 102*, 60–70.

45. Obata, A., Morimoto, K., Sato, H., et al. (2005). Effects of alcohol on hemodynamic and cardiovascular reaction in different genotypes. *Psychiatry Research, 139*, 65–72.

46. Peng, G. S., Yin, J. H., Wang, M. F., et al. (2002). Alcohol sensitivity in Taiwanese men with different alcohol and aldehyde dehydrogenase genotypes. *Journal of the Formosan Medical Association, 101*, 769–774.

47. Shin, H. Y., Shin, I. S., & Yoon, J. S. (2006). ALDH2 genotype-associated differences in the acute effects of alcohol on P300, psychomotor performance, and subjective response in healthy young Korean men: a double-blind placebo-controlled crossover study. *Human Psychopharmacology, 21*, 159–166.

48. Wall, T. L., & Ehlers, C. L. (1995). Acute effects of alcohol on P300 in Asians with different ALDH2 genotypes. *Alcoholism, Clinical and Experimental Research, 19*, 617–622.

49. Wall, T. L., Gallen, C. C., & Ehlers, C. L. (1993). Effects of alcohol on the EEG in Asian men with genetic variations of ALDH2. *Biological Psychiatry, 34*, 91–99.

50. Wall, T. L., Nemeroff, C. B., Ritchie, J. C., & Ehlers, C. L. (1994). Cortisol responses following placebo and alcohol in Asians with different ALDH2 genotypes. *Journal of Studies on Alcohol, 55*, 207–213.

51. Wall, T. L., Thomasson, H. R., Schuckit, M. A., & Ehlers, C. L. (1992). Subjective feelings of alcohol intoxication in Asians with genetic variations of ALDH2 alleles. *Alcoholism, Clinical and Experimental Research, 16*, 991–995.

52. Fromme, K., Stroot, E., & Kaplan, D. (1993). Comprehensive effects of alcohol: development and psychometric assessment of a new expectancy questionnaire. *Psychological Assessment, 5*, 19–26.

53. Goldman, M. S., Brown, S. A., Christiansen, B. A., & Smith, G. T. (1991). Alcoholism and memory: broadening the scope of alcohol-expectancy research. *Psychological Bulletin, 110*, 137–146.

54. Darkes, J., Greenbaum, P. E., & Goldman, M. S. (2004). Alcohol expectancy mediation of biopsychosocial risk: complex patterns of mediation. *Experimental and Clinical Psychopharmacology, 12*, 27–38.

55. Goldman, M. S., Darkes, J. & Del Boca, F. K. (1999). Expectancy mediation of biopsychosocial risk for alcohol use and alcoholism. In I. Kirsch (Ed.), *How Expectancies Shape Experience* (pp. 233–262). Washington, DC: American Psychological Association.

56. Nagoshi, C. T., Noll, R. T., & Wood, M. D. (1992). Alcohol expectancies and behavioral and emotional responses to placebo versus alcohol administration. *Alcoholism, Clinical and Experimental Research, 16*, 255–260.

57. Sher, K. J., Wood, M. D., Wood, P. K., & Raskin, G. (1996). Alcohol outcome expectancies and alcohol use: a latent variable cross-lagged panel study. *Journal of Abnormal Psychology, 105*, 561–574.

58. McCarthy, D. M., Wall, T. L., Brown, S. A., & Carr, L. G. (2000). Integrating biological and behavioral factors in alcohol use risk: the role of ALDH2 status and alcohol expectancies in a sample of Asian Americans. *Experimental and Clinical Psychopharmacology, 8*, 168–175.

59. McCarthy, D. M., Brown, S. A., Carr, L. G., & Wall, T. L. (2001). ALDH2 status, alcohol expectancies, and alcohol response: preliminary evidence for a mediation model. *Alcoholism, Clinical and Experimental Research, 25*, 1558–1563.

60. Hendershot, C. S., Neighbors, C., George, W. H., et al. (2009). ALDH2, ADH1B and alcohol expectancies: integrating genetic and learning perspectives. *Psychology of Addictive Behaviors, 23*, 452–463.

61. Hahn, C. Y., Huang, S. Y., Ko, H. C., et al. (2006). Acetaldehyde involvement in positive and negative alcohol expectancies in Han Chinese persons with alcoholism. *Archives of General Psychiatry, 63*, 817–823.

62. Higuchi, S., Muramatsu, T., Matsushita, S., Murayama, M., & Hayashida, M. (1996). Polymorphisms of ethanol-oxidizing enzymes in alcoholics with inactive ALDH2. *Human Genetics, 97*, 431–434.

63. Irons, D. E., McGue, M., Iacono, W. G., & Oetting, W. S. (2007). Mendelian randomization: a novel test of the gateway hypothesis and models of gene–environment interplay. *Development and Psychopathology, 19*, 1181–1195.

64. Luczak, S. E., Wall, T. L., Shea, S. H., Byun, S. M., & Carr, L. G. (2001). Binge drinking in Chinese, Korean, and White college students: genetic and ethnic group differences. *Psychology of Addictive Behaviors, 15,* 306–309.

65. Muramatsu, T., Wang, Z. C., Fang, Y. R., et al. (1995). Alcohol and aldehyde dehydrogenase genotypes and drinking behavior of Chinese living in Shanghai. *Human Genetics, 96,* 151–154.

66. Sun, F., Tsuritani, I., Honda, R., Ma, Z. Y., & Yamada, Y. (1999). Association of genetic polymorphisms of alcohol-metabolizing enzymes with excessive alcohol consumption in Japanese men. *Human Genetics, 105,* 295–300.

67. Sun, F., Tsuritani, I., & Yamada, Y. (2002). Contribution of genetic polymorphisms in ethanol-metabolizing enzymes to problem drinking behavior in middle-aged Japanese men. *Behavior Genetics, 32,* 229–236.

68. Takeshita, T., & Morimoto, K. (1999). Self-reported alcohol-associated symptoms and drinking behavior in three ALDH2 genotypes among Japanese university students. *Alcoholism, Clinical and Experimental Research, 23,* 1065–1069.

69. Tu, G. C., & Israel, Y. (1995). Alcohol consumption by orientals in North America is predicted largely by a single gene. *Behavior Genetics, 25,* 59–65.

70. Wall, T. L., Shea, S. H., Chan, K. K., & Carr, L. G. (2001). A genetic association with the development of alcohol and other substance use behavior in Asian Americans. *Journal of Abnormal Psychology, 110,* 173–178.

71. Heath, A. C., & Martin, N. G. (1988). Teenage alcohol use in the Australian twin register: genetic and social determinants of starting to drink. *Alcoholism, Clinical and Experimental Research, 12,* 735–741.

72. Koopmans, J. R., & Boomsma, D. I. (1996). Familial resemblances in alcohol use: genetic or cultural transmission? *Journal of Studies on Alcohol, 57,* 19–28.

73. Prescott, C. A., Hewitt, J. K., Heath, A. C., et al. (1994). Environmental and genetic influences on alcohol use in a volunteer sample of older twins. *Journal of Studies on Alcohol, 55,* 18–33.

74. Hendershot, C. S., Witkiewitz, K., George, W. H., et al. (2011). Evaluating a cognitive model of ALDH2 and drinking behavior. *Alcoholism, Clinical and Experimental Research, 35,* 91–98.

75. Wall, T. L., Horn, S. M., Johnson, M. L., Smith, T. L., & Carr, L. G. (2000). Hangover symptoms in Asian Americans with variations in the aldehyde dehydrogenase (ALDH2) gene. *Journal of Studies on Alcohol, 61,* 13–17.

76. Yokoyama, M., Yokoyama, A., Yokoyama, T., et al. (2005). Hangover susceptibility in relation to aldehyde dehydrogenase-2 genotype, alcohol flushing, and mean corpuscular volume in Japanese workers. *Alcoholism, Clinical and Experimental Research, 29,* 1165–1171.

77. Luczak, S. E., Shea, S. H., Hsueh, A. C., et al. (2006). ALDH2*2 is associated with a decreased likelihood of alcohol-induced blackouts in Asian American college students. *Journal of Studies on Alcohol, 67,* 349–353.

78. Chen, C. C., Lu, R. B., Chen, Y. C., et al. (1999). Interaction between the functional polymorphisms of the alcohol-metabolism genes in protection against alcoholism. *American Journal of Human Genetics, 65*, 795–807.

79. Luczak, S. E., Wall, T. L., Cook, T. A., Shea, S. H., & Carr, L. G. (2004). ALDH2 status and conduct disorder mediate the relationship between ethnicity and alcohol dependence in Chinese, Korean, and White American college students. *Journal of Abnormal Psychology, 113*, 271–278.

80. Iwahashi, K., Matsuo, Y., Suwaki, H., Nakamura, K., & Ichikawa, Y. (1995). CYP2E1 and ALDH2 genotypes and alcohol dependence in Japanese. *Alcoholism, Clinical and Experimental Research, 19*, 564–566.

81. Murayama, M., Matsushita, S., Muramatsu, T., & Higuchi, S. (1998). Clinical characteristics and disease course of alcoholics with inactive aldehyde dehydrogenase-2. *Alcoholism, Clinical and Experimental Research, 22*, 524–527.

82. Kimura, M., Miyakawa, T., Matsushita, S., So, M., & Higuchi, S. (2011). Gender differences in the effects of ADH1B and ALDH2 polymorphisms on alcoholism. *Alcoholism, Clinical and Experimental Research, 35*, 1923–1927.

83. Hendershot, C. S., MacPherson, L., Myers, M. G., Carr, L. G., & Wall, T. L. (2005). Psychosocial, cultural and genetic influences on alcohol use in Asian American youth. *Journal of Studies on Alcohol, 66*, 185–195.

84. Doran, N., Myers, M. G., Luczak, S. E., Carr, L. G., & Wall, T. L. (2007). Stability of heavy episodic drinking in Chinese- and Korean-American college students: effects of ALDH2 gene status and behavioral undercontrol. *Journal of Studies on Alcohol and Drugs, 68*, 789–797.

85. Heath, A. C. (1995). Genetic influences on drinking behavior in humans. In H. Begleiter & B. Kissin (Eds.), *The Genetics of Alcoholism* (pp. 82–121). New York: Oxford University Press.

86. Rose, R. J. (1998). A developmental behavior-genetic perspective on alcoholism risk. *Alcohol Health and Research World, 22*, 131–143.

87. Brennan, P., Lewis, S., Hashibe, M., et al. (2004). Pooled analysis of alcohol dehydrogenase genotypes and head and neck cancer: a HuGE review. *American Journal of Epidemiology, 159*, 1–16.

88. Brooks, P. J., Enoch, M. A., Goldman, D., Li, T. K., & Yokoyama, A. (2009). The alcohol flushing response: an unrecognized risk factor for esophageal cancer from alcohol consumption. *PLoS Medicine, 6*, e50.

89. Crabb, D. W., Matsumoto, M., Chang, D., & You, M. (2004). Overview of the role of alcohol dehydrogenase and aldehyde dehydrogenase and their variants in the genesis of alcohol-related pathology. *Proceedings of the Nutrition Society, 63*, 49–63.

90. Hao, P. P., Chen, Y. G., Wang, J. L., Wang, X. L., & Zhang, Y. (2011). Meta-analysis of aldehyde dehydrogenase 2 gene polymorphism and Alzheimer's disease in East Asians. *Canadian Journal of Neurological Sciences, 38*, 500–506.

91. Lewis, S. J., & Smith, G. D. (2005). Alcohol, ALDH2, and esophageal cancer: a meta-analysis which illustrates the potentials and limitations of a Mendelian randomization approach. *Cancer Epidemiology, Biomarkers & Prevention, 14*, 1967–1971.

92. Zintzaras, E., Stefanidis, I., Santos, M., & Vidal, F. (2006). Do alcohol-metabolizing enzyme gene polymorphisms increase the risk of alcoholism and alcoholic liver disease? *Hepatology, 43*, 352–361.

93. Yang, S. J., Yokoyama, A., Yokoyama, T., et al. (2010). Relationship between genetic polymorphisms of ALDH2 and ADH1B and esophageal cancer risk: a meta-analysis. *World Journal of Gastroenterology, 16*, 4210–4220.

94. Zhang, G. H., Mai, R. Q., & Huang, B. (2010). Meta-analysis of ADH1B and ALDH2 polymorphisms and esophageal cancer risk in China. *World Journal of Gastroenterology, 16*, 6020–6025.

95. Hurley, T. D., Edenberg, H., & Li, T.-K. (2002). The pharmacogenomics of alcoholism. In J. Licinio & M.-L. Wong (Eds.), *Pharmacogenomics: The Search for Individualized Therapies*. Weinheim, Germany: Wiley-VCH.

96. Lee, S. L., Hoog, J. O., & Yin, S. J. (2004). Functionality of allelic variations in human alcohol dehydrogenase gene family: assessment of a functional window for protection against alcoholism. *Pharmacogenetics, 14*, 725–732.

97. Duester, G., Farres, J., Felder, M. R., et al. (1999). Recommended nomenclature for the vertebrate alcohol dehydrogenase gene family. *Biochemical Pharmacology, 58*, 389–395.

98. Osier, M. V., Pakstis, A. J., Soodyall, H., Comas, D., Goldman, D., Odunsi, A., et al. (2002). A global perspective on genetic variation at the ADH genes reveals unusual patterns of linkage disequilibrium and diversity. *American Journal of Human Genetics, 71*, 84–99.

99. Goedde, H. W., Agarwal, D. P., Fritze, G., et al. (1992). Distribution of ADH2 and ALDH2 genotypes in different populations. *Human Genetics, 88*, 344–346.

100. Borras, E., Coutelle, C., Rosell, A., Fernandez-Muixi, F., Broch, M., Crosas, B., et al. (2000). Genetic polymorphisms of alcohol dehydrogenase in Europeans: the ADH2*2 allele decreases the risk for alcoholism and is associated with ADH3*1. *Hepatology, 31*, 984–989.

101. Macgregor, S., Lind, P. A., Bucholz, K. K., et al. (2009). Associations of ADH and ALDH2 gene variation with self report alcohol reactions, consumption and dependence: an integrated analysis. *Human Molecular Genetics, 18*, 580–593.

102. Ogurtsov, P. P., Garmash, I. V., Miandina, G. I., et al. (2001). Alcohol dehydrogenase ADH2–1 and ADH2–2 allelic isoforms in the Russian population correlate with type of alcoholic disease. *Addiction Biology, 6*, 377–383.

103. Sherva, R., Rice, J. P., Neuman, R. J., et al. (2009). Associations and interactions between SNPs in the alcohol metabolizing genes and alcoholism phenotypes in European Americans. *Alcoholism, Clinical and Experimental Research, 33*, 848–857.

104. Whitfield, J. B., Nightingale, B. N., Bucholz, K. K., et al. (1998). ADH genotypes and alcohol use and dependence in Europeans. *Alcoholism, Clinical and Experimental Research, 22*, 1463–1469.

105. Whitfield, J. B. (1997). Meta-analysis of the effects of alcohol dehydrogenase genotype on alcohol dependence and alcoholic liver disease. *Alcohol and Alcoholism, 32*, 613–619.

106. Whitfield, J. B. (2002). Alcohol dehydrogenase and alcohol dependence: variation in genotype-associated risk between populations. *American Journal of Human Genetics, 71*, 1247–1250, author reply 1250–1251.

107. Thomasson, H. R., Edenberg, H. J., Crabb, D. W., et al. (1991). Alcohol and aldehyde dehydrogenase genotypes and alcoholism in Chinese men. *American Journal of Human Genetics, 48*, 677–681.

108. Ehrig, T., & Li, T.-K. (1995). Metabolism of alcohol and metabolic consequences. In B. Tabakoff & P. L. Hoffman (Eds.), *Biological Aspects of Alcoholism: WHO Expert Series on Biological Psychiatry* (pp. 23–48). Seattle: Hogrefe & Huber.

109. Neumark, Y. D., Friedlander, Y., Durst, R., et al. (2004). Alcohol dehydrogenase polymorphisms influence alcohol-elimination rates in a male Jewish population. *Alcoholism, Clinical and Experimental Research, 28*, 10–14.

110. Yamamoto, K., Ueno, Y., Mizoi, Y., & Tatsuno, Y. (1993). Genetic polymorphism of alcohol and aldehyde dehydrogenase and the effects on alcohol metabolism. *Arukoru Kenkyu-to Yakubutsu Ison (Japanese Journal of Alcohol Studies and Drug Dependence), 28*, 13–25.

111. Yoshihara, E., Ameno, K., Nakamura, K., et al. (2000). The effects of the ALDH2*1/2, CYP2E1 C1/C2 and C/D genotypes on blood ethanol elimination. *Drug and Chemical Toxicology, 23*, 371–379.

112. Linneberg, A., Gonzalez-Quintela, A., Vidal, C., Jorgensen, T., Fenger, M., Hansen, T., et al. (2009). Genetic determinants of both ethanol and acetaldehyde metabolism influence hypersensitivity and drinking behavior among Scandinavians. *Clinical and Experimental Allergy, 40*, 123–140.

113. Carr, L. G., Foroud, T., Stewart, T., et al. (2002). Influence of ADH1B polymorphism on alcohol use and its subjective effects in a Jewish population. *American Journal of Medical Genetics, 112*, 138–143.

114. Shea, S. H., Wall, T. L., Carr, L. G., & Li, T. K. (2001). ADH2 and alcohol-related phenotypes in Ashkenazic Jewish American college students. *Behavior Genetics, 31*, 231–239.

115. Takeshita, T., Morimoto, K., Mao, X., Hashimoto, T., & Furuyama, J. (1994). Characterization of the three genotypes of low Km aldehyde dehydrogenase in a Japanese population. *Human Genetics, 94*, 217–223.

116. Duranceaux, N. C., Schuckit, M. A., Eng, M. Y., et al. (2006). Associations of variations in alcohol dehydrogenase genes with the level of response to alcohol in non-Asians. *Alcoholism, Clinical and Experimental Research, 30*, 1470–1478.

117. Hasin, D., Aharonovich, E., Liu, X., et al. (2002). Alcohol dependence symptoms and alcohol dehydrogenase 2 polymorphism: Israeli Ashkenazis, Sephardics, and recent Russian immigrants. *Alcoholism, Clinical and Experimental Research, 26,* 1315–1321.

118. Hasin, D., Aharonovich, E., Liu, X., et al. (2002). Alcohol and ADH2 in Israel: Ashkenazis, Sephardics, and recent Russian immigrants. *American Journal of Psychiatry, 159,* 1432–1434.

119. Higuchi, S., Matsushita, S., Imazeki, H., et al. (1994). Aldehyde dehydrogenase genotypes in Japanese alcoholics. *Lancet, 343,* 741–742.

120. Higuchi, S., Matsushita, S., Muramatsu, T., Murayama, M., & Hayashida, M. (1996). Alcohol and aldehyde dehydrogenase genotypes and drinking behavior in Japanese. *Alcoholism, Clinical and Experimental Research, 20,* 493–497.

121. Loew, M., Boeing, H., Sturmer, T., & Brenner, H. (2003). Relation among alcohol dehydrogenase 2 polymorphism, alcohol consumption, and levels of gamma-glutamyltransferase. *Alcohol, 29,* 131–135.

122. Neumark, Y. D., Friedlander, Y., Thomasson, H. R., & Li, T. K. (1998). Association of the ADH2*2 allele with reduced ethanol consumption in Jewish men in Israel: a pilot study. *Journal of Studies on Alcohol, 59,* 133–139.

123. Corbett, J., Saccone, N. L., Foroud, T., et al. (2005). A sex-adjusted and age-adjusted genome screen for nested alcohol dependence diagnoses. *Psychiatric Genetics, 15,* 25–30.

124. Reich, T., Edenberg, H. J., Goate, A., et al. (1998). Genome-wide search for genes affecting the risk for alcohol dependence. *American Journal of Medical Genetics, 81,* 207–215.

125. Saccone, N. L., Kwon, J. M., Corbett, J., et al. (2000). A genome screen of maximum number of drinks as an alcoholism phenotype. *American Journal of Medical Genetics, 96,* 632–637.

126. Williams, J. T., Begleiter, H., Porjesz, B., et al. (1999). Joint multipoint linkage analysis of multivariate qualitative and quantitative traits: II. alcoholism and event-related potentials. *American Journal of Human Genetics, 65,* 1148–1160.

127. Ehlers, C. L., Gilder, D. A., Wall, T. L., et al. (2004). Genomic screen for loci associated with alcohol dependence in Mission Indians. *American Journal of Medical Genetics. Part B, Neuropsychiatric Genetics, 129B,* 110–115.

128. Ehlers, C. L., & Wilhelmsen, K. C. (2005). Genomic scan for alcohol craving in Mission Indians. *Psychiatric Genetics, 15,* 71–75.

129. Long, J. C., Knowler, W. C., Hanson, R. L., et al. (1998). Evidence for genetic linkage to alcohol dependence on chromosomes 4 and 11 from an autosome-wide scan in an American Indian population. *American Journal of Medical Genetics, 81,* 216–221.

130. Prescott, C. A., Sullivan, P. F., Kuo, P. H., et al. (2006). Genomewide linkage study in the Irish affected sib pair study of alcohol dependence: evidence for a susceptibility region for symptoms of alcohol dependence on chromosome 4. *Molecular Psychiatry, 11,* 603–611.

131. Uhl, G. R., Liu, Q. R., Walther, D., Hess, J., & Naiman, D. (2001). Polysubstance abuse-vulnerability genes: genome scans for association, using 1,004 subjects and 1,494 single-nucleotide polymorphisms. *American Journal of Human Genetics, 69*, 1290–1300.

132. Chen, W. J., Loh, E. W., Hsu, Y. P., et al. (1996). Alcohol-metabolising genes and alcoholism among Taiwanese Han men: independent effect of ADH2, ADH3 and ALDH2. *British Journal of Psychiatry, 168*, 762–767.

133. Nakamura, K., Iwahashi, K., Matsuo, Y., et al. (1996). Characteristics of Japanese alcoholics with the atypical aldehyde dehydrogenase 2*2: I. a comparison of the genotypes of ALDH2, ADH2, ADH3, and cytochrome P-4502E1 between alcoholics and nonalcoholics. *Alcoholism, Clinical and Experimental Research, 20*, 52–55.

134. Konishi, T., Calvillo, M., Leng, A. S., et al. (2003). The ADH3*2 and CYP2E1 c2 alleles increase the risk of alcoholism in Mexican American men. *Experimental and Molecular Pathology, 74*, 183–189.

135. Mulligan, C. J., Robin, R. W., Osier, M. V., et al. (2003). Allelic variation at alcohol metabolism genes (ADH1B, ADH1C, ALDH2) and alcohol dependence in an American Indian population. *Human Genetics, 113*, 325–336.

136. Espinos, C., Sanchez, F., Ramirez, C., Juan, F., & Najera, C. (1997). Polymorphism of alcohol dehydrogenase genes in alcoholic and nonalcoholic individuals from Valencia (Spain). *Hereditas, 126*, 247–253.

137. Gilder, F. J., Hodgkinson, S., & Murray, R. M. (1993). ADH and ALDH genotype profiles in Caucasians with alcohol-related problems and controls. *Addiction, 88*, 383–388.

138. Chen, Y. C., Lu, R. B., Peng, G. S., et al. (1999). Alcohol metabolism and cardiovascular response in an alcoholic patient homozygous for the ALDH2*2 variant gene allele. *Alcoholism, Clinical and Experimental Research, 23*, 1853–1860.

139. Osier, M., Pakstis, A. J., Kidd, J. R., Lee, J.-F., Yin, S.-J., Ko, H.-C., et al. (1999). Linkage disequilibrium at the ADH2 and ADH3 loci and risk for alcoholism. *American Journal of Human Genetics, 64*, 1147–1157.

140. Luo, X., Kranzler, H. R., Zuo, L., et al. (2006). Diplotype trend regression analysis of the ADH gene cluster and the ALDH2 gene: multiple significant associations with alcohol dependence. *American Journal of Human Genetics, 78*, 973–987.

141. Choi, I. G., Son, H. G., Yang, B. H., et al. (2005). Scanning of genetic effects of alcohol metabolism gene (ADH1B and ADH1C) polymorphisms on the risk of alcoholism. *Human Mutation, 26*, 224–234.

142. Birley, A. J., Whitfield, J. B., Neale, M. C., et al. (2005). Genetic time-series analysis identifies a major QTL for in vivo alcohol metabolism not predicted by in vitro studies of structural protein polymorphism at the ADH1B or ADH1C loci. *Behavior Genetics, 35*, 509–524.

143. Bosron, W. F., Magnes, L. J., & Li, T. K. (1983). Human liver alcohol dehydrogenase: ADH Indianapolis results from genetic polymorphism at the ADH2 gene locus. *Biochemical Genetics, 21*, 735–744.

144. Wall, T. L., Carr, L. G., & Ehlers, C. L. (2003). Protective association of genetic variation in alcohol dehydrogenase with alcohol dependence in Native American Mission Indians. *American Journal of Psychiatry, 160*, 41–46.

145. Luo, X., Kranzler, H. R., Zuo, L., et al. (2006). ADH4 gene variation is associated with alcohol dependence and drug dependence in European Americans: results from HWD tests and case-control association studies. *Neuropsychopharmacology, 31*, 1085–1095.

146. Preuss, U. W., Ridinger, M., Rujescu, D., et al. (2011). Association of ADH4 genetic variants with alcohol dependence risk and related phenotypes: results from a larger multicenter association study. *Addiction Biology, 16*, 323–333.

147. Luo, X., Kranzler, H. R., Zuo, L., et al. (2007). Multiple ADH genes modulate risk for drug dependence in both African- and European-Americans. *Human Molecular Genetics, 16*, 380–390.

148. Osier, M. V., Lu, R. B., Pakstis, A. J., et al. (2004). Possible epistatic role of ADH7 in the protection against alcoholism. *American Journal of Medical Genetics. Part B, Neuropsychiatric Genetics, 126B*, 19–22.

149. Han, Y., Oota, H., Osier, M. V., et al. (2005). Considerable haplotype diversity within the 23kb encompassing the ADH7 gene. *Alcoholism, Clinical and Experimental Research, 29*, 2091–2100.

150. Gizer, I. R., Edenberg, H. J., Gilder, D. A., Wilhelmsen, K. C., & Ehlers, C. L. (2011). Association of alcohol dehydrogenase genes with alcohol-related phenotypes in a Native American community sample. *Alcoholism, Clinical and Experimental Research, 35*, 2008–2018.

151. Kuo, P. H., Kalsi, G., Prescott, C. A., et al. (2008). Association of ADH and ALDH genes with alcohol dependence in the Irish Affected Sib Pair Study of Alcohol Dependence (IASPSAD) sample. *Alcoholism, Clinical and Experimental Research, 32*, 785–795.

152. van Beek, J. H., Willemsen, G., de Moor, M. H., Hottenga, J. J., & Boomsma, D. I. (2009). Associations between ADH gene variants and alcohol phenotypes in Dutch adults. *Twin Research and Human Genetics, 13*, 30–42.

153. Ueshima, Y., Matsuda, Y., Tsutsumi, M., & Takada, A. (1993). Role of the aldehyde dehydrogenase-1 isozyme in the metabolism of acetaldehyde. *Alcohol and Alcoholism. Supplement, 1B*, 15–19.

154. Chan, A. W. (1986). Racial differences in alcohol sensitivity. *Alcohol and Alcoholism, 21*, 93–104.

155. Yoshida, A. (1993). Genetic polymorphisms of alcohol-metabolizing enzymes related to alcohol sensitivity and alcoholic diseases. In K. M. Lin, R. E. Poland, & G. Nakasaki (Eds.), *Psychopharmacology and Psychobiology of Ethnicity* (pp. 169–183). Washington, DC: American Psychiatric Association.

156. Ward, R. J., McPherson, A. J., Chow, C., et al. (1994). Identification and characterisation of alcohol-induced flushing in Caucasian subjects. *Alcohol and Alcoholism, 29*, 433–438.

157. Spence, J. P., Liang, T., Eriksson, C. J., et al. (2003). Evaluation of aldehyde dehydrogenase 1 promoter polymorphisms identified in human populations. *Alcoholism, Clinical and Experimental Research, 27*, 1389–1394.

158. Ehlers, C. L., Spence, J. P., Wall, T. L., Gilder, D. A., & Carr, L. G. (2004). Association of ALDH1 promoter polymorphisms with alcohol-related phenotypes in southwest California Indians. *Alcoholism, Clinical and Experimental Research, 28*, 1481–1486.

159. Lind, P. A., Eriksson, C. J., & Wilhelmsen, K. C. (2008). The role of aldehyde dehydrogenase-1 (ALDH1A1) polymorphisms in harmful alcohol consumption in a Finnish population. *Human Genomics, 3*, 24–35.

160. Dickson, P. A., James, M. R., Heath, A. C., et al. (2006). Effects of variation at the ALDH2 locus on alcohol metabolism, sensitivity, consumption, and dependence in Europeans. *Alcoholism, Clinical and Experimental Research, 30*, 1093–1100.

161. Mustavich, L. F., Miller, P., Kidd, K. K., & Zhao, H. (2010). Using a pharmacokinetic model to relate an individual's susceptibility to alcohol dependence to genotypes. *Human Heredity, 70*, 177–193.

4 Nicotine Metabolism as an Intermediate Phenotype

Meghan J. Chenoweth and Rachel F. Tyndale

The prevalence of cigarette smoking has plateaued in North America and Europe, where recent estimates suggest nearly a fifth of adults are current smokers [1–4]. Smoking causes approximately 20% of deaths in these geographic regions and remains the leading preventable cause of illness and early death [5–7]. In developing nations, the prevalence of cigarette smoking is increasing, and by 2030, more than 80% of the world's tobacco-related deaths are predicted to occur in these regions [8]. Although more than two-thirds of smokers express a desire to quit [6], the process of smoking cessation is arduous and only 3% of those making unaided quit attempts remain abstinent at 6 months [9].

4.1 Neurobiology of Nicotine Addiction

Nicotine is the main psychoactive compound in cigarette smoke and is responsible for the reinforcing properties of cigarette smoking, serving to both establish and maintain dependence [5]. Following cigarette smoke inhalation into the lungs, nicotine is rapidly absorbed into the circulatory system and transported to the brain within seconds, where it binds nicotinic acetylcholine receptors (nAChRs) [10]. The activation of nAChRs results in the downstream release of dopamine in a variety of brain regions including the shell of the nucleus accumbens, which is critical in mediating the rewarding effects of cigarette smoking [10]. Smokers adjust their patterns of cigarette use to achieve stable levels of nicotine in order to attain the desired effects and to avoid withdrawal [6, 10]; periods of abstinence result in the precipitation of nicotine withdrawal and craving symptoms [11]. These symptoms are alleviated upon reinstatement of smoking and can thus predict relapse [6]. The majority of smokers require a number of quit attempts before reaching sustained abstinence [10]; hence nicotine addiction is a chronic condition with an average duration of more than 20 years [12, 13].

4.2   Genetic Influences on Smoking

A combination of genetic and environmental factors influences the onset and main-tenance of smoking behaviors and the development of nicotine dependence. Twin studies reveal that genetics plays a substantial role in initiation, the development of dependence, level of dependence, amount smoked, continued smoking, and the ability to stop smoking [14–19]. There is some degree of overlap in the heritability of these phenotypes suggesting common genetic factors, in addition to unique genetic factors, mediate these behaviors [17]. Although environmental factors play a relatively minor role in cigarette smoking behaviors in adults, these factors may have a stronger influence on the smoking behaviors of adolescents/novice smokers [20–22]. The rate of nicotine metabolism in smokers is influenced by both genetic and environmental factors and can be considered an intermediate phenotype that modulates smoking behaviors in both adolescents and adults as discussed below.

4.3   Nicotine Metabolism as an Intermediate Phenotype

A variety of genetic and environmental factors, each with small individual effect sizes, contribute to complex neuropsychiatric disorders such as addiction, and individuals must display a threshold level of clinically overt signs in order to be positively diag-nosed [23, 24]. However, large interindividual variability exists regarding the clinical features of these disorders; thus intermediate phenotypes with clearer and narrower features are a powerful tool to study the causes of diseases that have complex genetic and environmental components [23, 25, 26]. Therefore, the intermediate phenotype approach, in unraveling causal links between genetic factors and disease [27], may provide greater mechanistic insight into the etiology of complex neuropsychiatric disease. Moreover, the evaluation of intermediate phenotypes may allow for the greater identification of at-risk individuals.

The majority (~80%) of nicotine is inactivated to cotinine, a reaction primarily catalyzed by cytochrome P450 2A6 (CYP2A6). CYP2A6 is responsible for more than 90% of the inactivation of nicotine to cotinine and 100% of the conversion of cotinine to 3'-hydroxycotinine [28, 29] (see figure 4.1). Thus the metabolism of nicotine accounts for the majority of nicotine removal from the body. Two surrogates of nico-tine clearance, *CYP2A6* genotype and the 3'-hydroxycotinine/cotinine ratio (also known as the nicotine metabolite ratio; NMR), have been utilized to study associations between nicotine metabolism rates and smoking behaviors.

The metabolism of nicotine can be considered an intermediate phenotype that mediates the relationship between genetic risk factors and the susceptibility to nico-tine addiction. Additive genetic factors play a major role in mediating both total nico-tine clearance (59%) and the nicotine metabolite ratio in plasma and urine (67% and

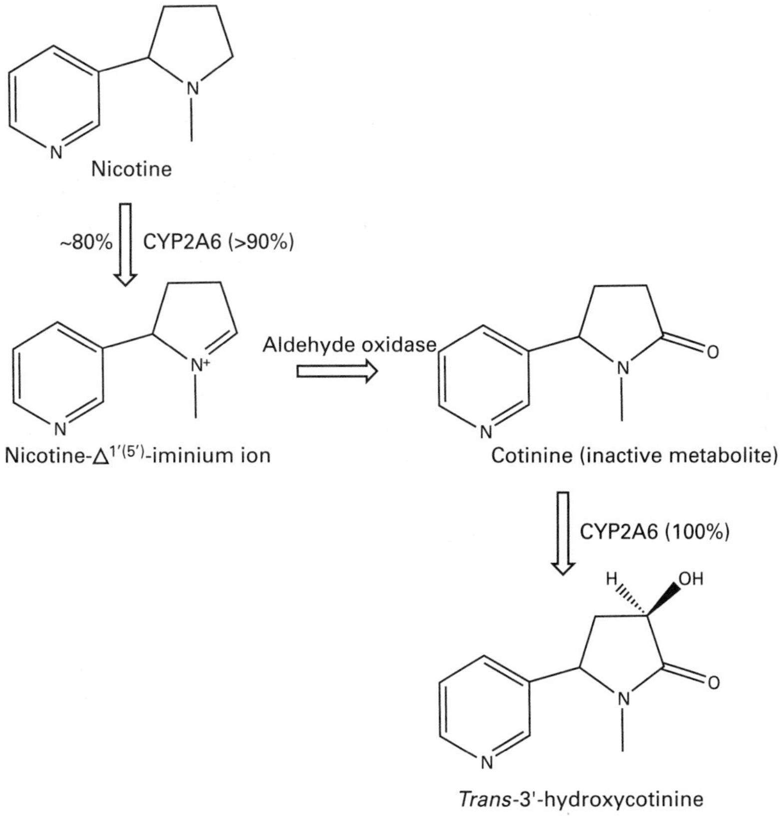

Figure 4.1

Nicotine metabolism: Approximately 80% of inhaled nicotine is inactivated to cotinine, in a two-step reaction catalyzed by cytochrome P450 2A6 (CYP2A6) and cytoplasmic aldehyde oxidase. CYP2A6 is responsible for more than 90% of the conversion of nicotine to nicotine-$\Delta^{1'(5')}$-iminium ion, which is the rate-limiting step of the reaction [160]. Nicotine-$\Delta^{1'(5')}$-iminium ion is then rapidly converted to cotinine by aldehyde oxidase [161]. Cotinine is subsequently metabolized to *trans*-3′-hydroxycotinine, in a reaction mediated exclusively by CYP2A6. The 3′-hydroxycotinine/cotinine ratio (or nicotine metabolite ratio) serves as a proxy for both CYP2A6 activity and total nicotine clearance. (Chemical structures were created using ChemDraw software, PerkinElmer, Cambridge, MA.)

47%, respectively) [30, 31]. In addition, variation in the rates of nicotine metabolism is associated with a number of smoking-related behaviors [32–35].

4.3.1 Nicotine Metabolism Rate Approximated by CYP2A6 Genotype

A variety of studies have investigated the impact of genetic factors on smoking in adults and have demonstrated significant associations with multiple smoking behaviors for variation in the gene encoding CYP2A6. Specifically, variation in CYP2A6 contributes to differences in current smoking status, cigarette consumption and puff volume, level of dependence, duration of smoking, likelihood of being a former versus current smoker, and likelihood of cessation in adults [33, 34, 36–40].

The highly polymorphic CYP2A6 gene consists of 9 exons and 8 introns and encodes 494 amino acids [41]. To date, 37 unique numbered CYP2A6 alleles have been characterized, many of which are single nucleotide polymorphisms (SNPs), or the result of nucleotide insertions and deletions. CYP2A6 gene duplications, conversions, and hybrid alleles have also been described (http://www.cypalleles.ki.se/cyp2a6.htm). A number of the CYP2A6 variant alleles meaningfully alter CYP2A6 enzymatic activity in vitro and in vivo [42–45] and thus are of particular importance for nicotine metabolism and smoking behaviors.

Four CYP2A6 alleles (CYP2A6*2, CYP2A6*4, CYP2A6*9, and CYP2A6*12) have received widespread attention in studies examining the association between CYP2A6 variation and a number of smoking-related outcome measures in Caucasians [33, 35, 36, 38, 46, 47]. Each of these alleles is present in Caucasian populations at relatively high frequencies (between ~1% for CYP2A6*4 to ~8% for CYP2A6*9) [33, 36, 47], and each encodes a CYP2A6 enzyme with reduced or inactive function. Specifically, CYP2A6*2 and CYP2A6*4 both code for inactive CYP2A6 enzymes. CYP2A6*2 is the consequence of a nonsynonymous SNP in the third exon of the CYP2A6 gene and encodes an inactive CYP2A6 enzyme due to the lack of incorporation of a heme molecule [48]. Similarly, individuals possessing CYP2A6*4 also do not express functional CYP2A6 enzyme as this variant results in a CYP2A6 gene deletion likely stemming from unequal crossover events within the CYP2A gene cluster [49]. Conversely, CYP2A6*9 and CYP2A6*12 are both decrease-of-function alleles. CYP2A6*9 is a SNP in the TATA box of the CYP2A6 gene, resulting in reduced promoter activity and decreased production of CYP2A6 mRNA [50]. The CYP2A6*12 variant is thought to result from an unequal crossover event between the CYP2A6 and the nonfunctional CYP2A7 pseudogene in intron 2, resulting in a decrease in both the protein and activity level of CYP2A6 as demonstrated in vitro [45] and in vivo [44].

A number of additional CYP2A6 alleles reducing CYP2A6 enzymatic function have been characterized, including CYP2A6*7, CYP2A6*10, CYP2A6*17, CYP2A6*20, CYP2A6*23, CYP2A6*25, CYP2A6*26, CYP2A6*27, and CYP2A6*35 [42, 51–54].

Expression of each of these variants is associated with reduced or inactive CYP2A6 function in vitro and/or in vivo [42, 51–54].

When investigating associations between *CYP2A6* genotype and smoking behaviors, individuals are grouped into predicted CYP2A6 enzymatic activity groups based on their *CYP2A6* genotype [33, 36, 46, 47]. Individuals having one copy of a reduced function variant, as determined by pharmacokinetic studies, are expected to have 75% to 80% of normal CYP2A6 activity whereas individuals expressing one copy of a inactive function variant or two copies of a reduced function variant are predicted to have approximately 50% of normal CYP2A6 enzymatic activity [44].

Large interindividual variability in the rate of nicotine metabolism exists within and between populations [42]. The frequency of *CYP2A6* variants in populations varies widely by ethnicity [42]. For example, *CYP2A6*17* is found at high frequency in African American subjects but rarely in Caucasian or East Asian populations [16]. In contrast, the frequencies of *CYP2A6*4* and *CYP2A6*9* are relatively lower in Caucasians and African Americans compared to East Asian populations [36, 42, 51]. Overall, given the variation in the prevalence of these variant alleles among different ethnic groups, it is not surprising that there is a lower prevalence of *CYP2A6* reduced metabolizers (≤75% CYP2A6 activity as predicted by *CYP2A6* genotype) in Caucasian populations (~10%–25%), relative to East Asian populations (~50%) [35, 42, 46, 47]. This is exemplified by overall slower rates of nicotine metabolism in East Asians compared to Caucasians [42, 55]. Since smoking rates are generally higher in East Asian countries relative to Western countries, especially among men [1, 2, 56, 57], the effect of slower nicotine metabolism on smoking behaviors is likely to have a greater relative impact in East Asian populations compared to Caucasian populations.

4.3.2 Nicotine Metabolism Rate Measured by Phenotype

Although *CYP2A6* genotype is often used to predict an individual's capacity to metabolize nicotine, it does not capture all sources of variation in CYP2A6 and nicotine metabolism rates [30]. Even among individuals classified as normal nicotine metabolizers, rates of nicotine clearance are widely variable [30]; thus it is important to also include alternative approaches that account for both genetic and environmental influences on nicotine metabolism [58]. Environmental factors including the use of certain medications may meaningfully alter nicotine metabolism rates [58]. For example, the antipsoriasis drug methoxsalen, a CYP2A6 inhibitor, is associated with reduced nicotine metabolism [59], whereas nicotine metabolism is higher in women compared to men [58], and even higher in women taking estrogen-containing oral contraceptives [60]. Moreover, dietary factors may also significantly alter nicotine metabolism rates, as broccoli consumption is associated with increased CYP2A6 enzymatic activity [61].

As previously mentioned, CYP2A6 is responsible for more than 90% of the inactivation of nicotine into cotinine and mediates 100% of the conversion of cotinine into 3'hydroxycotinine [28, 29]. The 3'hydroxycotinine/cotinine ratio (the NMR), is a phenotypic measure of the nicotine metabolism rate that has been validated as a phenotypic marker of CYP2A6 activity [62, reviewed in 63]. The utility of this phenotypic measure stems from the long elimination half-life of cotinine (approximately 13–19 hours), and the formation dependence of 3'hydroxycotinine [64]. The NMR is calculable from plasma, saliva, and urine and has strong concordance with CYP2A6 genotype [33, 54, 65]. Although using the NMR has many advantages, some of the influences on the NMR are transient (e.g., diet, drugs) and may confound the interpretation of intrinsic nicotine metabolism rates [66]. However, the NMR correlates well with total nicotine clearance and is more closely related than CYP2A6 genotype to total nicotine clearance [62, 67] (see figure 4.2).

4.4   Role of Nicotine Metabolism in Smoking Behaviors and Dependence

Variation in the rate of nicotine metabolism, due to its relationship with nicotine clearance, is associated with a variety of smoking-related outcomes. The following sections will describe the current state of knowledge regarding the impact of the rate of nicotine metabolism on smoking behaviors and nicotine addiction (summarized in figure 4.3). In addition, the association between nicotine metabolism, CYP2A6 genotype, and the risk for a variety of smoking-related diseases will be discussed.

4.4.1   Consumption and Nicotine Dependence

Level of Consumption
Differences in the level of cigarette consumption, as a function of predicted CYP2A6 metabolic group, have been observed in both adults and adolescents [35, 46]. In adults, these changes manifest not only in terms of the number of cigarettes smoked per day but also in terms of smoking topography, which is another quantifiable measure of smoking behavior [38]. In addition, a positive correlation exists between the rate of nicotine clearance and the total daily dose of nicotine acquired from cigarette smoking [55].

Number of Cigarettes
Smokers expressing reduced or inactive CYP2A6 variants (*2, *4, *9, and *12) smoked a mean of 5.7 fewer cigarettes daily (20.2 cigarettes) compared to smokers with normal CYP2A6 activity (25.9 cigarettes) among current controls from a lung cancer case–control study in Caucasians [35]. Moreover, CYP2A6 slow metabolizers (≤50% CYP2A6 enzymatic activity as predicted by genotype) smoked a mean of 7 fewer cigarettes per

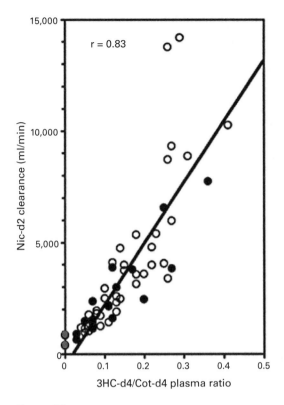

Figure 4.2

Correlation between total nicotine clearance and 3'-hydroxycotinine/cotinine ratio: Correlation between total nicotine clearance and the plasma 3'-hydroxycotinine/cotinine ratio four hours following the administration of an oral solution of deuterium-labeled nicotine (2 mg d2 for nicotine clearance) and cotinine (10 mg d4 for the ratio) to healthy volunteers. The filled black circles represent smokers, and the empty circles represent nonsmokers. The filled gray circles represent nonsmokers homozygous for the *CYP2A6*4* allele (0% CYP2A6 enzymatic activity as predicted by *CYP2A6* genotype and ratio).

From Dempsey, D., Tutka, P., Jacob, P., III, Allen, F., Schoedel, K., Tyndale, R. F., and Benowitz, N. L., Nicotine metabolite ratio as an index of cytochrome P450 2A6 metabolic activity, *Clinical Pharmacology and Therapeutics*, 2004. *76*(1), 64–72.

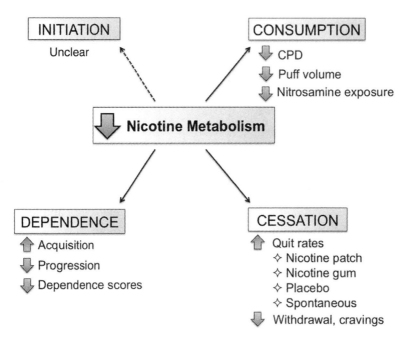

Figure 4.3
Nicotine metabolism is an intermediate phenotype that modulates smoking behaviors: The rate
of nicotine metabolism, measured by *CYP2A6* genotype and the nicotine metabolite ratio, is
associated with a number of smoking behaviors on the tobacco use continuum. While the role
of nicotine metabolism in smoking initiation (i.e., puffing on a cigarette for the first time) is
currently unclear, a decreased rate of nicotine metabolism is associated with a number of other
smoking behaviors. CPD, cigarettes smoked per day.

day compared to *CYP2A6* normal metabolizers in a community-based sample of Caucasian adult smokers [36]. This finding was also replicated in a population of Caucasian
treatment-seeking adults, where individuals with *CYP2A6* slow activity smoked a mean
of 4 fewer cigarettes per day (20 cigarettes per day) at baseline relative to *CYP2A6*
normal metabolizers (24 cigarettes per day) [33]. The association between *CYP2A6*
reduced metabolism and lower cigarette consumption also manifests in Japanese [34]
and Chinese [68] smokers. In Japanese smokers, *CYP2A6* slow metabolizers (≤50%
CYP2A6 activity as predicted by *CYP2A6* genotype) smoked 10 to 14 fewer cigarettes
per day relative to *CYP2A6* normal metabolizers [34], and among Chinese smokers,
CYP2A6 poor metabolizers (<25% CYP2A6 activity as predicted by *CYP2A6* genotype)
reported lower cigarette consumption compared to *CYP2A6* normal metabolizers (OR
[odds ratio] = 0.49) [68].

The NMR also correlated (r = 0.33; p = 0.005) with cigarette consumption in a
predominantly Caucasian population of heavy smokers [69] and was positively associ-

ated (p = 0.04) with cigarette consumption in another population of European heavy smokers that previously participated in a smoking cessation trial [67]. These findings were replicated in a pooled analysis of pretreatment data derived from three smoking cessation clinical trials [70–72] comprising over 1,000 participants, where faster nicotine metabolizers (top three NMR quartiles) smoked more cigarettes per day relative to slower nicotine metabolizers (lowest NMR quartile) [73].

In a cohort of Caucasian adolescents, there was also a trend toward a 40% to 56% decrease in cigarette consumption among *CYP2A6* reduced metabolizers, relative to *CYP2A6* normal metabolizers [46]. In a separate cohort of Caucasian adolescents, ~55% decreased weekly cigarette consumption (p = 0.04) was noted among *CYP2A6* reduced metabolizers relative to *CYP2A6* normal metabolizers [47].

Despite the observed association between reduced nicotine metabolism and lower cigarette consumption in a variety of adult and adolescent populations, these findings are not replicated in all light-smoking (≤10 cigarettes per day) populations [54, 74]. For example, African American light smokers did not alter total cigarette consumption to compensate for differences in the rate of nicotine metabolism, as measured by either *CYP2A6* genotype or the NMR [54]. In addition, the NMR was not correlated with cigarette consumption in adolescent light smokers [74]. Among light smokers, cigarettes per day may be a relatively weaker indicator of smoking, and individuals with slower nicotine metabolism may acquire lower nicotine doses through reducing puff volume [38, 75], described in the following section.

In general, there is a need for utilizing alternative biomarkers of consumption to more accurately assess the total nicotine dose achieved by smokers. Only modest associations are observed between cigarette consumption and biomarkers of cigarette smoke exposure [54, 76, 77]. For example, in an African American light-smoking population, exhaled CO and plasma cotinine levels were only weakly correlated (r ~ 0.31–0.37) with self-reported cigarette consumption [77]. Overall, urinary total nicotine equivalents (TNE), which is the molar sum of nicotine and its major metabolites, may be a more optimal biomarker for approximating total cigarette smoke exposure in smokers [76, 78, 79].

Topography

Both *CYP2A6* genotype and the NMR are associated with alterations in smoking topography, whereby different nicotine doses can be acquired from an equivalent number of cigarettes through changes in puff volume [38, 75]. In a sample of treatment-seeking smokers who smoked more than 20 cigarettes per day, *CYP2A6* slow metabolizers (≤50% CYP2A6 activity as predicted by genotype) puffed significantly less deeply on cigarettes relative to *CYP2A6* normal metabolizers, despite there being no significant difference in the number of puffs taken [38]. Similarly, heavy smokers in the slowest (first) NMR quartile had a significantly reduced total puff volume relative to heavy smokers in the third and fourth NMR quartiles [75].

Nicotine Dependence

Acquisition and Progression

Up to 90% of adult smokers surveyed report initiating smoking by the end of their teenage years [80]; thus adolescence is a critical period for the acquisition of smoking behaviors. To date, the association between *CYP2A6* variation and risk for acquisition of and progression in nicotine dependence has been studied prospectively in two adolescent cohorts [46, 47]. An investigation of the risk of tobacco dependence among adolescent ever-inhalers in the Nicotine Dependence in Teens (NDIT) cohort revealed the hazard ratio for conversion to International Classification of Diseases (10th rev.; ICD-10) tobacco dependence was 2.8 for slowest (one to two copies of *2 or *4) metabolizers compared to normal metabolizers [46]. Having symptoms of depression was also associated with conversion to tobacco dependence among *CYP2A6* normal metabolizers; however the association was more modest among *CYP2A6* slowest metabolizers [81].

Although *CYP2A6* slowest metabolizers were more likely to convert to ICD-10 [46] or modified Fagerström Tolerance Questionnaire (mFTQ) [82] nicotine dependence compared to *CYP2A6* normal metabolizers, data from the Georgetown Adolescent Tobacco Research (GATOR) study suggested reduced metabolizers progress in mFTQ dependence levels more slowly than normal metabolizers [47]. Although having a positive initial smoking experience predicts the future development of nicotine dependence [83, 84], positive initial smoking experiences did not appear to be responsible for mediating the association between *CYP2A6* genotype and risk for, or progression in, nicotine dependence [46, 47]. Nevertheless, the findings from NDIT and GATOR provide the framework for a novel, integrated hypothesis regarding the relationship between *CYP2A6* and nicotine dependence: while *CYP2A6* reduced metabolizers acquire nicotine dependence before *CYP2A6* normal metabolizers, they stabilize in the course of onset sooner, at lower levels of nicotine dependence and smoking quantities [82].

In late adolescence (age 18), *CYP2A6* reduced metabolism is also associated with a greater likelihood of being a current smoker (OR = 2.23 for current vs. ex-smoker) [85], consistent with the findings from NDIT [46]. However, their slower progression in nicotine dependence relative to normal *CYP2A6* metabolizers [47] may translate into increased quit rates in later adolescence and early adulthood, resulting in the lower observed prevalence of adult smokers (both *Diagnostic and Statistical Manual of Mental Disorders* [4th ed.] dependent and nondependent) with reduced *CYP2A6* metabolism [36] (see below).

The NMR also positively correlates with self-described level of addiction in adolescent light smokers (consuming ≤6 cigarettes per day), such that individuals with faster nicotine metabolism report greater levels of self-described nicotine addiction [74].

However, the NMR was not significantly correlated with nicotine dependence as measured by the Hooked on Nicotine Checklist or mFTQ [74].

Nicotine Dependence Scores in Adults

Among Caucasian heavy smokers, FTND (Fagerström Test for Nicotine Dependence) scores were significantly higher in *CYP2A6* normal metabolizers (5.1) relative to *CYP2A6* reduced metabolizers (4.2) [35]. However, the NMR did not correlate with FTND scores in a community-based sample of heavy smokers [69].

Several clinical studies examining potential associations between the rate of nicotine metabolism and level of dependence have also found no significant relationship [33, 54, 67, 69, 86]. In a clinical trial investigating the efficacy of transdermal nicotine for smoking cessation, neither NMR [86] nor *CYP2A6* genotype [33] were significantly associated with FTND scores. However, one aspect of the FTND, cigarettes consumed per day, was significantly associated with *CYP2A6* genotype [33]. In addition, neither pretreatment NMR nor *CYP2A6* genotype was associated with FTND scores in a clinical trial involving nicotine gum [54].

While the FTND captures daily cigarette consumption and time to first cigarette, both of which are associated with *CYP2A6* [33], the FTND also includes variables less likely to be altered by nicotine metabolism, such as the ability to refrain from smoking in places where it is forbidden [87]. Conversely, the rate of nicotine metabolism can modulate withdrawal and craving scores [32, 74]; however these elements are not captured by the FTND. Future studies assessing nicotine dependence using measures in addition to the FTND scale may reveal significant associations.

4.4.2 Cessation

Variation in the rate of nicotine metabolism also predicts patient response to smoking cessation pharmacotherapy in clinical trials. In addition, altered nicotine metabolism is associated with spontaneous quit rates in observational studies. These findings are discussed in detail below.

Clinical Trials

Clinical trials involving smoking cessation pharmacotherapies demonstrate significant associations between variation in nicotine metabolism and therapeutic response to the nicotine patch [32, 86]. In a predominantly Caucasian population, pretreatment NMR, among individuals receiving 21 mg transdermal nicotine, was significantly associated with treatment response at both end of treatment and follow-up. At end of treatment, 46% of smokers in the lowest NMR quartile achieved abstinence, compared to only 28% of smokers in the fastest NMR quartile. Interestingly, this difference persisted at six months follow-up, where 30% of individuals in the slowest NMR quartile successfully quit relative to only 11% in the fastest NMR quartile [32]. Lerman et al.

also examined whether an association exists between the rate of nicotine metabolism and efficacy of nicotine nasal spray but found no significant association between NMR quartile and treatment response [32]. However, the rate of nicotine metabolism was associated with the degree of nicotine nasal spray usage, such that CYP2A6 normal metabolizers used nearly six more doses per day relative to CYP2A6 slow metabolizers (p < 0.02) [33].

A follow-up investigation involving a larger, predominantly Caucasian study population reexamined whether pretreatment NMR predicts therapeutic response to nicotine patch treatment [86]. All patients received 21 mg nicotine patch treatment for eight weeks, and end-of-treatment quit rates were significantly associated with pretreatment NMR [86]. Forty-two percent of individuals in the slowest NMR quartile achieved abstinence, compared to only 28% of individuals in the fastest NMR quartile, validating the previous findings of Lerman et al. [32]. The quit rate observed among slow nicotine metabolizers is on par with that achieved by the newest smoking cessation drug, varenicline [86]. Thus, variation in nicotine metabolism rates as assessed by pretreatment NMR serves as a predictor of successful smoking abstinence achieved with transdermal nicotine therapy.

In a smoking cessation trial in African American light smokers randomized to receive either nicotine gum or placebo, both CYP2A6 genotype and pretreatment NMR predicted overall cessation rates [54]. While there was no overall effect of nicotine gum treatment on cessation outcome [88], individuals in the slowest NMR quartile were more likely to quit at the end of follow-up relative to individuals in faster NMR quartiles, and there was a nonsignificant trend toward increased quit rates at both end of treatment and follow-up among CYP2A6 slow metabolizers relative to CYP2A6 normal and intermediate metabolizers [54]. In addition, a significant twofold increase in abstinence rates on nicotine gum was observed among CYP2A6 slow metabolizers relative to CYP2A6 normal and intermediate metabolizers [54].

In a placebo-controlled smoking cessation clinical trial involving bupropion, individuals with faster pretreatment NMR that received bupropion had higher end-of-treatment quit rates compared to those that received placebo (34% vs. 10%, respectively) [89]. In addition, the relationship among fastest nicotine metabolizers remained significant at six-month follow-up, when 27% of individuals that received bupropion were abstinent relative to only 8% of individuals who received placebo [89]. Of note, the NMR effect was primarily in the placebo arm, whereby 32% of individuals in the slowest NMR quartile quit smoking at end of treatment relative to only 10% of individuals in the fastest NMR quartile, with little difference in bupropion response (32% vs. 34%) [89]. As a result, treatment with bupropion did not significantly enhance end-of-treatment cessation rates among slow metabolizers: individuals in the slowest NMR quartile had equal quit rates on placebo and bupropion (32%) [89]. Overall, the effect of NMR on quitting smoking while on placebo pill [89] is similar to that while on placebo gum [54], suggesting slow metabolizers have better quit rates in general.

Observational Studies

Observational studies have also found significant associations between *CYP2A6* genotype and smoking cessation. Adults expressing the inactive *CYP2A6*2* allele experienced a shorter duration of smoking and increased probability of quitting relative to individuals homozygous for the wild type *CYP2A6*1* allele [40], and *CYP2A6* reduced metabolizers were less likely to be current smokers (OR = 0.52) [36]. However, these findings were not replicated in an epidemiological study of Southern Chinese [68]. A recent study in Caucasian adolescent smokers also demonstrated a significantly increased likelihood of smoking cessation among *CYP2A6* slow metabolizers (~≤50% CYP2A6 activity) relative to normal metabolizers (~100% CYP2A6 activity; OR = 2.3), suggesting *CYP2A6* variation influences smoking cessation even in novice smokers [90].

Withdrawal and Cravings

The rate of nicotine metabolism is also associated with the severity of both withdrawal and craving symptoms in smokers. Adults with higher *CYP2A6* activity (*CYP2A6*1/*1*, *1/*9*, *1/*4*, *9/*9*) reported more severe withdrawal symptoms compared to adults with lower *CYP2A6* activity (*CYP2A6*4/*9*, *4/*4*) [91], consistent with the lower cessation rates observed for faster nicotine metabolizers compared to slower nicotine metabolizers [32]. Rubinstein et al. [74] examined the effect of NMR on withdrawal symptoms in adolescent smokers. Individuals in the fastest NMR quartile reported greater withdrawal symptoms after 24-hour abstinence compared to individuals in slower NMR quartiles, even after adjusting for cigarette consumption [74].

Among abstinent adults receiving nicotine patch treatment for one week, craving intensity increased linearly with increasing quartile of nicotine patch–derived NMR [32]. This suggests that faster nicotine metabolizers experienced more intense cravings compared to slower nicotine metabolizers even while abstinent and on patch, which is an important consideration for smoking cessation interventions as craving associated with abstinence can predict relapse in adults [92].

Differences in smoking cue-evoked brain activity in faster versus slower nicotine metabolizers, measured by *CYP2A6* genotype or NMR, have also been documented [93] and may contribute to smoking relapse. Among young adult smokers, faster metabolizers demonstrated increased responding to visual smoking cues (versus control cues) in several brain reward regions, relative to slower metabolizers, as measured using functional magnetic resonance imaging (fMRI) [93]. This increase in cue-evoked responding may contribute to the relatively higher rate of smoking relapse observed among faster metabolizers relative to slower metabolizers.

4.4.3   Gene–Gene Interactions

CYP2A6 and CYP2B6

Cytochrome P450 2B6 (CYP2B6) may contribute to nicotine metabolism but possesses a lower affinity for nicotine and metabolizes nicotine at a slower rate than CYP2A6 in human liver microsomes [94]. Some studies have implicated a significant role for *CYP2B6* genetic variation in altering NMR [67] and nicotine clearance [95]. Interestingly, Ring et al. [95] observed a larger effect of *CYP2B6* variation on nicotine clearance in reduced *CYP2A6* metabolizers compared to normal *CYP2A6* metabolizers. However, when controlling for *CYP2A6* genotype, virtually no effect of *CYP2B6* variation on baseline NMR [96] and in vitro nicotine C-oxidation activity [97] was observed. In addition, the *CYP2B6*6* variant was associated with poorer abstinence rates in the placebo arm of a bupropion smoking cessation trial [98]; however it is not clear whether this effect was mediated by alterations in nicotine metabolism and/or smoking behaviors.

The apparent effects of *CYP2B6* genetic variation on nicotine metabolism and/or smoking behaviors may occur via coregulation with *CYP2A6,* as the two genes may share a 5′ regulatory region due to their close proximity and opposed transcriptional start sites [97]. In support of this notion, correlations between CYP2B6 and CYP2A6 mRNA [99] and protein [97, 100] levels have been observed. Alternatively, *CYP2B6* may be associated with nicotine metabolism through genetic linkage disequilibrium (LD) with *CYP2A6*. This is plausible since *CYP2A6* and *CYP2B6* are within close proximity to one another (<150 kb apart) in a common gene cluster on chromosome 19 [101]. While no statistically significant effect of LD was observed on in vivo nicotine clearance [95] or on in vitro nicotine metabolism in human liver microsomes [97], a large sequencing study found some evidence of LD between several variants in *CYP2B6* and the *CYP2A6*12* allele [102]. In addition, evidence of strong LD between *CYP2B6*6* and *CYP2A6*9* has been reported [67]. Thus, the role of *CYP2B6* genetic variation in altered nicotine metabolism, and resulting smoking behaviors, requires further clarification.

CYP2A6 and CHRNA5–CHRNA3–CHRNB4 Cluster

Several genome-wide association studies identifying susceptibility loci associated with smoking amount, nicotine dependence, lung cancer risk, and response to smoking cessation therapy have found significant associations with variation in a region of chromosome 15 containing the α5, α3, and β4 subunit genes of the nAChR [103–107]. A recent meta-analysis investigating the impact of variation in the *CHRNA5–CHRNA3–CHRNB4* cluster found significant associations for level of cigarette consumption and lung cancer risk [108]. Because variation in *CYP2A6* is also associated with smoking behaviors [33, 34, 36, 38] and lung cancer risk [109–112], it is logical to investigate the combined effect of variation in nicotine and nitrosamine pharmacokinetic genes (e.g., *CYP2A6*) and nicotine and nitrosamine pharmacodynamic genes (e.g., *CHRNA5–CHRNA3–CHRNB4*) on these outcome measures. In a case–control lung cancer study in adult smokers [35], *CYP2A6* appeared to play a relatively larger role in modulating smoking behaviors and nicotine dependence while *CHRNA5–CHRNA3–CHRNB4* was more important in influencing lung cancer risk. Normal *CYP2A6* metabolizers smoked about six more cigarettes per day and had significantly higher FTND scores than *CYP2A6* reduced metabolizers while those with the *CHRNA5–CHRNA3–CHRNB4* rs1051730 AA genotype smoked about one more cigarette per day (not significant) and did not have significantly higher FTND scores relative to those with the CHRNA5–CHRNA3–CHRNB4 rs1051730 GG/GA genotype [35]. Although *CHRNA5–CHRNA3–CHRNB4* played only a small role in modulating smoking behaviors, *CHRNA5–CHRNA3–CHRNB4* AA individuals were at significantly greater risk for lung cancer relative to *CHRNA5–CHRNA3–CHRNB4* GG/GA individuals (OR = 1.57). In contrast, there was a modest, nonsignificant increase in lung cancer risk among *CYP2A6* normal metabolizers relative to *CYP2A6* reduced metabolizers (OR = 1.26), except in those smoking less than or equal to 20 cigarettes per day [35]. In examining the effects of *CYP2A6* and *CHRNA5–CHRNA3–CHRNB4* in combination, individuals in the high-risk genotype group (*CYP2A6* normal metabolism and *CHRNA5–CHRNA3–CHRNB4* AA) smoked a mean of 7.1 more cigarettes per day (27.9 vs. 20.8) and had greater lung cancer risk (OR = 2.03) relative to individuals with neither risk genotype [35]. This suggests that genetic variation in *CYP2A6* and *CHRNA5–CHRNA3–CHRNB4* independently and additively combines to modulate both cigarette consumption and lung cancer risk [35]. A follow-up investigation in the same study population revealed a potential role for *CYP2B6* variation as a risk factor for lung cancer, independently of *CYP2A6* and *CHRNA5–CHRNA3–CHRNB4*, where the *CYP2B6*1/*1 + CYP2B6*1/*6* genotype group displayed a trend toward increased risk for lung cancer (OR = 1.25; vs. *CYP2B6*6/*6* genotype group) [113]. Variation in *CYP2B6* is thought to modulate lung cancer risk through the differential activation of tobacco-specific nitrosamines, rather than via alterations in cigarette consumption [113].

An independent and combined effect of variation in *CYP2A6* and *CHRNA5–CHRNA3–CHRNB4* on smoking behavior has also been demonstrated in adolescents [114]. *CYP2A6* reduced metabolizers with two copies of the nonsynonymous *CHRNA5* SNP rs16969968 (AA) were at an increased risk of regular smoking relative to *CHRNA5* GG/GA individuals expressing *CYP2A6*1* (OR > 4) [114].

Thus, it is important to investigate the individual and combined impact of variation in nicotine metabolism genes and nicotine receptor genes on smoking behaviors and the risk for tobacco-related diseases, as both appear to contribute singly and additively to modulate smoking behaviors and disease risk.

4.5   Impact of Smoking and Nicotine Metabolism on Disease

4.5.1   Cancer

Cigarette smoking increases the risk of developing several different types of cancers including, but not limited to, those of the oral cavity, lung, bladder, cervix, stomach, and kidney [115]. A variety of epidemiological studies have shown cigarette smoke exposure to be the main causal factor in the etiology of lung cancer, and lung cancer accounts for the highest rate of cancer-associated mortality worldwide [115]. Up to 90% of lung cancer cases are attributable to cigarette smoking [115, 116]. In addition, lung cancer risk increases with increasing levels of cigarette consumption and duration of smoking [117]. Exposure to secondhand smoke as well as the use of smokeless tobacco products such as chewing tobacco also increase cancer risk, with the latter causing mostly oral cancer [115].

More than 60 carcinogens have been identified in cigarette smoke, including polycyclic aromatic hydrocarbons, aromatic amines, and tobacco-specific nitrosamines such as 4-(methylnitrosamino)-1-(3-pyridyl)-1-butanone (NNK) and N'-nitrosonornicotine (NNN) [115, 118]. These tobacco-specific nitrosamines are present at considerable levels in both cigarette smoke and a variety of smokeless tobacco products [118]. CYP2A6 is involved in the metabolic activation of both NNN and NNK [119, 120]. Thus, in addition to its role as the principal enzyme mediating nicotine inactivation, CYP2A6 contributes to the activation of procarcinogens. Importantly, altered CYP2A6 metabolism may influence the risk for lung cancer directly, as well as indirectly through the modulation of smoking behaviors [35]. Interestingly, polymorphisms in *CYP2A6* resulting in reduced CYP2A6 enzymatic activity are associated with reduced risk of lung cancer [35, 109–112]. Conversely, normal *CYP2A6* activity is associated with greater lung cancer risk, even after adjusting for cigarette consumption [110] and cigarette pack years [35], a measure of total active cigarette smoke exposure. This suggests *CYP2A6* may directly contribute to lung cancer risk independently of consumption, perhaps via procarcinogen activation [35]. Moreover, the relative lung cancer risk associated with *CYP2A6* may be magnified in people smoking 20 or fewer cigarettes per day, suggesting greater relative genetic risk for lung cancer at lower levels of cigarette consumption [35].

In addition to significant protective effects of reduced *CYP2A6* metabolism on lung cancer risk, reduced *CYP2A6* metabolism also protects against head and neck squamous cell carcinoma (HNSCC) [121], colorectal cancer [122], oral cancer [123], and upper aerodigestive tract cancer [124].

4.5.2   Cardiovascular Disease

Cigarette smoking is also associated with a number of adverse cardiovascular outcomes including peripheral vascular disease, aortic aneurysm, stroke, coronary heart disease, and myocardial infarction [125–127]. Although smoking indirectly influences cardiovascular disease risk by modulating levels of other risk factors, there is also a direct association between smoking and cardiovascular disease [125]. After adjusting for differences in other risk factors between smokers and nonsmokers, an independent association between smoking and risk for cardiovascular disease remains [125].

A variety of studies have implicated a role for cigarette smoke in vascular endothelial cell dysfunction and accelerated atherosclerosis [128–131], which are major antecedents of coronary heart disease [132]. Importantly, the risk for cardiovascular disease attributable to smoking is positively associated with heaviness of smoking and smoking duration [125, 133–135]. Even a low level of smoking (one to four cigarettes per day) increases the risk for coronary heart disease [136] and death from ischemic heart disease [135]. In addition, passive cigarette smoke exposure is associated with a greater risk of coronary heart disease among nonsmokers, and the risk increases with increasing level and duration of exposure [137].

To date, only one study has examined a potential role for altered nicotine metabolism/*CYP2A6* genotype as a modulator of cardiovascular disease risk in smokers [138]. Consistent with the positive association between cigarette smoke exposure and risk for cardiovascular disease, heavier smokers (more than 15 cigarettes per day) had an increased risk of hypertension relative to lighter smokers (15 or fewer cigarettes per day) (OR = 1.59), and this relationship was moderated by *CYP2A6* [138]. Specifically, the risk for hypertension was highest among heavier smokers with normal *CYP2A6* metabolism, relative to lighter smokers with slower *CYP2A6* metabolism (≤50% CYP2A6 enzymatic activity as predicted by genotype) (OR = 2.74) [138].

4.5.3 Obesity

Cigarette smoking is negatively associated with total body weight as smokers tend to present with lower body mass index (BMI) than nonsmokers [139, 140]. The rate of nicotine metabolism, as measured by NMR, is also negatively associated with BMI among smokers [31, 54, 141]; however the mechanisms governing this relationship remain to be clarified. Differences in BMI may change the volume of distribution, metabolism, or renal clearance of nicotine's metabolites, or it may alter the regulation of the CYP2A6 enzyme [77].

Smokers gain body weight upon quitting smoking [142]. In a smoking cessation trial, individuals who were continuously abstinent for one year had gained an average of 5.90 kg while individuals who had continuously smoked for one year had gained an average of only 1.09 kg [142]. The ability of nicotine to reduce food intake has been demonstrated in animal models, and cigarette smoking is associated with appetite suppression in humans [143, 144]. Nicotinic drugs decrease food intake in mice via β4 subunit-containing nicotinic acetylcholine receptors, in a process that involves activation of pro-opiomelanocortin neurons [144]. Activation of pro-opiomelanocortin neurons in the arcuate nucleus is thought to decrease food intake as well as increase energy expenditure [145], and disruption of pro-opiomelanocortin neurons promotes obesity in both humans and animals [146, 147].

While smoking appears to be associated with lower total body weight, it is associated with increased waist circumference and waist-to-hip ratio [148, 149]. A prospective investigation in five birth cohorts of Finnish twins revealed a significant positive association between smoking in adolescence and future development of abdominal adiposity [150]. After adjusting for confounders, smoking at least ten cigarettes per day between the ages of 16 and 18 was associated with an approximately 30% increased risk of developing abdominal obesity in adulthood [150]. In addition, the level of cigarette consumption appears to alter risk for abdominal obesity, such that heavier smokers (more than 15 cigarettes per day) are at increased risk of abdominal obesity compared to lighter smokers (15 or fewer cigarettes per day) (OR = 1.57) [151]. Importantly, *CYP2A6* has been shown to moderate the association between heavy smoking and abdominal obesity [151]. Among individuals with poor *CYP2A6* metabolism (less than 25% of the activity of normal metabolizers [68]), heavy smokers were at increased risk of abdominal obesity compared to lighter smokers (OR = 3.90) [151].

These findings hold important implications for public health as abdominal obesity is a major risk factor for type 2 diabetes (T2DM) and a number of adverse cardiovascular outcomes, including congestive heart failure, myocardial infarction, and stroke [152, 153]. Together smoking and abdominal obesity promote insulin resistance and altered secretion of proinflammatory cytokines, the latter of which are thought to cause endothelial dysfunction and the subsequent development of cardiovascular disease [153].

4.5.4   Diabetes

Cigarette smoking is associated with the development of T2DM [154], consistent with the increased level of insulin resistance observed among smokers relative to nonsmokers [155]. A recent meta-analysis comprising over one million subjects found a pooled adjusted relative risk of 1.44 for developing T2DM among current smokers relative to nonsmokers [154]. In addition, heavy smoking (more than 20 cigarettes per day) increases the risk of T2DM relative to light smoking (10 or fewer cigarettes per day), with an odds ratio of 1.75 [156]. Interestingly, *CYP2A6* genotype modulated the interaction between heavy smoking and risk for T2DM, such that heavier-smoking individuals with poor *CYP2A6* metabolism (≤25% CYP2A6 enzymatic activity as predicted by genotype) were at greater risk of having T2DM compared to lighter-smoking individuals with normal *CYP2A6* metabolism (OR = 8.54) [156]. This increased risk for T2DM among heavy-smoking individuals with poor *CYP2A6* metabolism may be due to greater exposure of the pancreas to nicotine as a result of higher circulating nicotine levels among heavy-smoking poor metabolizers [156]. Nicotine exposure was previously shown to cause impaired function and apoptosis of beta cells in rodent models [157, 158], and a reduction in beta cell mass is a hallmark of T2DM in humans [159].

4.6 Conclusions and Significance

Nicotine addiction is a complex, multifactorial neuropsychiatric disorder with numerous genetic and environmental risk factors contributing to its etiology. The utility of an intermediate phenotype approach to aid our understanding of the causal pathways underpinning genetic susceptibility to nicotine addiction is evident. As an intermediate phenotype, the rate of nicotine metabolism has a strong genetic component and varies widely between individuals, influencing a number of smoking-related behaviors including cigarette consumption, dependence, and the ability to stop smoking. Moreover, variation in nicotine metabolism contributes to the risk among smokers for a variety of illnesses, such as lung cancer, cardiovascular disease, abdominal obesity, and diabetes.

Future studies investigating the individual and combined impact of variation in *CYP2A6* and genes encoding other putative targets involved in nicotine pharmacokinetics and pharmacodynamics may reveal additional pharmacologically relevant intermediate phenotypes and gene interactions that may offer new insights into the complexities of nicotine addiction. Advancements in this field will enhance our current understanding of the mechanisms governing the progression from experimental smoking to nicotine addiction, interindividual variation in the success of smoking cessation pharmacotherapies, and differences among people in their ability to stop smoking unaided. This in turn may inform novel tobacco control programs and improve personalized treatment interventions, with the eventual goal of reducing the high morbidity, mortality, and societal burden associated with cigarette smoking.

Acknowledgments

We acknowledge the support of a University Endowed Chair in Addictions (RFT), Canadian Institutes of Health Research (CIHR) Frederick Banting and Charles Best Canada Graduate Scholarship Doctoral Award (MJC), CIHR grant MOP86471 and TMH-109787, and National Institutes of Health grant DA 020830 (PGRN), the Campbell Family Mental Health Research Institute at the Centre for Addiction and Mental Health (CAMH), CAMH, the CAMH foundation, the Canada Foundation for Innovation (#20289 and #16014) and the Ontario Ministry of Research and Innovation.

Conflict of Interest

Dr. R. F. Tyndale has consulted for Novartis and McNeil. No support was provided by either of these companies for the writing of this chapter, nor did any members of the companies review the chapter.

References

1. Centers for Disease Control and Prevention. (2011). Current cigarette smoking prevalence among working adults—United States, 2004–2010. *MMWR. Morbidity and Mortality Weekly Report, 60*, 1305–1309.

2. Health Canada. *Canadian Tobacco Use Monitoring Survey (CTUMS).* http://www.hc-sc.gc.ca/hc-ps/tobac-tabac/research-recherche/stat/index-eng.php, 1999–2010.

3. Office for National Statistics UK. (2011). *Smoking and drinking among adults.* General Lifestyle Survey, 2009.

4. World Health Organization. *Tobacco control database.* http://www.euro.who.int/en/what-we-do/data-and-evidence/databases, 1994–2005.

5. Benowitz, N. L. (2008). Clinical pharmacology of nicotine: implications for understanding, preventing, and treating tobacco addiction. *Clinical Pharmacology and Therapeutics, 83*, 531–541.

6. Benowitz, N. L. (2010). Nicotine addiction. *New England Journal of Medicine, 362*, 2295–2303.

7. Davy, M. (2006). Time and generational trends in smoking among men and women in Great Britain, 1972–2004/05. *Health Statistics Quarterly*, (32), 35–43.

8. World Health Organization. (2008). *Report on the global tobacco epidemic.* http://www.who.int/tobacco/mpower/mpower_report_full_2008.pdf.

9. Hughes, J. R., Gulliver, S. B., Fenwick, J. W., et al. (1992). Smoking cessation among self-quitters. *Health Psychology, 11*, 331–334.

10. Benowitz, N. L. (2008). Neurobiology of nicotine addiction: implications for smoking cessation treatment. *American Journal of Medicine, 121*(4 Suppl 1), S3–S10.

11. Hughes, J. R., & Hatsukami, D. (1986). Signs and symptoms of tobacco withdrawal. *Archives of General Psychiatry, 43*, 289–294.

12. Burns, E. K., Levinson, A. H., Lezotte, D., et al. (2007). Differences in smoking duration between Latinos and Anglos. *Nicotine & Tobacco Research, 9*, 731–737.

13. Flanders, W. D., Lally, C. A., Zhu, B. P., et al. (2003). Lung cancer mortality in relation to age, duration of smoking, and daily cigarette consumption: results from Cancer Prevention Study II. *Cancer Research, 63*, 6556–6562.

14. Vink, J. M., Willemsen, G., & Boomsma, D. I. (2005). Heritability of smoking initiation and nicotine dependence. *Behavior Genetics, 35*, 397–406.

15. Tyndale, R. F., & Sellers, E. M. (2002). Genetic variation in CYP2A6-mediated nicotine metabolism alters smoking behavior. *Therapeutic Drug Monitoring, 24*, 163–171.

16. Fukami, T., Nakajima, M., Yoshida, R., et al. (2004). A novel polymorphism of human CYP2A6 gene CYP2A6*17 has an amino acid substitution (V365M) that decreases enzymatic activity in vitro and in vivo. *Clinical Pharmacology and Therapeutics, 76*, 519–527.

17. Maes, H. H., Sullivan, P. F., Bulik, C. M., et al. (2004). A twin study of genetic and environmental influences on tobacco initiation, regular tobacco use and nicotine dependence. *Psychological Medicine, 34*, 1251–1261.

18. Broms, U., Silventoinen, K., Madden, P. A., et al. (2006). Genetic architecture of smoking behavior: a study of Finnish adult twins. *Twin Research and Human Genetics, 9*, 64–72.

19. Vink, J. M., Beem, A. L., Posthuma, D., et al. (2004). Linkage analysis of smoking initiation and quantity in Dutch sibling pairs. *Pharmacogenomics Journal, 4*, 274–282.

20. Boomsma, D. I., Koopmans, J. R., Van Doornen, L. J., et al. (1994). Genetic and social influences on starting to smoke: a study of Dutch adolescent twins and their parents. *Addiction, 89*, 219–226.

21. McGue, M., Elkins, I., & Iacono, W. G. (2000). Genetic and environmental influences on adolescent substance use and abuse. *American Journal of Medical Genetics, 96*, 671–677.

22. White, V. M., Hopper, J. L., Wearing, A. J., et al. (2003). The role of genes in tobacco smoking during adolescence and young adulthood: a multivariate behaviour genetic investigation. *Addiction, 98*, 1087–1100.

23. Cannon, T. D., & Keller, M. C. (2006). Endophenotypes in the genetic analyses of mental disorders. *Annual Review of Clinical Psychology, 2*, 267–290.

24. Bearden, C. E., Reus, V. I., & Freimer, N. B. (2004). Why genetic investigation of psychiatric disorders is so difficult. *Current Opinion in Genetics & Development, 14*, 280–286.

25. Meyer-Lindenberg, A., & Weinberger, D. R. (2006). Intermediate phenotypes and genetic mechanisms of psychiatric disorders. *Nature Reviews. Neuroscience, 7*, 818–827.

26. Rasetti, R., & Weinberger, D. R. (2011). Intermediate phenotypes in psychiatric disorders. *Current Opinion in Genetics & Development, 21*, 340–348.

27. Bearden, C. E., & Freimer, N. B. (2006). Endophenotypes for psychiatric disorders: ready for primetime? *Trends in Genetics, 22*, 306–313.

28. Messina, E. S., Tyndale, R. F., & Sellers, E. M. (1997). A major role for CYP2A6 in nicotine C-oxidation by human liver microsomes. *Journal of Pharmacology and Experimental Therapeutics, 282*, 1608–1614.

29. Nakajima, M., Yamamoto, T., Nunoya, K., et al. (1996). Characterization of CYP2A6 involved in 3'-hydroxylation of cotinine in human liver microsomes. *Journal of Pharmacology and Experimental Therapeutics, 277*, 1010–1015.

30. Swan, G. E., Benowitz, N. L., Lessov, C. N., et al. (2005). Nicotine metabolism: the impact of CYP2A6 on estimates of additive genetic influence. *Pharmacogenetics and Genomics, 15*, 115–125.

31. Swan, G. E., Lessov-Schlaggar, C. N., Bergen, A. W., et al. (2009). Genetic and environmental influences on the ratio of 3' hydroxycotinine to cotinine in plasma and urine. *Pharmacogenetics and Genomics, 19*, 388–398.

32. Lerman, C., Tyndale, R., Patterson, F., et al. (2006). Nicotine metabolite ratio predicts efficacy of transdermal nicotine for smoking cessation. *Clinical Pharmacology and Therapeutics, 79*, 600–608.

33. Malaiyandi, V., Lerman, C., Benowitz, N. L., et al. (2006). Impact of CYP2A6 genotype on pretreatment smoking behaviour and nicotine levels from and usage of nicotine replacement therapy. *Molecular Psychiatry, 11*, 400–409.

34. Minematsu, N., Nakamura, H., Furuuchi, M., et al. (2006). Limitation of cigarette consumption by CYP2A6*4, *7 and *9 polymorphisms. *European Respiratory Journal, 27*, 289–292.

35. Wassenaar, C. A., Dong, Q., Wei, Q., et al. (2011). Relationship between CYP2A6 and CHRNA5–CHRNA3–CHRNB4 variation and smoking behaviors and lung cancer risk. *Journal of the National Cancer Institute, 103*, 1342–1346.

36. Schoedel, K. A., Hoffmann, E. B., Rao, Y., et al. (2004). Ethnic variation in CYP2A6 and association of genetically slow nicotine metabolism and smoking in adult Caucasians. *Pharmacogenetics, 14*, 615–626.

37. Khokhar, J. Y., Ferguson, C. S., Zhu, A. Z., et al. (2010). Pharmacogenetics of drug dependence: role of gene variations in susceptibility and treatment. *Annual Review of Pharmacology and Toxicology, 50*, 39–61.

38. Strasser, A. A., Malaiyandi, V., Hoffmann, E., et al. (2007). An association of CYP2A6 genotype and smoking topography. *Nicotine & Tobacco Research, 9*, 511–518.

39. Wassenaar, C. A., Dong, Q., Wei, Q., et al. (2011). Relationship between CYP2A6 and CHRNA5–CHRNA3–CHRNB4 variation and smoking behaviors and lung cancer risk. *Journal of the National Cancer Institute, 103*, 1342–1346.

40. Gu, D. F., Hinks, L. J., Morton, N. E., et al. (2000). The use of long PCR to confirm three common alleles at the CYP2A6 locus and the relationship between genotype and smoking habit. *Annals of Human Genetics, 64*(Pt 5), 383–390.

41. Yamano, S., Nagata, K., Yamazoe, Y., et al. (1989). cDNA and deduced amino acid sequences of human P450 IIA3 (CYP2A3). *Nucleic Acids Research, 17*, 4888.

42. Nakajima, M., Fukami, T., Yamanaka, H., et al. (2006). Comprehensive evaluation of variability in nicotine metabolism and CYP2A6 polymorphic alleles in four ethnic populations. *Clinical Pharmacology and Therapeutics, 80*, 282–297.

43. Mwenifumbo, J. C., Zhou, Q., Benowitz, N. L., et al. (2010). New CYP2A6 gene deletion and conversion variants in a population of Black African descent. *Pharmacogenomics, 11*, 189–198.

44. Benowitz, N. L., Swan, G. E., Jacob, P., III, et al. (2006). CYP2A6 genotype and the metabolism and disposition kinetics of nicotine. *Clinical Pharmacology and Therapeutics, 80*, 457–467.

45. Oscarson, M., McLellan, R. A., Asp, V., et al. (2002). Characterization of a novel CYP2A7/CYP2A6 hybrid allele (CYP2A6*12) that causes reduced CYP2A6 activity. *Human Mutation, 20*, 275–283.

46. O'Loughlin, J., Paradis, G., Kim, W., et al. (2004). Genetically decreased CYP2A6 and the risk of tobacco dependence: a prospective study of novice smokers. *Tobacco Control, 13*, 422–428.

47. Audrain-McGovern, J., Al Koudsi, N., Rodriguez, D., et al. (2007). The role of CYP2A6 in the emergence of nicotine dependence in adolescents. *Pediatrics, 119*, e264–e274.

48. Yamano, S., Tatsuno, J., & Gonzalez, F. J. (1990). The CYP2A3 gene product catalyzes coumarin 7-hydroxylation in human liver microsomes. *Biochemistry, 29*, 1322–1329.

49. Nunoya, K., Yokoi, T., Takahashi, Y., et al. (1999). Homologous unequal cross-over within the human CYP2A gene cluster as a mechanism for the deletion of the entire CYP2A6 gene associated with the poor metabolizer phenotype. *Journal of Biochemistry, 126*, 402–407.

50. Pitarque, M., von Richter, O., Oke, B., et al. (2001). Identification of a single nucleotide polymorphism in the TATA box of the CYP2A6 gene: impairment of its promoter activity. *Biochemical and Biophysical Research Communications, 284*, 455–460.

51. Mwenifumbo, J. C., Al Koudsi, N., Ho, M. K., et al. (2008). Novel and established CYP2A6 alleles impair in vivo nicotine metabolism in a population of Black African descent. *Human Mutation, 29*, 679–688.

52. Al Koudsi, N., Ahluwalia, J. S., Lin, S. K., et al. (2009). A novel CYP2A6 allele (CYP2A6*35) resulting in an amino-acid substitution (Asn438Tyr) is associated with lower CYP2A6 activity in vivo. *Pharmacogenomics Journal, 9*, 274–282.

53. Ho, M. K., Mwenifumbo, J. C., Zhao, B., et al. (2008). A novel CYP2A6 allele, CYP2A6*23, impairs enzyme function in vitro and in vivo and decreases smoking in a population of Black-African descent. *Pharmacogenetics and Genomics, 18*, 67–75.

54. Ho, M. K., Mwenifumbo, J. C., Al Koudsi, N., et al. (2009). Association of nicotine metabolite ratio and CYP2A6 genotype with smoking cessation treatment in African-American light smokers. *Clinical Pharmacology and Therapeutics, 85*, 635–643.

55. Benowitz, N. L., Perez-Stable, E. J., Herrera, B., et al. (2002). Slower metabolism and reduced intake of nicotine from cigarette smoking in Chinese-Americans. *Journal of the National Cancer Institute, 94*, 108–115.

56. Fagerstrom, K., Nakamura, M., Cho, H. J., et al. (2010). Varenicline treatment for smoking cessation in Asian populations: a pooled analysis of placebo-controlled trials conducted in six Asian countries. *Current Medical Research and Opinion, 26*, 2165–2173.

57. World Health Organization. (2010). *Global Adult Tobacco Survey* http://www.who.int/tobacco/ surveillance/en_tfi_china_gats_factsheet_2010.pdf.

58. Benowitz, N. L., Hukkanen, J., & Jacob, P., III. (2009). Nicotine chemistry, metabolism, kinetics and biomarkers. In J. E. Henningfield, E. D. London, & S. Pogun (Eds.), *Handbook of Experimental Pharmacology*: Vol. 192. Nicotine Psychopharmacology (pp. 29–60). Berlin: Springer-Verlag.

59. Sellers, E. M., Ramamoorthy, Y., Zeman, M. V., et al. (2003). The effect of methoxsalen on nicotine and 4-(methylnitrosamino)-1-(3-pyridyl)-1-butanone (NNK) metabolism in vivo. *Nicotine & Tobacco Research, 5*, 891–899.

60. Benowitz, N. L., Lessov-Schlaggar, C. N., Swan, G. E., et al. (2006). Female sex and oral contraceptive use accelerate nicotine metabolism. *Clinical Pharmacology and Therapeutics, 79*, 480–488.

61. Hakooz, N., & Hamdan, I. (2007). Effects of dietary broccoli on human in vivo caffeine metabolism: a pilot study on a group of Jordanian volunteers. *Current Drug Metabolism, 8*, 9–15.

62. Dempsey, D., Tutka, P., Jacob, P., III, et al. (2004). Nicotine metabolite ratio as an index of cytochrome P450 2A6 metabolic activity. *Clinical Pharmacology and Therapeutics, 76*, 64–72.

63. Bough, K. J., Lerman, C., Rose, J. E., McClernon, F. J., Kenny, J. P., Tyndale, R. F., et al. (in press). Biomarkers for smoking cessation. *Clinical Pharmacology and Therapeutics*.

64. Benowitz, N. L., & Jacob, P., III. (1994). Metabolism of nicotine to cotinine studied by a dual stable isotope method. *Clinical Pharmacology and Therapeutics, 56*, 483–493.

65. Malaiyandi, V., Goodz, S. D., Sellers, E. M., et al. (2006). CYP2A6 genotype, phenotype, and the use of nicotine metabolites as biomarkers during ad libitum smoking. *Cancer Epidemiology, Biomarkers & Prevention, 15*, 1812–1819.

66. Epstein, R. S. (2008). What's needed for personalized therapy in smoking cessation. *Clinical Pharmacology and Therapeutics, 84*, 309–310.

67. Johnstone, E., Benowitz, N., Cargill, A., et al. (2006). Determinants of the rate of nicotine metabolism and effects on smoking behavior. *Clinical Pharmacology and Therapeutics, 80*, 319–330.

68. Liu, T., David, S. P., Tyndale, R. F., et al. (2011). Associations of CYP2A6 genotype with smoking behaviors in southern China. *Addiction, 106*, 985–994.

69. Benowitz, N. L., Pomerleau, O. F., Pomerleau, C. S., et al. (2003). Nicotine metabolite ratio as a predictor of cigarette consumption. *Nicotine & Tobacco Research, 5*, 621–624.

70. Lerman, C., Kaufmann, V., Rukstalis, M., et al. (2004). Individualizing nicotine replacement therapy for the treatment of tobacco dependence: a randomized trial. *Annals of Internal Medicine,* *140,* 426–433.

71. Collins, B. N., Wileyto, E. P., Patterson, F., et al. (2004). Gender differences in smoking cessation in a placebo-controlled trial of bupropion with behavioral counseling. *Nicotine & Tobacco Research, 6,* 27–37.

72. Schnoll, R. A., Patterson, F., Wileyto, E. P., et al. (2010). Effectiveness of extended-duration transdermal nicotine therapy: a randomized trial. *Annals of Internal Medicine, 152,* 144–151.

73. Falcone, M., Jepson, C., Benowitz, N., et al. (2011). Association of the nicotine metabolite ratio and CHRNA5/CHRNA3 polymorphisms with smoking rate among treatment-seeking smokers. *Nicotine & Tobacco Research, 13,* 498–503.

74. Rubinstein, M. L., Benowitz, N. L., Auerback, G. M., et al. (2008). Rate of nicotine metabolism and withdrawal symptoms in adolescent light smokers. *Pediatrics, 122,* e643–e647.

75. Strasser, A. A., Benowitz, N. L., Pinto, A. G., et al. (2011). Nicotine metabolite ratio predicts smoking topography and carcinogen biomarker level. *Cancer Epidemiology, Biomarkers & Prevention, 20,* 234–238.

76. Benowitz, N. L., Dains, K. M., Dempsey, D., et al. (2011). Racial differences in the relationship between number of cigarettes smoked and nicotine and carcinogen exposure. *Nicotine & Tobacco Research, 13,* 772–783.

77. Ho, M. K., Faseru, B., Choi, W. S., et al. (2009). Utility and relationships of biomarkers of smoking in African-American light smokers. *Cancer Epidemiology, Biomarkers & Prevention, 18,* 3426–3434.

78. Wang, J., Liang, Q., Mendes, P., et al. (2011). Is 24h nicotine equivalents a surrogate for smoke exposure based on its relationship with other biomarkers of exposure? *Biomarkers, 16,* 144–154.

79. Le Marchand, L., et al. (2008). Smokers with the CHRNA lung cancer-associated variants are exposed to higher levels of nicotine equivalents and a carcinogenic tobacco-specific nitrosamine. *Cancer Research, 68,* 9137–9140.

80. Health Canada. (2003). *Smoking in Canada: an overview.* http://www.hc-sc.gc.ca/hc-ps/alt_formats/ hecs-sesc/pdf/tobac-tabac/research-recherche/stat/_ctums-esutc_fs-if/2003-smok-fum-eng.pdf.

81. Karp, I., O'Loughlin, J., Hanley, J., et al. (2006). Risk factors for tobacco dependence in adolescent smokers. *Tobacco Control, 15,* 199–204.

82. Al Koudsi, N., O'Loughlin, J., Rodriguez, D., et al. (2010). The genetic aspects of nicotine metabolism and their impact on adolescent nicotine dependence. *Journal of Pediatric Biochemistry, 1,* 19.

83. Hu, M. C., Griesler, P., Schaffran, C., et al. (2011). Risk and protective factors for nicotine dependence in adolescence. *Journal of Child Psychology and Psychiatry, and Allied Disciplines, 52,* 1063–1072.

84. Rodriguez, D., & Audrain-McGovern, J. (2004). Construct validity analysis of the early smoking experience questionnaire for adolescents. *Addictive Behaviors, 29*, 1053–1057.

85. Huang, S., Cook, D. G., Hinks, L. J., et al. (2005). CYP2A6, MAOA, DBH, DRD4, and 5HT2A genotypes, smoking behaviour and cotinine levels in 1518 UK adolescents. *Pharmacogenetics and Genomics, 15*, 839–850.

86. Schnoll, R. A., Patterson, F., Wileyto, E. P., et al. (2009). Nicotine metabolic rate predicts successful smoking cessation with transdermal nicotine: a validation study. *Pharmacology, Biochemistry, and Behavior, 92*, 6–11.

87. Payne, T. J., Smith, P. O., McCracken, L. M., et al. (1994). Assessing nicotine dependence: a comparison of the Fagerstrom Tolerance Questionnaire (FTQ) with the Fagerstrom Test for Nicotine Dependence (FTND) in a clinical sample. *Addictive Behaviors, 19*, 307–317.

88. Ahluwalia, J. S., Okuyemi, K., Nollen, N., et al. (2006). The effects of nicotine gum and counseling among African American light smokers: a 2 × 2 factorial design. design. *Addiction, 101*, 883–891.

89. Patterson, F., Schnoll, R. A., Wileyto, E. P., et al. (2008). Toward personalized therapy for smoking cessation: a randomized placebo-controlled trial of bupropion. *Clinical Pharmacology and Therapeutics, 84*, 320–325.

90. Chenoweth, M. J., O'Loughlin, J., Sylvestre, M., & Tyndale, R. F. (2013). CYP2A6 slow nicotine metabolism is associated with increased quitting by adolescent smokers. *Pharmacogenetics and Genomics, 23*, 232–235.

91. Kubota, T., Nakajima-Taniguchi, C., Fukuda, T., et al. (2006). CYP2A6 polymorphisms are associated with nicotine dependence and influence withdrawal symptoms in smoking cessation. *Pharmacogenomics Journal, 6*, 115–119.

92. Swan, G. E., Ward, M. M., & Jack, L. M. (1996). Abstinence effects as predictors of 28-day relapse in smokers. *Addictive Behaviors, 21*, 481–490.

93. Tang, D. W., Mroziewicz, M., Fellows, L. K., Tyndale, R. F., & Dagher, A. (2012). Genetic variation in CYP2A6 predicts neural response to smoking cues as measured using fMRI. *NeuroImage, 60*, 2136–2143.

94. Yamazaki, H., Inoue, K., Hashimoto, M., et al. (1999). Roles of CYP2A6 and CYP2B6 in nicotine C-oxidation by human liver microsomes. *Archives of Toxicology, 73*, 65–70.

95. Ring, H. Z., Valdes, A. M., Nishita, D. M., et al. (2007). Gene-gene interactions between CYP2B6 and CYP2A6 in nicotine metabolism. *Pharmacogenetics and Genomics, 17*, 1007–1015.

96. Lee, A. M., Jepson, C., Shields, P. G., et al. (2007). CYP2B6 genotype does not alter nicotine metabolism, plasma levels, or abstinence with nicotine replacement therapy. *Cancer Epidemiology, Biomarkers & Prevention, 16*, 1312–1314.

97. Al Koudsi, N., & Tyndale, R. F. (2010). Hepatic CYP2B6 is altered by genetic, physiologic, and environmental factors but plays little role in nicotine metabolism. *Xenobiotica, 40*, 381–392.

98. Lee, A. M., Jepson, C., Hoffmann, E., et al. (2007). CYP2B6 genotype alters abstinence rates in a bupropion smoking cessation trial. *Biological Psychiatry, 62,* 635–641.

99. Miles, J. S., Bickmore, W., Brook, J. D., et al. (1989). Close linkage of the human cytochrome P450IIA and P450IIB gene subfamilies: implications for the assignment of substrate specificity. *Nucleic Acids Research, 17,* 2907–2917.

100. Forrester, L. M., Henderson, C. J., Glancey, M. J., et al. (1992). Relative expression of cytochrome P450 isoenzymes in human liver and association with the metabolism of drugs and xenobiotics. *Biochemical Journal, 281*(Pt 2), 359–368.

101. Hoffman, S. M., Nelson, D. R., & Keeney, D. S. (2001). Organization, structure and evolution of the CYP2 gene cluster on human chromosome 19. *Pharmacogenetics, 11,* 687–698.

102. Haberl, M., Anwald, B., Klein, K., et al. (2005). Three haplotypes associated with CYP2A6 phenotypes in Caucasians. *Pharmacogenetics and Genomics, 15,* 609–624.

103. Sarginson, J. E., Killen, J. D., Lazzeroni, L. C., et al. (2011). Markers in the 15q24 nicotinic receptor subunit gene cluster (CHRNA5–A3-B4) predict severity of nicotine addiction and response to smoking cessation therapy. *American Journal of Medical Genetics. Part B, Neuropsychiatric Genetics, 156B,* 275–284.

104. Hung, R. J., McKay, J. D., Gaborieau, V., et al. (2008). A susceptibility locus for lung cancer maps to nicotinic acetylcholine receptor subunit genes on 15q25. *Nature, 452,* 633–637.

105. Thorgeirsson, T. E., Geller, F., Sulem, P., et al. (2008). A variant associated with nicotine dependence, lung cancer and peripheral arterial disease. *Nature, 452,* 638–642.

106. Amos, C. I., Wu, X., Broderick, P., et al. (2008). Genome-wide association scan of tag SNPs identifies a susceptibility locus for lung cancer at 15q25.1. *Nature Genetics, 40,* 616–622.

107. Spitz, M. R., Amos, C. I., Dong, Q., et al. (2008). The CHRNA5–A3 region on chromosome 15q24–25.1 is a risk factor both for nicotine dependence and for lung cancer. *Journal of the National Cancer Institute, 100,* 1552–1556.

108. Saccone, N. L., Culverhouse, R. C., Schwantes-An, T. H., et al. (2010). Multiple independent loci at chromosome 15q25.1 affect smoking quantity: a meta-analysis and comparison with lung cancer and COPD. *PLOS Genetics, 6,* e1001053.

109. Tamaki, Y., Arai, T., Sugimura, H., et al. (2011). Association between cancer risk and drug metabolizing enzyme gene (CYP2A6, CYP2A13, CYP4B1, SULT1A1, GSTM1, and GSTT1) polymorphisms in Japanese cases of lung cancer. *Drug Metabolism and Pharmacokinetics, 26,* 516–522.

110. Fujieda, M., Yamazaki, H., Saito, T., et al. (2004). Evaluation of CYP2A6 genetic polymorphisms as determinants of smoking behavior and tobacco-related lung cancer risk in male Japanese smokers. *Carcinogenesis, 25,* 2451–2458.

111. Ariyoshi, N., Miyamoto, M., Umetsu, Y., et al. (2002). Genetic polymorphism of CYP2A6 gene and tobacco-induced lung cancer risk in male smokers. *Cancer Epidemiology, Biomarkers & Prevention, 11,* 890–894.

112. Gemignani, F., Landi, S., Szeszenia-Dabrowska, N., et al. (2007). Development of lung cancer before the age of 50: the role of xenobiotic metabolizing genes. *Carcinogenesis, 28,* 1287–1293.

113. Wassenaar, C. A., Dong, Q., Amos, C. I., Spitz, M. R., & Tyndale, R. F. (Accepted). Pilot study of CYP2B6 genetic variation to explore the contribution of nitrosamine activation to lung carcinogenesis. *International Journal of Molecular Sciences.*

114. Rodriguez, S., Cook, D. G., Gaunt, T. R., et al. (2011). Combined analysis of CHRNA5, CHRNA3 and CYP2A6 in relation to adolescent smoking behaviour. *Journal of Psychopharmacology, 25,* 915–923.

115. Hecht, S. S. (2003). Tobacco carcinogens, their biomarkers and tobacco-induced cancer. *Nature Reviews. Cancer, 3,* 733–744.

116. Haiman, C. A., Stram, D. O., Wilkens, L. R., et al. (2006). Ethnic and racial differences in the smoking-related risk of lung cancer. *New England Journal of Medicine, 354,* 333–342.

117. Rachet, B., Siemiatycki, J., Abrahamowicz, M., et al. (2004). A flexible modeling approach to estimating the component effects of smoking behavior on lung cancer. *Journal of Clinical Epidemiology, 57,* 1076–1085.

118. Hecht, S. S. (2008). Progress and challenges in selected areas of tobacco carcinogenesis. *Chemical Research in Toxicology, 21,* 160–171.

119. Patten, C. J., Smith, T. J., Friesen, M. J., et al. (1997). Evidence for cytochrome P450 2A6 and 3A4 as major catalysts for N'-nitrosonornicotine alpha-hydroxylation by human liver microsomes. *Carcinogenesis, 18,* 1623–1630.

120. Rossini, A., de Almeida Simao, T., Albano, R. M., et al. (2008). CYP2A6 polymorphisms and risk for tobacco-related cancers. *Pharmacogenomics, 9,* 1737–1752.

121. Ruwali, M., Pant, M. C., Shah, P. P., et al. (2009). Polymorphism in cytochrome P450 2A6 and glutathione S-transferase P1 modifies head and neck cancer risk and treatment outcome. *Mutation Research, 669,* 36–41.

122. Sachse, C., Smith, G., Wilkie, M. J., et al. (2002). A pharmacogenetic study to investigate the role of dietary carcinogens in the etiology of colorectal cancer. *Carcinogenesis, 23,* 1839–1849.

123. Topcu, Z., Chiba, I., Fujieda, M., et al. (2002). CYP2A6 gene deletion reduces oral cancer risk in betel quid chewers in Sri Lanka. *Carcinogenesis, 23,* 595–598.

124. Canova, C., Hashibe, M., Simonato, L., et al. (2009). Genetic associations of 115 polymorphisms with cancers of the upper aerodigestive tract across 10 European countries: the ARCAGE project. *Cancer Research, 69,* 2956–2965.

125. Burns, D. M. (2003). Epidemiology of smoking-induced cardiovascular disease. *Progress in Cardiovascular Diseases, 46,* 11–29.

126. Bonita, R., Duncan, J., Truelsen, T., et al. (1999). Passive smoking as well as active smoking increases the risk of acute stroke. *Tobacco Control, 8,* 156–160.

127. Mahonen, M. S., McElduff, P., Dobson, A. J., et al. (2004). Current smoking and the risk of non-fatal myocardial infarction in the WHO MONICA Project populations. *Tobacco Control, 13*, 244–250.

128. Raij, L., DeMaster, E. G., & Jaimes, E. A. (2001). Cigarette smoke-induced endothelium dysfunction: role of superoxide anion. *Journal of Hypertension, 19*, 891–897.

129. Noronha-Dutra, A. A., Epperlein, M. M., & Woolf, N. (1993). Effect of cigarette smoking on cultured human endothelial cells. *Cardiovascular Research, 27*, 774–778.

130. Bernhard, D., Pfister, G., Huck, C. W., et al. (2003). Disruption of vascular endothelial homeostasis by tobacco smoke: impact on atherosclerosis. *FASEB Journal, 17*, 2302–2304.

131. Celermajer, D. S., Adams, M. R., Clarkson, P., et al. (1996). Passive smoking and impaired endothelium-dependent arterial dilatation in healthy young adults. *New England Journal of Medicine, 334*, 150–154.

132. Hansson, G. K. (2005). Inflammation, atherosclerosis, and coronary artery disease. *New England Journal of Medicine, 352*, 1685–1695.

133. Miettinen, O. S., Neff, R. K., & Jick, H. (1976). Cigarette-smoking and nonfatal myocardial infarction: rate ratio in relation to age, sex and predisposing conditions. *American Journal of Epidemiology, 103*, 30–36.

134. Al-Delaimy, W. K., Manson, J. E., Solomon, C. G., et al. (2002). Smoking and risk of coronary heart disease among women with type 2 diabetes mellitus. *Archives of Internal Medicine, 162*, 273–279.

135. Bjartveit, K., & Tverdal, A. (2005). Health consequences of smoking 1–4 cigarettes per day. *Tobacco Control, 14*, 315–320.

136. Rosengren, A., Wilhelmsen, L., & Wedel, H. (1992). Coronary heart disease, cancer and mortality in male middle-aged light smokers. *Journal of Internal Medicine, 231*, 357–362.

137. He, J., Vupputuri, S., Allen, K., et al. (1999). Passive smoking and the risk of coronary heart disease–a meta-analysis of epidemiologic studies. *New England Journal of Medicine, 340*, 920–926.

138. Liu, T., Tyndale, R. F., David, S. P., et al. (2013). Association between daily cigarette consumption and hypertension moderated by CYP2A6 genotypes in Chinese male current smokers. *Journal of Human Hypertension, 27*, 24–30.

139. Akbartabartoori, M., Lean, M. E., & Hankey, C. R. (2005). Relationships between cigarette smoking, body size and body shape. *International Journal of Obesity, 29*, 236–243.

140. Albanes, D., Jones, D. Y., Micozzi, M. S., et al. (1987). Associations between smoking and body weight in the US population: analysis of NHANES II. *American Journal of Public Health, 77*, 439–444.

141. Mooney, M. E., Li, Z. Z., Murphy, S. E., et al. (2008). Stability of the nicotine metabolite ratio in ad libitum and reducing smokers. *Cancer Epidemiology, Biomarkers & Prevention, 17*, 1396–1400.

142. Klesges, R. C., Winders, S. E., Meyers, A. W., et al. (1997). How much weight gain occurs following smoking cessation? A comparison of weight gain using both continuous and point prevalence abstinence. *Journal of Consulting and Clinical Psychology*, *65*, 286–291.

143. Perkins, K. A., Epstein, L. H., Stiller, R. L., et al. (1991). Acute effects of nicotine on hunger and caloric intake in smokers and nonsmokers. *Psychopharmacology*, *103*, 103–109.

144. Mineur, Y. S., Abizaid, A., Rao, Y., et al. (2011). Nicotine decreases food intake through activation of POMC neurons. *Science*, *332*, 1330–1332.

145. Barsh, G. S., & Schwartz, M. W. (2002). Genetic approaches to studying energy balance: perception and integration. *Nature Reviews. Genetics*, *3*, 589–600.

146. Smart, J. L., & Low, M. J. (2003). Lack of proopiomelanocortin peptides results in obesity and defective adrenal function but normal melanocyte pigmentation in the murine C57BL/6 genetic background. *Annals of the New York Academy of Sciences*, *994*, 202–210.

147. Krude, H., Biebermann, H., Schnabel, D., et al. (2003). Obesity due to proopiomelanocortin deficiency: three new cases and treatment trials with thyroid hormone and ACTH4–10. *Journal of Clinical Endocrinology and Metabolism*, *88*, 4633–4640.

148. Bamia, C., Trichopoulou, A., Lenas, D., et al. (2004). Tobacco smoking in relation to body fat mass and distribution in a general population sample. *International Journal of Obesity and Related Metabolic Disorders*, *28*, 1091–1096.

149. Canoy, D., Wareham, N., Luben, R., et al. (2005). Cigarette smoking and fat distribution in 21,828 British men and women: a population-based study. *Obesity Research*, *13*, 1466–1475.

150. Saarni, S. E., Pietilainen, K., Kantonen, S., et al. (2009). Association of smoking in adolescence with abdominal obesity in adulthood: a follow-up study of 5 birth cohorts of Finnish twins. *American Journal of Public Health*, *99*, 348–354.

151. Liu, T., David, S. P., Tyndale, R. F., et al. (2012). Relationship between amounts of daily cigarette consumption and abdominal obesity moderated by CYP2A6 genotypes in Chinese male current smokers. *Annals of Behavioral Medicine*, *43*, 253–261.

152. Yusuf, S., Hawken, S., Ounpuu, S., et al. (2004). Effect of potentially modifiable risk factors associated with myocardial infarction in 52 countries (the INTERHEART study): case-control study. *Lancet*, *364*, 937–952.

153. Van Gaal, L. F., Mertens, I. L., & De Block, C. E. (2006). Mechanisms linking obesity with cardiovascular disease. *Nature*, *444*, 875–880.

154. Willi, C., Bodenmann, P., Ghali, W. A., et al. (2007). Active smoking and the risk of type 2 diabetes: a systematic review and meta-analysis. *Journal of the American Medical Association*, *298*, 2654–2664.

155. Facchini, F. S., Hollenbeck, C. B., Jeppesen, J., et al. (1992). Insulin resistance and cigarette smoking. *Lancet*, *339*, 1128–1130.

156. Liu, T., Chen, W., David, S. P., et al. (2011). Interaction between heavy smoking and CYP2A6 genotypes on type 2 diabetes and its possible pathways. *European Journal of Endocrinology, 165*, 961–967.

157. Bruin, J. E., Petre, M. A., Raha, S., et al. (2008). Fetal and neonatal nicotine exposure in Wistar rats causes progressive pancreatic mitochondrial damage and beta cell dysfunction. *PLoS ONE, 3*, e3371.

158. Bruin, J. E., Gerstein, H. C., Morrison, K. M., et al. (2008). Increased pancreatic beta-cell apoptosis following fetal and neonatal exposure to nicotine is mediated via the mitochondria. *Toxicological Sciences, 103*, 362–370.

159. Butler, A. E., Janson, J., Bonner-Weir, S., et al. (2003). Beta-cell deficit and increased beta-cell apoptosis in humans with type 2 diabetes. *Diabetes, 52*, 102–110.

160. Murphy, P. J. (1973). Enzymatic oxidation of nicotine to nicotine 1'(5') iminium ion: a newly discovered intermediate in the metabolism of nicotine. *Journal of Biological Chemistry, 248*, 2796–2800.

161. Brandange, S., & Lindblom, L. (1979). The enzyme "aldehyde oxidase" is an iminium oxidase: reaction with nicotine delta 1'(5') iminium ion. *Biochemical and Biophysical Research Communications, 91*, 991–996.

5 Subjective Responses to Alcohol as an Endophenotype: Implications for Alcoholism Etiology and Treatment Development

Lara A. Ray and Markus Heilig

Alcoholism is a complex psychiatric disorder marked by the interplay between genetic and environmental risk and protective factors [1]. A multitude of pathways may lead to a common outcome of heavy drinking resulting in the development of alcoholism. These intraindividual pathways of risk include poor impulse control and impulsive decision making, externalizing psychopathology, and alleviation of mood and/or anxiety symptoms. Another risk pathway for alcoholism is indexed by subjective responses to the neuropharmacological effects of alcohol [2]. Research has demonstrated that individuals who are less sensitive to the aversive effects of alcohol (i.e., sedative and unpleasant effects] are in turn more likely to drink heavily and, ultimately, to develop an alcohol use disorder [3–5]. More recent findings have shown that greater sensitivity to the stimulant and reinforcing effects of alcohol may also predispose individuals to binge drinking and alcohol-related problems [6].

The pharmacological effects of alcohol involve both stimulant and sedative effects with the former being more prominent during the ascending limb of the blood alcohol curve (BAC) and the latter being most salient during the descending limb [7]. Therefore, subjective responses to alcohol represent the interplay between both rewarding and aversive behavioral effects [8], which in sum and over the course of repeated alcohol exposure function as a determinant of alcohol intake and related alcoholism risk. Since subjective responses to alcohol represent a fairly discrete pathway of vulnerability to alcoholism, they offer unique opportunities for translational science in addiction [9]. From a behavioral genetics perspective, subjective response to alcohol represents a potential endophenotype for alcoholism [2, 10].

Understanding genetic and neurobiological underpinnings of individual differences in subjective responses to alcohol may identify individual susceptibility factors for alcoholism and inform the development of personalized treatments that can effectively target specific pathophysiological mechanisms that are related to the individual risk factors. To that end, the objective of this chapter is threefold. First, it will review the literature on subjective responses to alcohol with a focus on its application to alcoholism etiology and evaluate the evidence supporting the use of subjective

responses as an endophenotype. Second, it will discuss the application of this phenotype to treatment development, including pharmacogenetics. Third, it will discuss
conceptual and methodological issues associated with the application of endophenotypes, such as subjective responses to alcohol, to translational science in addiction.
Specifically, we will discuss issues of endophenotype specificity across psychiatric
disorders and improvements in statistical power afforded by endophenotype-driven
approaches.

5.1 Subjective Response to Alcohol and Alcoholism Etiology

Individuals vary widely in their subjective experience of the pharmacological and
neurobehavioral effects of alcohol upon its consumption. Pharmacological effects
focus on the cellular and physiological effects of alcohol while subjective experiences
focus on an individual's self-reported perceptions of the effects of the substance.
While some individuals may be more or less sensitive to the positively reinforcing
and stimulant effects of alcohol, others report higher sensitivity to the aversive sedative effects. Recent research, primarily from alcohol administration studies, has documented the substantial variability in individuals' subjective responses to alcohol and
has shown that differences in these subjective experiences may play a significant
role in the predisposition to alcohol use and misuse (e.g., [4]). Importantly, recent
studies have shown that subjective response to alcohol represents a heritable phenotype [11, 12].

Schuckit and colleagues produced the early seminal work on the assessment of
self-reported subjective response to alcohol by measuring self-reported subjective
intoxication during alcohol administration sessions (i.e., alcohol challenge) (e.g., [13]).
In Schuckit's studies, the primary measure of subjective responses to alcohol is the
Subjective High Assessment Scale (SHAS). The SHAS consists of various positive and
negative mood-related adjectives, in addition to a single-item ad hoc scale of "feeling
high." Principal-components analysis of the SHAS suggested that the "maximum ter-
rible feelings" construct loaded into a first factor and accounted for 46% of the total
variance [14], thereby suggesting that the SHAS may be most sensitive to the unpleasant effects of alcohol. In fact, a factor analysis found that the SHAS is most strongly
correlated with measures of alcohol-induced sedation [8]. Perhaps the most compelling evidence that subjective responses to alcohol predict alcohol use and misuse
comes from a longitudinal study of sons of alcohol-dependent probands and controls,
suggesting that individuals who demonstrated low response to alcohol in the laboratory (measured by the SHAS) were more likely to develop alcoholism at follow-up [4].

More recent work has suggested that alcohol's pharmacological effects may be
biphasic in nature [7, 15–18]. It has been well documented that when blood alcohol
levels are rising, alcohol produces robust stimulatory and other pleasurable subjective

effects. Conversely, when blood alcohol levels are declining, alcohol's effects are largely sedative and unpleasant. This conceptualization of the effects of alcohol argues for the construct of subjective responses to be further parsed out into stimulant and sedative effects. Indeed, the Biphasic Alcohol Effects Scale [17] has been developed to directly assess the stimulant and sedative aspects of intoxication in alcohol administration studies. When subjective responses to alcohol are divided into stimulant and sedative effects, studies have shown that greater alcohol-induced stimulation and reinforcement are associated with increased alcohol consumption [19] whereas greater subjective experiences of the sedative and unpleasant effects of alcohol are associated with decreased alcohol use [20, 21].

An important effort toward resolving discrepancies in the alcohol administration literature comes from the work of Newlin and Thomson [22]. In the context of their review of alcohol challenge studies of sons of alcohol-dependent parents and controls, they proposed the differentiator model (DM) for understanding psychobiological responses to alcohol as a function of family history. This model proposes that responses to alcohol may be accentuated during the rising BAC (i.e., acute sensitization) and attenuated during the falling BAC (i.e., acute tolerance). The authors propose that sons of alcohol-dependent individuals may be both more sensitive to the rewarding effects of alcohol during the rising limb of the BAC and less sensitive to the unpleasant effects of alcohol when BAC is dropping. Acute tolerance and acute sensitization occur within session and represent a useful way to capture the "snapshot" of alcohol's effects obtained in a single administration session. This model has influenced efforts to parse out the phenotype of subjective response to alcohol into rewarding (primarily during the rising limb of BAC) and unpleasant (most salient during the descending limb of BAC).

This approach is somewhat consistent with the psychomotor stimulant theory of addictions, which posits that the stimulatory and rewarding effects of addictive substances, including alcohol, share a common underlying biological mechanism and that individuals who experience greater alcohol-induced reward are thought to be more likely to develop alcohol problems [23]. A recent meta-analysis [24] contrasting predictions from the low level of response model (LLRM) [4] and the DM [22] found support for both models and argued that they describe two sets of distinct phenotypic risk, with different etiological implications for alcoholism. In fact, the field is moving toward a paradigm shift in understanding subjective response to alcohol as a pathway of risk. Specifically, there is increasing recognition that the positive hedonic and sedative effects underlying responses to alcohol may be distinct and subserved by specific brain regions, neurotransmitters, and functional circuitry. Further, it has been recommended that subjective response to alcohol be specified by (a) the response being measured, (b) the amount and rate of alcohol administered, (c) BAC and whether in the ascending or descending limb, and (d) other potential risk factors under

investigation [25]. Together, these recommendations are in line with the conceptualization of subjective response to alcohol as a useful yet complex behavioral phenotype
which challenges the field to reach more standard assessment methods and reporting
conventions. To that end, we will next discuss subjective responses to alcohol as
endophenotypes for alcoholism.

5.1.1 Subjective Responses as Endophenotypes for Alcoholism

Genetic association studies typically rely on diagnostic phenotypes such as alcohol
abuse or dependence, which are influenced by many different genetic as well as environmental factors. Given the noted heterogeneity of diagnostic phenotypes, it has
become increasingly important to identify more specific and narrowly defined behavioral phenotypes (i.e., intermediate phenotypes or "endophenotypes") that are related
to the larger disorder [26]. Endophenotypes are thought to facilitate research in the
neurobiology of psychiatric disorders by being more homogenous and proximal to
the underlying genetic variation than the broader, more heterogenous diagnostic
phenotype [26, 27]. The identification of more narrow neurobehavioral phenotypes,
or endophenotypes, for alcoholism has received increased attention [10, 28] as is the
case for most psychiatric disorders [27, 29]. A good endophenotype should be narrowly
defined, readily identifiable, and related to the disorder of interest. When used correctly, endophenotypes for psychiatric disorders are thought to increase the power to
detect genetic risk for a given disorder (see discussion of statistical power). While there
is great interest in identifying potential endophenotypes for a range of complex phenotypes, including psychiatric disorders, it is important to systematically evaluate
phenotypes as to whether or not they constitute an adequate, and potentially useful,
endophenotype for the disorder of interest. To that end, Gottesman and Gould [27]
have put forth specific criteria for evaluating endophenotypes, which are the following: (1) specificity, (2) state independence, (3) heritability, (4) familial association, (5)
cosegregation, and (6) biological and clinical plausibility. Next we review each of these
criteria as they apply to subjective responses to alcohol.

Specificity refers to the expectation that the endophenotype is more strongly associated with the disease of interest than with other psychiatric disorders. Given that
subjective responses to alcohol are directly dependent upon alcohol consumption and
its associated pharmacological effects, the argument can also be made that this phenotype is highly specific to the disorder of interest. It is recognized that in addition
to the pharmacodynamic effects of alcohol, expectancies for the effects of alcohol and
alcohol cues also play a role in subjective effects of alcohol [30]. To address expectancy
effect, intravenous alcohol administration paradigms have been developed [31, 32],
which may result in a "cleaner" and more pharmacologically based endophenotype.

The argument can be made that similar mechanisms of pharmacological response
may be in play for substances other than alcohol. Patterns of cross-tolerance between

alcohol and benzodiazepines, for example, have been widely documented in the animal and human literature (e.g., [33]). Moreover, common neurotransmitter systems and pathways are involved in the reinforcing effects of multiple substances, as is the case for the dopaminergic and opioidergic systems, for example [23, 34, 35]. In summary, responses to alcohol may be an alcohol-specific phenotype, although future research is needed to determine its specificity in relation to other substances of abuse and their common reward pathways.

State independence refers to the phenotype's being stable and not simply a reflection of the disease process. As discussed above, Schuckit and colleagues found that subjective responses to alcohol measured in the laboratory before the development of alcohol-related problems predict the development of alcohol use disorders at eight-year follow-up [4]. That is true even in the case of individuals who were relatively alcohol naive at the time of the alcohol challenge. Significantly less is known about the state independence of alcohol's hedonic effects and its association to alcohol use disorders, but some data suggest that these effects become progressively attenuated as alcohol use becomes heavy and individuals develop alcohol use disorders [36, 37].

Heritability is an important criterion for evaluating endophenotypes and represents the degree to which the phenotype is influenced by genetic variance. Ideal endophenotypes are highly heritable, and one would wish for a highly heritable endophenotype very much in the same way we expect our behavioral measures to be reliable, so as to reduce "noise." Subjective responses to alcohol, measured in an experimental twin study, had a heritability estimate of 60% [12]. A more recent laboratory study of the offspring of fathers who completed an alcohol challenge 20 years earlier revealed a significant positive parent–offspring association for subjective feelings of intoxication and body sway, among family history positive individuals [38]. Although not providing direct evidence of heritability, this study is consistent with prior reports of the genetic influences on these endophenotypes and provides support for its reliability.

Familial association refers to the expectation that the endophenotype will be more prevalent among relatives of affected probands as compared to controls. There is substantial evidence that subjective responses to alcohol are influenced by family history of alcoholism (e.g., [39–41]), and a recent study has shown that the number of alcohol-dependent relatives was significantly associated with subjective response to alcohol in the laboratory [38]. However, as reviewed by Newlin and Thomson [22], the broader literature on alcohol challenge studies with sons of alcohol-dependent parents and controls is mixed with regard to several behavioral and biological markers.

Cosegregation is the expectation that the endophenotype will be more prevalent among the affected relatives compared to the unaffected relatives of ill probands. To date, there have been no animal or human studies of cosegregation patterns for alcohol endophenotypes.

Biological and clinical plausibility refers to the assumption that the endophenotype will bear a conceptual relationship to the disorder of interest. Subjective responses to alcohol are conceptually related to the clinical construct of alcoholism in that individuals who are more sensitive to the rewarding and positive effects of alcohol are thought to crave alcohol more and to drink more [6, 19, 23] whereas sensitivity to the unpleasant effects of alcohol may deter alcohol use and serve as a protective factor [42, 43]. From a biological standpoint, there is ample evidence that subjective responses to alcohol are informative regarding neurobiological and genetic factors underlying alcoholism.

In sum, subjective responses to alcohol meet several of the criteria for evaluating an endophenotype. However, most of the available evidence comes from studies using the SHAS, which, as discussed above, loads most strongly on the sedative effects of alcohol [8]. Therefore, additional work capturing the reinforcing and stimulant effects of alcohol and carefully examining its utility as an endophenotype for alcoholism is warranted.

5.1.2 Genetics of Subjective Responses to Alcohol

There are multiple neurotransmitter systems underlying the subjective effects of alcohol. Given the complexity of the neurobiological effects of alcohol, we have argued for considering subjective response to alcohol a moving target [44, 45]. In that regard, the endogenous opioid system has been implicated in the pathophysiology of alcoholism as it modulates the reinforcing effects of alcohol via activation of opioid receptors in the ventral tegmental area and nucleus accumbens, which enhances extracellular concentrations of dopamine in the mesolimbic pathway [46–50]. In light of the implication of endogenous opioids in alcohol-induced reward, several genetic association studies have focused on allelic variation in the mu-opioid receptor (OPRM1) gene as a plausible candidate gene for several alcoholism-related phenotypes.

In particular, a single nucleotide polymorphism (SNP) of *OPRM1*, the Asn40Asp SNP (rs17799971),[1] has received significant attention. This variant results in an amino acid change from asparagine to aspartic acid and eliminates a putative glycosylation site in the N-terminal extracellular loop of the receptor protein. Despite numerous attempts, the molecular consequences of this substitution remain unclear. The initial study that identified this variant reported it to result in a modestly increased receptor binding affinity for β-endorphin, consistent with a straightforward gain-of-function role of the mutation [34]. Subsequent studies from the same group [51] or others [49, 52] have, however, not replicated these findings. Furthermore, the minor allele at this locus (Asp40) has been associated with lower levels of receptor expression [52]. While numerous studies demonstrate this variant to be functional, it is presently unclear whether it is a gain-of-function or a loss-of-function mutation. Available data suggest that this may in fact vary depending on the mu-opioid receptor ligand exerting receptor activation, and in fact both can be observed under different conditions. For instance, studies of alcohol effects presumably rely on the actions of endogenous opioid peptides released in response to alcohol, and numerous studies with alcohol administration reviewed below consistently suggest the minor allele to be a gain-of-function variant. The same is found when pain thresholds, once again presumably reflecting endogenous opioid activity, are examined [53]. However, the opposite is found with administration of the exogenous mu-opioid receptor agonist morphine, both for human analgesia, and analgesia as well as inhibition of intracellular Ca^{2++} responses in humanized mice [54].

A number of human laboratory studies employing behavioral pharmacology and experimental designs have examined the effect of the Asn40Asp SNP in a host of alcohol-related phenotypes, including subjective response to alcohol. These studies offer an advantage over case–control genetic association studies in that experimental phenotypes are putatively closer to the neurobiology of alcoholism [2, 55] and laboratory designs can more easily control extraneous variables. Experimental approaches allow for manipulation and assessment of critical components of alcoholism that are more proximal to the underlying neurobiology of the disorder, such as mechanisms of alcohol-induced reward ("liking") and craving ("wanting") [28]. Given that alcohol dependence is highly heterogeneous in terms of both clinical phenomenology and causal pathways, more focused studies may be more effective at parsing discrete translational phenotypes and their underlying genetic bases. Furthermore, an often overlooked aspect is that studies that utilize subjects preselected by genotype and using quantitative traits can be analyzed within a framework of general linear models, which inherently has greater power than frequency-based approaches used with categorical traits such as presence or absence of a diagnosis (see below). Ultimately, however, the utility of experimental studies hinges upon their ability to effectively inform clinical practice, such as disorder etiology and treatment outcome.

A series of studies focusing on behavioral mechanisms of alcohol reinforcement (i.e., the hedonic subjective response to alcohol) have shown that compared to Asn40 homozygotes, Asp40 carriers report greater subjective reinforcement from alcohol consumption in the laboratory [32] and in the natural environment, measured using ecological momentary assessment (EMA) methods [56]. These studies are consistent with the hypothesized role of endogenous opioids as mediators of the reinforcing effects of alcohol and suggest that favorable phenotypes to probe for the effect of the Asn40Asp SNP may involve assays of the rewarding effects of alcohol. A self-report study in a sample of American Indians showed that Asp40 carriers reported expecting a more intense subjective response to alcohol (e.g., feeling buzzed, dizzy, drunk) as compared to Asn40 homozygotes, which in turn was correlated with lower alcohol use [57] although it remains unclear whether these responses were primarily stimulant or sedative.

A recent report combined a placebo-controlled intravenous alcohol administration with positron-emission tomography (PET) methodology to examine the striatal dopamine response to alcohol in social-drinking male Asp40 carriers and Asn40 homozygotes [49]. Using 11C-raclopride, this study showed that Asp40 carriers displayed a more potent striatal dopamine response to alcohol compared to individuals who were homozygous for the Asn40 allele. Further, a study by Filbey et al. [58] employed a novel functional magnetic resonance imaging (fMRI) based alcohol taste-cue paradigm to elicit alcohol craving and to activate the mesocorticolimbic circuitry underlying the phenotypic expression of craving using a sample of heavy drinkers [58, 59]. This study compared Asp40 carriers to Asn40 homozygotes on hemodynamic response to the alcohol taste-cue task and found that Asp40 carriers had greater fMRI blood-oxygen-level-dependent response in the mesocorticolimbic areas (i.e., ventral striatum, ventromedial prefrontal cortex, and orbitofrontal cortex) before and after a priming dose of alcohol relative to control cues and as compared to Asn40 homozygotes [59]. Together, these studies provide support for the biological plausibility of this polymorphism as a determinant of alcohol-induced reward, both in terms of hemodynamic response and dopamine release in the striatum following alcohol consumption.

In summary, a number of experimental paradigms have examined the effects of this polymorphism on subjective responses to alcohol. Results suggest that this variant may be associated with greater alcohol-induced reward, which is in turn consistent with the role of the opioidergic system in the hedonic properties, or liking, of alcohol as well as natural rewards [60]. Inconsistencies in the literature may be associated with factors such as sample characteristics (e.g., heavy drinkers vs. alcohol-dependent samples), which in turn may serve as a proxy for stages of alcohol-related problems. To the extent that this polymorphism is related to the reinforcing effects of alcohol, current neurobiological theories of addiction suggest that positive reinforcement is most salient early in the transition from heavy drinking to dependence whereas late stage alcoholism is characterized primarily by negative reinforcement processes [47, 61]. Nevertheless, such models of addiction have not been sufficiently translated to human samples, and the transition from positive to negative reinforcement remains poorly understood in clinical populations. On balance, the Asn40Asp SNP of the OPRM1 has been implicated in subjective response to alcohol, particularly sensitivity to the rewarding effects of alcohol.

While the OPRM1 gene may be involved in the stimulant and rewarding effects of alcohol, genetic markers underlying the sedative subjective effects of alcohol have not been clearly characterized. The sedative and anxiolytic effects of alcohol, more prominent during the descending limb, are thought to be mediated by gamma-aminobutyric acid (GABA) neurotransmission [62, 63]. A number of genetic association studies have examined candidate genes for level of response to alcohol, as defined by Schuckit and colleagues (e.g., [4, 64]). Results to date have not been entirely consistent, but there is some support for the role of variation in the $GABA_A\alpha_6$ and serotonin transporter genes in the subjective response to alcohol in the laboratory [64–66]. In a review of the literature on genes contributing to low level of response to alcohol, Schuckit, Smith, and Kalmijn [67] highlighted several candidate gene findings with possible pharmacodynamic effects on the level of response to alcohol and alcoholism risk. These findings included genes relating to alterations in the GABAergic system, with a particular focus on $GABA_A$ and its subunits, and genes involved in second messenger systems (e.g., G proteins and protein kinases) [67].

Recent studies have found some support for the involvement of genes coding for $GABA_A$ and its subunits on measures of subjective responses to alcohol, particularly with regard to its sedative effects (indexed by the LLRM) [68, 69]. Specifically, studies have examined a cluster of $GABA_A$ receptor genes (GABRA1, GABRA6, GABRB2, and GABRG2) on chromosome 5q, and while some results suggested significant associations to drinking behaviors, such as level of response to alcohol, history of blackouts, age of first drunkenness, and alcoholism [70, 71], others found no association between $GABA_A$ receptor genes and alcohol dependence [72, 73]. Importantly, a study by Covault et al. [74] suggested that markers in the 5′-region of the GABRG1 gene, which encodes the $GABA_A$ receptor γ-1 subunit, are in linkage disequilibrium (LD) with markers in the GABRA2 gene, which is adjacent to the GABRG1 gene and also located in chromosome 4p. Moreover, markers in the 5′-region of the GABRG1 gene showed associations with alcoholism in two samples of individuals of European ancestry [74]. A recent study of LD patterns of GABRG1 and GABRA2 across various populations showed further support for intergenic LD in five different populations, including European American [75]. A study of subclinical heavy drinkers found that a SNP of the GABRG1 gene (rs1497571) was associated with level of response to alcohol and drinking patterns [68]. These studies highlight the importance of genetic variation in $GABA_A$ receptor genes to the sedative effects of alcohol (LLRM approach), drinking patterns, and to the alcoholism phenotype more broadly. Next we discuss how these findings may be applied to treatment development and pharmacogenetics in particular.

5.2 Subjective Response to Alcohol and Alcoholism Treatment

Several studies have suggested that naltrexone, an opioid receptor antagonist, exerts its clinical effects, at least in part, by blocking the positively reinforcing effects of alcohol. Specifically, naltrexone alters subjective responses to alcohol by dampening feelings of alcohol-induced stimulation [76–78] and alcohol "high" [79], decreasing ratings of liking and enjoyment of the alcohol intoxication [77, 80] and increasing self-reported fatigue, tension, and confusion [81]. Naltrexone's effects are thought to be mediated through the blockade of opioid receptors, which in turn are associated with the positively reinforcing effects of alcohol upon consumption [35]. Although naltrexone is not entirely selective for any of the opioid receptor subtypes, the mu-opioid receptor subtype encoded by the OPRM1 gene is thought to be its primary target at clinically used doses. Given that naltrexone attenuates alcohol-induced reinforcement and that the Asn40Asp SNP of the OPRM1 gene is associated with a greater degree of positive reinforcement from alcohol, it stands to reason that individuals with the Asp40 allele may be more responsive to the behavioral and clinical effects of naltrexone.

The first pharmacogenetic reports of the Asn40Asp SNP were from studies of naloxone, an opioid antagonist often used to provide a pharmacological challenge of the hypothalamic–pituitary–adrenal (HPA) axis. Because the HPA axis in humans is under tonic mu-opioid inhibition [82], the degree of HPA-axis activation following a challenge with mu-opioid antagonists serves as a gauge of the potency with which molecules in this class block their target. These challenge studies found that the Asp40 allele was associated with greater cortisol response to opioid receptor blockade via naloxone [83]. The effects of this SNP on cortisol response to naloxone were independently replicated [84], and shortly after, the first report of naltrexone pharmacogenetics was published [85]. Oslin et al. [85], in a combined retrospective analysis of three separate clinical trials, showed that the Asn40Asp polymorphism was associated with clinical response to naltrexone among alcohol-dependent patients. The results were such that individuals with at least one copy of the Asp40 allele had lower relapse rates and longer time to return to heavy drinking when treated with naltrexone, as compared to Asn40 homozygotes [85]. In fact, Asp40 carriers were the only ones to benefit from naltrexone, while no significant difference between naltrexone and placebo was observed in subjects homozygous for the major Asn40 allele. As these findings emerged, influential reviews have highlighted the potential clinical utility of this polymorphism as a pharmacogenetic predictor of response to naltrexone, one of only three pharmacotherapies approved by the FDA for the treatment of alcoholism [86, 87]. Importantly, the Asn40Asp SNP was advanced as a potential genetic moderator of clinical response to naltrexone in the Combining Medications and Behavioral Interventions for Alcoholism (COMBINE) study, a large multisite trial allowing for analysis of clinically meaningful pharmacogenetic effects [88]. As with other studies, the COMBINE project allowed for a retrospective analysis of OPRM1 effects given that no prospective genotyping or treatment matching was conducted.

The COMBINE study found that carriers of the Asp40 allele receiving naltrexone plus medication management (MM; a minimal form of behavioral intervention less likely to obscure purely pharmacological treatment effects) reported a significantly greater decrease in heavy drinking days, compared to homozygotes for the Asn40 allele. A total of 87% of carriers of the Asp40 allele were classified as having a good clinical outcome to naltrexone plus MM while only 55% of individuals who were homozygous for the Asn40 allele were classified as good responders, suggesting a clinically meaningful predictive utility for this polymorphism [89]. These findings were upheld in haplotype-based analyses of the COMBINE study data set, which implicated the Asn40Asp SNP as the single *OPRM1* locus predictive of naltrexone response regarding the good clinical outcome variable [90].

Additional studies have examined the Asn40Asp SNP and naltrexone pharmacogenetics. A controlled laboratory study of naltrexone found that Asp40 carriers reported greater naltrexone-induced blunting of alcohol "high" as compared to placebo and to individuals who were homozygous for the Asn40 allele [77]. This study suggested a biobehavioral mechanism by which naltrexone may be differentially effective among Asp40 carriers, such that these individuals may be more sensitive to the reinforcing effects of alcohol and, in turn, more responsive to the dampening of alcohol "high" afforded by naltrexone. Secondary analyses from this laboratory study found that naltrexone selectively elevated GABAergic neurosteroid levels among Asp40 carriers, indicating a potential neurosteroid contribution to the differential effects of naltrexone among Asp40 carriers [91]. A recent placebo-controlled study has replicated the naltrexone-induced blunting of alcohol "euphoria" among Asp40 carriers after a priming dose of alcohol in a sample of social drinkers [92]. These effects were only seen in females and were not extended to a progressive ratio paradigm, such that naltrexone did not attenuate motivation to work for additional alcoholic beverages.

Given the known minor allele frequency imbalance across ethnic groups, extension of these findings to diverse samples is warranted. To that end, a placebo-controlled study of naltrexone among Korean alcohol-dependent patients found that, among treatment adherent individuals, Asp40 carriers reported longer time to relapse compared to Asn40 homozygotes [93]. These results only reached statistical significance when medication-noncompliant patients were excluded from the analysis. As with the preclinical literature discussed below, there is evidence of sex-specific effects of this polymorphism in a Korean sample, such that the Asp40 allele was overrepresented in women with alcohol dependence versus controls, but not in males [94]. This is relevant as some clinical trials have reported sex-specific effects, with males showing a better clinical response to naltrexone [95, 96], although these differences may be a function of reduced sample size and/or the end point drinking variable selection [97].

A laboratory-based study of naloxone comparing individuals of European ancestry to those of Asian ancestry found that the effects of the Asn40Asp polymorphism on naloxone-induced HPA-axis activation were restricted to individuals of European ancestry [98].

More recently, a placebo-controlled laboratory study of naltrexone among heavy drinkers of Asian descent found that Asp40 carriers experienced greater alcohol-induced sedation, subjective intoxication, and lower alcohol craving on naltrexone as compared to placebo and to Asn40 homozygotes. These results were maintained when controlling for *ALDH2* (rs671) and *ADH1B* (rs1229984) markers and when examining the three levels of OPRM1 genotype, thereby supporting an OPRM1 gene dose response [99]. These findings extend previous studies of naltrexone pharmacogenetics to individuals of Asian descent, an ethnic group more likely to carry the minor allele putatively associated with improved biobehavioral and clinical response to this medication.

A reanalysis of the COMBINE study data set focusing exclusively on African American participants did not support the efficacy of naltrexone [100]. However, the power to detect this effect was low, increasing the risk of type II error. Moreover, the lack of naltrexone effect might be explained by the low Asp40 allele frequency among African Americans (approximately 7% in the COMBINE study). Ancestry-specific effects remain a critical area of investigation for naltrexone pharmacogenetics, as well as etiological studies. While it remains unclear how alcohol-dependent African Americans, and those in other ethnic groups, respond to naltrexone overall, population effects and allele frequency considerations might be of particular importance as the field of pharmacogenetics (and genomics) progresses. These issues may have important implications for health disparities in the era of personalized medicine [101].

Pharmacogenetic studies of naltrexone and *OPRM1* are far from conclusive, and nonreplications have been reported. A large clinical study of male veterans did not find support for the moderating effect of Asn40Asp polymorphism on clinical response to naltrexone [102]. In the COMBINE study, patients who received a more extensive psychosocial intervention showed no effect of naltrexone and no *OPRM1* pharmacogenetic interaction, suggesting that robust psychosocial interventions may obscure pure pharmacological, as well as pharmacogenetic, effects. Another plausible explanation for the mixed findings is that the effect size of this pharmacogenetic interaction is small. There is increasing recognition in behavioral genetics that psychiatric disorders may be due to multiple genes of small effect size [103]. Applying a similar polygenic framework to pharmacogenetic studies would suggest a relatively small effect size for this particular pharmacogenetic finding, which in turn can account for the mixed findings.

Results of behavioral/laboratory studies of naltrexone pharmacogenetics have not been completely consistent. For instance, a placebo-controlled study of non-treatment seekers found that Asp40 carriers treated with naltrexone reported *greater* cue-induced craving for alcohol than Asn40 homozygotes and as compared to placebo [104]. Further analyses of the same sample failed to support a pharmacogenetic effect on measures of alcohol use and urge to drink in the natural environment, using EMA [105]. Similar null findings were reported for measures of cue reactivity in a mixed sample comprised of both non-treatment-seeking and treatment-seeking alcohol-dependent individuals [106]. A small neuroimaging study examined the pharmacogenetics of naltrexone on delay discounting and did not support the pharmacogenetic effect of the Asn40Asp SNP of *OPRM1* on impulsive decision making as well as activation of its underlying neurocircuitry [107]. Such null findings highlight the need to cautiously evaluate the empirical evidence before pharmacogenetic prescriptions can be made regarding naltrexone for alcoholism. As highlighted by Gelernter et al. [102], attention to additional opioid genes such as those encoding kappa and delta receptors, which are also targeted by naltrexone although to a lesser degree than mu receptors, represents an important avenue for future research.

In summary, the empirical data provide some evidence that the Asn40Asp SNP is associated with a differential subjective response to alcohol and is a predictor of the clinical response to naltrexone. These findings have been met with considerable enthusiasm as well as a healthy level of skepticism. It appears that considerable work remains to be done before the promise of targeted therapies may be realized for naltrexone. Nevertheless, the biological and clinical plausibility of this line of research is rather compelling in the case of naltrexone and the Asn40Asp SNP of *OPRM1*.

Beyond the pharmacogenetic studies of naltrexone, subjective responses to alcohol have been studied in the laboratory in the context of medication development for alcoholism (for a review, see [55]). Some medications with potential for treating alcoholism have been found to attenuate alcohol-induced reward or to potentiate the aversive effects of alcohol. For example, a pilot study of quetiapine in alcohol-dependent individuals found that it reduced subjective response to alcohol and, in particular, alcohol-induced sedation during placebo-controlled alcohol administration [108]. Further, a study of varenicline, a partial nicotinic receptor agonist, found that it potentiated the negative (sedative, unpleasant) subjective effects of alcohol, as compared to placebo [109]. Together, these studies highlight the clinical utility of assessing subjective responses to alcohol as a treatment target for alcoholism. Specifically, some effective medications for alcoholism may reduce motivation to drink by "blocking the buzz" [9, 110], or, in other words, attenuating the positively reinforcing effects of alcohol. Other medications may work by potentiating the aversive and sedative effects of alcohol or attenuating negative affective states that ultimately emerge in the absence of alcohol.

5.3 Methodological Issues and Future Directions

Alcohol administration studies have a long and rich history in the field of alcoholism. These studies have allowed for the identification of specific pathways of risk, namely the qualitative and quantitative nature of an individual's subjective responses to alcohol. Specifically, two sets of distinct phenotypic risk have emerged from the literature, each with different etiological and treatment implications. The first risk pathway conferred by subjective response to alcohol is that of greater sensitivity to the hedonic and stimulant effects of alcohol. The second related pathway of risk consists of lower sensitivity to the aversive and sedative effects of alcohol intoxication. While both alcohol reward and aversion co-occur within a drinking episode and change over the course of one's drinking history, the underlying neurotransmitter systems involved in the rewarding and sedative effects of alcohol appear to differ sufficiently to warrant further phenotypic refinement of subjective responses into these two dimensions. As has been recently argued in factor-analytic [8], quantitative meta-analysis [24], literature reviews [2, 111], and opinion papers [25, 112], the field has come to recognize subjective responses to alcohol as a multidimensional construct, most prominently marked by dimensions of reward and punishment. Future studies in the field would benefit from a more clear characterization of the specific dimension of response being measured, as well as considerations of alcohol dosage, limb of intoxication, and risk factors under study [25]. Moreover, sample characterization that takes into account alcohol involvement and stages of alcoholism, when present, is more likely to advance translational models of disorder etiology and treatment [113].

Another issue stemming from the conceptualization of subjective responses to alcohol is whether or not they represent endophenotypes that are relevant for alcoholism. If that is the case, it becomes important to ask the question of what advantages and shortcomings this category of phenotypes may have for alcoholism research. Based on the literature reviewed above, subjective responses to alcohol appear to meet a priori criteria for an endophenotype. While meeting the criteria for an endophenotype represents a meaningful index of this purported risk pathway, a number of issues remain to be resolved with regard to the utility of endophenotypes for the study of psychiatric disorders, broadly, and alcoholism in particular.

One such issue involves the notion that endophenotypes would improve statistical power, over diagnostic phenotypes, to detect genetic effects. This issue is often raised in theory and not addressed empirically. Quantifying the "gain of power" afforded by endophenotypes is critical to setting acceptable sample size requirements. This is particularly true given that alcohol administration is a costly and time-consuming process. For the sake of illustrating the point of statistical power, we have taken the OPRM1 findings as an example. We estimated effect sizes for four studies, all reporting significant findings. The first study is a traditional case and control association study with the diagnostic phenotype of alcohol dependence [114], the second study is a human laboratory investigation with the phenotype of subjective response to alcohol (i.e., stimulant/rewarding effects) [32], the third study is a PET investigation quantifying dopamine output in the striatum following alcohol administration [49], and the fourth study is a preclinical investigation of dopaminergic response to alcohol in mice [49]. As can be seen in figure 5.1, effect size estimates, a necessary requirement for power analysis, suggest that as we move away from the diagnostic phenotype and closer to the neurobiology of alcohol response, the effect sizes increase. This is the equivalent of "cranking up the microscope" from a diagnostic phenotype, to a behav-

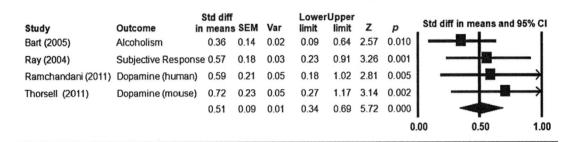

Study	Outcome	Std diff in means	SEM	Var	Lower limit	Upper limit	Z	p
Bart (2005)	Alcoholism	0.36	0.14	0.02	0.09	0.64	2.57	0.010
Ray (2004)	Subjective Response	0.57	0.18	0.03	0.23	0.91	3.26	0.001
Ramchandani (2011)	Dopamine (human)	0.59	0.21	0.05	0.18	1.02	2.81	0.005
Thorsell (2011)	Dopamine (mouse)	0.72	0.23	0.05	0.27	1.17	3.14	0.002
		0.51	0.09	0.01	0.34	0.69	5.72	0.000

Figure 5.1
Effect size estimates for studies of the A118G SNP of the OPRM1 gene across alcoholism phenotypes. All studies reported significant genetic findings, and findings were in the same direction,
such that Asp40 carriers scored higher than Asn40 homozygotes. Std Diff, standard difference;
SEM, standard error of the mean; CI, confidence interval.

ioral and pharmacological phenotype (subjective response to alcohol), to a more
biologically based brain measure in humans, and ending in the controlled animal
model. Further studies that can effectively quantify the gain in statistical power
afforded by a host of biological phenotypes, including neuroimaging paradigms, are
clearly needed to inform endophenotype-based research of psychiatric disorders.

Another issue is that of phenotypic specificity. Even though subjective responses
to alcohol are clearly linked to alcohol pharmacology and therefore appear to be substance specific, there is ample recognition of a common reward pathway by which
multiple drugs of abuse exert their positively reinforcing effects [115, 116]. Moreover,
quantitative genetic studies have demonstrated that the risk for substance use disorders is mostly shared across disorders, as opposed to being substance specific [117,
118]. Nevertheless, endophenotype-based research by definition narrows the risk to a
distinct pathway, which may in fact be more substance and disorder specific. A conceptual analysis of endophentoypes in psychiatry has highlighted several related issues
such as the bidirectional relationship between endophenotype and disorder, such that
the risk for the disorder is only partially mediated by the endophenotype [29]. In other
words, the endophenotype accounts for only a proportion of the variance in the disorder phenotype. Relatedly, not all of the variance in the endophenotype will be
shared with the disorder. These arguments have been elegantly described using path
analytic models [29] and suggest the need for more integrative and broader sets of
multivariate genetic models across psychiatric disorders.

In conclusion, this chapter reviewed the literature on subjective responses to
alcohol as they advance etiological and treatment models for alcohol dependence. It
is argued that while subjective responses to alcohol meet several of the criteria for an

endophenotype, the recent refinement of this construct into a multidimensional construct marked by both reward and punishment has yet to be incorporated in etiological and treatment models. The combination of insights from the human laboratory with clinical neuroscience paradigms, such as neuroimaging and preclinical models, represents a promising avenue for translational science in the field of alcoholism. Finally, a number of methodological and conceptual issues were highlighted with regard to the implementation of endophenotype-driven research in psychiatry, and alcoholism in particular. In conclusion, while alcohol administrations have had a long and productive history in alcoholism research, the next steps in this line of inquiry clearly call for further integration of this paradigm within the neuroscience literature. Such efforts include translational and reverse translational approaches that can more adequately capture the complex nature of subjective responses to alcohol and their biological bases.

Note

1. Note that although this SNP is referred to in the literature, as well as this manuscript, as the Asn40Asp (or the A118G SNP), this designation has been recently updated in the public bioinformatics databases (ABI, NCBI, HapMap) as it has been determined that the mu-opioid receptor may contain an additional 62 amino acids. The new designation of this SNP on the NCBI Human Genome Assembly 36 is Asn102Asp (or A355G).

References

1. Prescott, C. A., & Kendler, K. S. (1999). Genetic and environmental contributions to alcohol abuse and dependence in a population-based sample of male twins. *American Journal of Psychiatry*, *156*, 34–40.

2. Ray, L. A., Mackillop, J., & Monti, P. M. (2010). Subjective responses to alcohol consumption as endophenotypes: advancing behavioral genetics in etiological and treatment models of alcoholism. *Substance Use & Misuse*, *45*, 1742–1765.

3. Schuckit, M. A. (1984). Subjective responses to alcohol in sons of alcoholics and control subjects. *Archives of General Psychiatry*, *41*, 879–884.

4. Schuckit, M. A., & Smith, T. L. (1996). An 8-year follow-up of 450 sons of alcoholic and control subjects. *Archives of General Psychiatry*, *53*, 202–210.

5. Schuckit, M. A., & Smith, T. L. (2011). Onset and course of alcoholism over 25 years in middle class men. *Drug and Alcohol Dependence*, *113*, 21–28.

6. King, A. C., de Wit, H., McNamara, P. J., & Cao, D. (2011). Rewarding, stimulant, and sedative alcohol responses and relationship to future binge drinking. *Archives of General Psychiatry*, *68*, 389–399.

7. Martin, C. S., Earleywine, M., Musty, R. E., Perrine, M. W., & Swift, R. M. (1993). Development and validation of the Biphasic Alcohol Effects Scale. *Alcoholism, Clinical and Experimental Research*, *17*, 140–146.

8. Ray, L. A., MacKillop, J., Leventhal, A., & Hutchison, K. E. (2009). Catching the alcohol buzz: an examination of the latent factor structure of subjective intoxication. *Alcoholism, Clinical and Experimental Research*, *33*, 2154–2161.

9. Heilig, M., Thorsell, A., Sommer, W. H., et al. (2009). Translating the neuroscience of alcoholism into clinical treatments: from blocking the buzz to curing the blues. *Neuroscience and Biobehavioral Reviews*, *35*, 334–344.

10. Hines, L. M., Ray, L., Hutchison, K., & Tabakoff, B. (2005). Alcoholism: the dissection for endophenotypes. *Dialogues in Clinical Neuroscience*, *7*, 153–163.

11. Heath, A. C., & Martin, N. G. (1991). Intoxication after an acute dose of alcohol: an assessment of its association with alcohol consumption patterns by using twin data. *Alcoholism, Clinical and Experimental Research*, *15*, 122–128.

12. Viken, R. J., Rose, R. J., Morzorati, S. L., Christian, J. C., & Li, T. K. (2003). Subjective intoxication in response to alcohol challenge: heritability and covariation with personality, breath alcohol level, and drinking history. *Alcoholism, Clinical and Experimental Research*, *27*, 795–803.

13. Schuckit, M. A. (1980). Self-rating of alcohol intoxication by young men with and without family histories of alcoholism. *Journal of Studies on Alcohol*, *41*, 242–249.

14. Schuckit, M. A. (1985). Ethanol-induced changes in body sway in men at high alcoholism risk. *Archives of General Psychiatry*, *42*, 375–379.

15. Earleywine, M. (1994). Confirming the factor structure of the anticipated biphasic alcohol effects scale. *Alcoholism, Clinical and Experimental Research*, *18*, 861–866.

16. Earleywine, M. (1994). Anticipated biphasic effects of alcohol vary with risk for alcoholism: a preliminary report. *Alcoholism, Clinical and Experimental Research*, *18*, 711–714.

17. Earleywine, M., & Martin, C. S. (1993). Anticipated stimulant and sedative effects of alcohol vary with dosage and limb of the blood alcohol curve. *Alcoholism, Clinical and Experimental Research*, *17*, 135–139.

18. Erblich, J., Earleywine, M., Erblich, B., & Bovbjerg, D. H. (2003). Biphasic stimulant and sedative effects of ethanol: are children of alcoholics really different? *Addictive Behaviors*, *28*, 1129–1139.

19. Lewis, M. J., & June, H. L. (1990). Neurobehavioral studies of ethanol reward and activation. *Alcohol*, *7*, 213–219.

20. Leigh, B. C. (1987). Beliefs about the effects of alcohol on self and others. *Journal of Studies on Alcohol*, *48*, 467–475.

21. O'Malley, S. S., & Maisto, S. A. (1984). Factors affecting the perception of intoxication: dose, tolerance, and setting. *Addictive Behaviors*, *9*, 111–120.

22. Newlin, D. B., & Thomson, J. B. (1990). Alcohol challenge with sons of alcoholics: a critical review and analysis. *Psychological Bulletin, 108*, 383–402.

23. Wise, R. A., & Bozarth, M. A. (1987). A psychomotor stimulant theory of addiction. *Psychological Review, 94*, 469–492.

24. Quinn, P. D., & Fromme, K. (2011). Subjective response to alcohol challenge: a quantitative review. *Alcoholism, Clinical and Experimental Research, 35*, 1759–1770.

25. King, A. C., Roche, D. J., & Rueger, S. Y. (2011). Subjective responses to alcohol: a paradigm shift may be brewing. *Alcoholism, Clinical and Experimental Research, 35*, 1726–1728.

26. Gottesman, I., & Shields, J. (1972). *Schizophrenia and Genetics: A Twin Study Vantage Point.* London: Academic.

27. Gottesman, I. I., & Gould, T. D. (2003). The endophenotype concept in psychiatry: etymology and strategic intentions. *American Journal of Psychiatry, 160*, 636–645.

28. Ducci, F., & Goldman, D. (2008). Genetic approaches to addiction: genes and alcohol. *Addiction, 103*, 1414–1428.

29. Kendler, K. S., & Neale, M. C. (2010). Endophenotype: a conceptual analysis. *Molecular Psychiatry, 15*, 789–797.

30. Hasking, P. A., & Oei, T. P. (2008). Incorporating coping into an expectancy framework for explaining drinking behaviour. *Current Drug Abuse Reviews, 1*, 20–35.

31. Ramchandani, V. A., O'Connor, S., Neumark, Y., et al. (2006). The alcohol clamp: applications, challenges, and new directions—an RSA 2004 symposium summary. *Alcoholism, Clinical and Experimental Research, 30*, 155–164.

32. Ray, L. A., & Hutchison, K. E. (2004). A polymorphism of the mu-opioid receptor gene (OPRM1) and sensitivity to the effects of alcohol in humans. *Alcoholism, Clinical and Experimental Research, 28*, 1789–1795.

33. Toki, S., Saito, T., Nabeshima, A., et al. (1996). Changes in GABAA receptor function and cross-tolerance to ethanol in diazepam-dependent rats. *Alcoholism, Clinical and Experimental Research, 20*, 40A–44A.

34. Bond, C., LaForge, K. S., Tian, M., et al. (1998). Single-nucleotide polymorphism in the human mu opioid receptor gene alters beta-endorphin binding and activity: possible implications for opiate addiction. *Proceedings of the National Academy of Sciences of the United States of America, 95*, 9608–9613.

35. Herz, A. (1997). Endogenous opioid systems and alcohol addiction. *Psychopharmacology, 129*, 99–111.

36. Gilman, J. M., Ramchandani, V. A., Davis, M. B., Bjork, J. M., & Hommer, D. W. (2008). Why we like to drink: a functional magnetic resonance imaging study of the rewarding and anxiolytic effects of alcohol. *Journal of Neuroscience, 28*, 4583–4591.

37. Gilman, J. M., Ramchandani, V. A., Crouss, T., & Hommer, D. W. (2012). Subjective and neural responses to intravenous alcohol in young adults with light and heavy drinking patterns. *Neuropsychopharmacology, 37*, 467–477.

38. Schuckit, M. A., Smith, T. L., Kalmijn, J., & Danko, G. P. (2005). A cross-generational comparison of alcohol challenges at about age 20 in 40 father–offspring pairs. *Alcoholism, Clinical and Experimental Research, 29*, 1921–1927.

39. Conrod, P. J., Peterson, J. B., Pihl, R. O., & Mankowski, S. (1997). Biphasic effects of alcohol on heart rate are influenced by alcoholic family history and rate of alcohol ingestion. *Alcoholism, Clinical and Experimental Research, 21*, 140–149.

40. Pollock, V. E. (1992). Meta-analysis of subjective sensitivity to alcohol in sons of alcoholics. *American Journal of Psychiatry, 149*, 1534–1538.

41. Schuckit, M. A., & Gold, E. O. (1988). A simultaneous evaluation of multiple markers of ethanol/placebo challenges in sons of alcoholics and controls. *Archives of General Psychiatry, 45*, 211–216.

42. Duranceaux, N. C., Schuckit, M. A., Luczak, S. E., et al. (2008). Ethnic differences in level of response to alcohol between Chinese Americans and Korean Americans. *Journal of Studies on Alcohol and Drugs, 69*, 227–234.

43. Eng, M. Y., Luczak, S. E., & Wall, T. L. (2007). ALDH2, ADH1B, and ADH1C genotypes in Asians: a literature review. *Alcohol Research & Health, 30*, 22–27.

44. Ray, L. A., Hutchison, K. E., MacKillop, J., et al. (2008). Effects of naltrexone during the descending limb of the blood alcohol curve. *American Journal on Addictions, 17*, 257–264.

45. Heilig, M., & Egli, M. (2006). Pharmacological treatment of alcohol dependence: target symptoms and target mechanisms. *Pharmacology & Therapeutics, 111*, 855–876.

46. Gianoulakis, C. (2009). Endogenous opioids and addiction to alcohol and other drugs of abuse. *Current Topics in Medicinal Chemistry, 9*, 999–1015.

47. Koob, G. F., & Kreek, M. J. (2007). Stress, dysregulation of drug reward pathways, and the transition to drug dependence. *American Journal of Psychiatry, 164*, 1149–1159.

48. Kreek, M. J. (1996). Opiates, opioids and addiction. *Molecular Psychiatry, 1*, 232–254.

49. Ramchandani, V. A., Umhau, J., Pavon, F. J., et al. (2011). A genetic determinant of the striatal dopamine response to alcohol in men. *Molecular Psychiatry, 16*, 809–817.

50. Mitchell, J. M., O'Neil, J. P., Janabi, M. et al. (2012). Alcohol consumption induces endogenous opioid release in the human orbitofrontal cortex and nucleus accumbens. *Science Translational Medicine, 4*, 116ra6.

51. Kroslak, T., Laforge, K. S., Gianotti, R. J., et al. (2007). The single nucleotide polymorphism A118G alters functional properties of the human mu opioid receptor. *Journal of Neurochemistry, 103*, 77–87.

52. Zhang, Y., Wang, D., Johnson, A. D., Papp, A. C., & Sadee, W. (2005). Allelic expression imbalance of human mu opioid receptor (OPRM1) caused by variant A118G. *Journal of Biological Chemistry, 280*, 32618–32624.

53. Fillingim, R. B., Kaplan, L., Staud, R., et al. (2005). The A118G single nucleotide polymorphism of the mu-opioid receptor gene (OPRM1) is associated with pressure pain sensitivity in humans. *Journal of Pain, 6*, 159–167.

54. Lotsch, J., & Geisslinger, G. (2006). Current evidence for a genetic modulation of the response to analgesics. *Pain, 121*, 1–5.

55. Ray, L. A., Hutchison, K. E., & Tartter, M. (2010). Application of human laboratory models to pharmacotherapy development for alcohol dependence. *Current Pharmaceutical Design, 16*, 2149–2158.

56. Ray, L. A., Miranda, R., Jr., Tidey, J. W., et al. (2010). Polymorphisms of the mu-opioid receptor and dopamine D4 receptor genes and subjective responses to alcohol in the natural environment. *Journal of Abnormal Psychology, 119*, 115–125.

57. Ehlers, C. L., Lind, P. A., & Wilhelmsen, K. C. (2008). Association between single nucleotide polymorphisms in the mu opioid receptor gene (OPRM1) and self-reported responses to alcohol in American Indians. *BMC Medical Genetics, 9*, 35.

58. Filbey, F. M., Claus, E., Audette, A. R., et al. (2008). Exposure to the taste of alcohol elicits activation of the mesocorticolimbic neurocircuitry. *Neuropsychopharmacology, 33*, 1391–1401.

59. Filbey, F. M., Ray, L., Smolen, A., et al. (2008). Differential neural response to alcohol priming and alcohol taste cues is associated with DRD4 VNTR and OPRM1 genotypes. *Alcoholism, Clinical and Experimental Research, 32*, 1113–1123.

60. Robinson, T. E., & Berridge, K. C. (1993). The neural basis of drug craving: an incentive-sensitization theory of addiction. *Brain Research. Brain Research Reviews, 18*, 247–291.

61. Heilig, M., Egli, M., Crabbe, J. C., & Becker, H. C. (2010). Acute withdrawal, protracted abstinence and negative affect in alcoholism: are they linked? *Addiction Biology, 15*, 169–184.

62. Buck, K. J. (1996). Molecular genetic analysis of the role of GABAergic systems in the behavioral and cellular actions of alcohol. *Behavior Genetics, 26*, 313–323.

63. Grobin, A. C., Matthews, D. B., Devaud, L. L., & Morrow, A. L. (1998). The role of GABA(A) receptors in the acute and chronic effects of ethanol. *Psychopharmacology, 139*, 2–19.

64. Schuckit, M. A., Mazzanti, C., Smith, T. L., et al. (1999). Selective genotyping for the role of 5-HT2A, 5-HT2C, and GABA alpha 6 receptors and the serotonin transporter in the level of response to alcohol: a pilot study. *Biological Psychiatry, 45*, 647–651.

65. Corbin, W. R., Fromme, K., & Bergeson, S. E. (2006). Preliminary data on the association among the serotonin transporter polymorphism, subjective alcohol experiences, and drinking behavior. *Journal of Studies on Alcohol, 67*, 5–13.

66. Fromme, K., de Wit, H., Hutchison, K. E., et al. (2004). Biological and behavioral markers of alcohol sensitivity. *Alcoholism, Clinical and Experimental Research, 28,* 247–256.

67. Schuckit, M. A., Smith, T. L., & Kalmijn, J. (2004). The search for genes contributing to the low level of response to alcohol: patterns of findings across studies. *Alcoholism, Clinical and Experimental Research, 28,* 1449–1458.

68. Ray, L. A., & Hutchison, K. E. (2009). Associations among GABRG1, level of response to alcohol, and drinking behaviors. *Alcoholism, Clinical and Experimental Research, 33,* 1382–1390.

69. Roh, S., Matsushita, S., Hara, S., et al. (2011). Role of GABRA2 in moderating subjective responses to alcohol. *Alcoholism, Clinical and Experimental Research, 35,* 400–407.

70. Dick, D. M., Plunkett, J., Wetherill, L. F., et al. (2006). Association between GABRA1 and drinking behaviors in the collaborative study on the genetics of alcoholism sample. *Alcoholism, Clinical and Experimental Research, 30,* 1101–1110.

71. Radel, M., Vallejo, R. L., Iwata, N., et al. (2005). Haplotype-based localization of an alcohol dependence gene to the 5q34 {gamma}-aminobutyric acid type A gene cluster. *Archives of General Psychiatry, 62,* 47–55.

72. Dick, D. M., Edenberg, H. J., Xuei, X., et al. (2005). No association of the GABAA receptor genes on chromosome 5 with alcoholism in the collaborative study on the genetics of alcoholism sample. *American Journal of Medical Genetics. Part B, Neuropsychiatric Genetics, 132B,* 24–28.

73. Sander, T., Ball, D., Murray, R., et al. (1999). Association analysis of sequence variants of GABA(A) alpha6, beta2, and gamma2 gene cluster and alcohol dependence. *Alcoholism, Clinical and Experimental Research, 23,* 427–431.

74. Covault, J., Gelernter, J., Jensen, K., Anton, R., & Kranzler, H. R. (2008). Markers in the 5'-region of GABRG1 associate to alcohol dependence and are in linkage disequilibrium with markers in the adjacent GABRA2 gene. *Neuropsychopharmacology, 33,* 837–848.

75. Ittiwut, C., Listman, J., Mutirangura, A., et al. (2008). Interpopulation linkage disequilibrium patterns of GABRA2 and GABRG1 genes at the GABA cluster locus on human chromosome 4. *Genomics, 91,* 61–69.

76. Drobes, D. J., Anton, R. F., Thomas, S. E., & Voronin, K. (2004). Effects of naltrexone and nalmefene on subjective response to alcohol among non-treatment-seeking alcoholics and social drinkers. *Alcoholism, Clinical and Experimental Research, 28,* 1362–1370.

77. Ray, L. A., & Hutchison, K. E. (2007). Effects of naltrexone on alcohol sensitivity and genetic moderators of medication response: a double-blind placebo-controlled study. *Archives of General Psychiatry, 64,* 1069–1077.

78. Swift, R. M., Whelihan, W., Kuznetsov, O., Buongiorno, G., & Hsuing, H. (1994). Naltrexone-induced alterations in human ethanol intoxication. *American Journal of Psychiatry, 151,* 1463–1467.

79. Volpicelli, J. R., Watson, N. T., King, A. C., Sherman, C. E., & O'Brien, C. P. (1995). Effect of naltrexone on alcohol "high" in alcoholics. *American Journal of Psychiatry*, *152*, 613–615.

80. McCaul, M. E., Wand, G. S., Stauffer, R., Lee, S. M., & Rohde, C. A. (2001). Naltrexone dampens ethanol-induced cardiovascular and hypothalamic–pituitary–adrenal axis activation. *Neuropsychopharmacology*, *25*, 537–547.

81. King, A. C., Volpicelli, J. R., Frazer, A., & O'Brien, C. P. (1997). Effect of naltrexone on subjective alcohol response in subjects at high and low risk for future alcohol dependence. *Psychopharmacology*, *129*, 15–22.

82. Kreek, M. J., LaForge, K. S., & Butelman, E. (2002). Pharmacotherapy of addictions. *Nature Reviews. Drug Discovery*, *1*, 710–726.

83. Wand, G. S., McCaul, M., Yang, X., et al. (2002). The mu-opioid receptor gene polymorphism (A118G) alters HPA axis activation induced by opioid receptor blockade. *Neuropsychopharmacology*, *26*, 106–114.

84. Hernandez-Avila, C. A., Wand, G., Luo, X., Gelernter, J., & Kranzler, H. R. (2003). Association between the cortisol response to opioid blockade and the Asn40Asp polymorphism at the mu-opioid receptor locus (OPRM1). *American Journal of Medical Genetics. Part B, Neuropsychiatric Genetics*, *118B*, 60–65.

85. Oslin, D. W., Berrettini, W., Kranzler, H. R., et al. (2003). A functional polymorphism of the mu-opioid receptor gene is associated with naltrexone response in alcohol-dependent patients. *Neuropsychopharmacology*, *28*, 1546–1552.

86. Oslin, D. W., Berrettini, W. H., & O'Brien, C. P. (2006). Targeting treatments for alcohol dependence: the pharmacogenetics of naltrexone. *Addiction Biology*, *11*, 397–403.

87. Oroszi, G., & Goldman, D. (2004). Alcoholism: genes and mechanisms. *Pharmacogenomics*, *5*, 1037–1048.

88. Goldman, D., Oroszi, G., O'Malley, S., & Anton, R. (2005). COMBINE genetics study: the pharmacogenetics of alcoholism treatment response: genes and mechanisms. *Journal of Studies on Alcohol* (Suppl 15), 56–64, discussion 33.

89. Anton, R. F., Oroszi, G., O'Malley, S., et al. (2008). An evaluation of mu-opioid receptor (OPRM1) as a predictor of naltrexone response in the treatment of alcohol dependence: results from the Combined Pharmacotherapies and Behavioral Interventions for Alcohol Dependence (COMBINE) study. *Archives of General Psychiatry*, *65*, 135–144.

90. Oroszi, G., Anton, R. F., O'Malley, S., et al. (2008). OPRM1 Asn40Asp predicts response to naltrexone treatment: a haplotype-based approach. *Alcoholism, Clinical and Experimental Research*, *33*, 383–393.

91. Ray, L. A., Hutchison, K. E., Ashenhurst, J. R., & Morrow, A. L. (2010). Naltrexone selectively elevates GABAergic neuroactive steroid levels in heavy drinkers with the ASP40 allele of the OPRM1 gene: a pilot investigation. *Alcoholism, Clinical and Experimental Research*, *34*, 1479–1487.

92. Setiawan, E., Pihl, R. O., Cox, S. M., et al. (2011). The effect of naltrexone on alcohol's stimulant properties and self-administration behavior in social drinkers: influence of gender and genotype. *Alcoholism, Clinical and Experimental Research*, *35*, 1134–1141.

93. Kim, S. G., Kim, C. M., Choi, S. W., et al. (2009). A micro opioid receptor gene polymorphism (A118G) and naltrexone treatment response in adherent Korean alcohol-dependent patients. *Psychopharmacology*, *201*, 611–618.

94. Kim, S. G. (2009). Gender differences in the genetic risk for alcohol dependence—the results of a pharmacogenetic study in Korean alcoholics. *Nihon Arukoru. Yakubutsu Igakkai Zasshi*, *44*, 680–685.

95. Garbutt, J. C., Kranzler, H. R., O'Malley, S. S., et al. (2005). Efficacy and tolerability of long-acting injectable naltrexone for alcohol dependence: a randomized controlled trial. *Journal of the American Medical Association*, *293*, 1617–1625.

96. Kranzler, H. R., Tennen, H., Armeli, S., et al. (2009). Targeted naltrexone for problem drinkers. *Journal of Clinical Psychopharmacology*, *29*, 350–357.

97. Baros, A. M., Latham, P. K., & Anton, R. F. (2008). Naltrexone and cognitive behavioral therapy for the treatment of alcohol dependence: do sex differences exist? *Alcoholism, Clinical and Experimental Research*, *32*, 771–776.

98. Hernandez-Avila, C. A., Covault, J., Wand, G., et al. (2007). Population-specific effects of the Asn40Asp polymorphism at the mu-opioid receptor gene (OPRM1) on HPA-axis activation. *Pharmacogenetics and Genomics*, *17*, 1031–1038.

99. Ray, L. A., Bujarski, S., Chin, P. F., & Miotto, K. (2012). Pharmacogenetics of naltrexone in Asian Americans: a randomized placebo-controlled laboratory study. *Neuropsychopharmacology*, *37*, 445–455.

100. Ray, L. A., & Oslin, D. W. (2009). Naltrexone for the treatment of alcohol dependence among African Americans: results from the COMBINE Study. *Drug and Alcohol Dependence*, *105*, 256–258.

101. Tate, S. K., & Goldstein, D. B. (2004). Will tomorrow's medicines work for everyone? *Nature Genetics*, *36*, S34–S42.

102. Gelernter, J., Gueorguieva, R., Kranzler, H. R., et al. (2007). Opioid receptor gene (OPRM1, OPRK1, and OPRD1) variants and response to naltrexone treatment for alcohol dependence: results from the VA Cooperative Study. *Alcoholism, Clinical and Experimental Research*, *31*, 555–563.

103. Plomin, R., Haworth, C. M., & Davis, O. S. (2009). Common disorders are quantitative traits. *Nature Reviews. Genetics*, *10*, 872–878.

104. McGeary, J. E., Monti, P. M., Rohsenow, D. J., et al. (2006). Genetic moderators of naltrexone's effects on alcohol cue reactivity. *Alcoholism, Clinical and Experimental Research*, *30*, 1288–1296.

105. Tidey, J. W., Monti, P. M., Rohsenow, D. J., et al. (2008). Moderators of naltrexone's effects on drinking, urge, and alcohol effects in non-treatment-seeking heavy drinkers in the natural environment. *Alcoholism, Clinical and Experimental Research, 32,* 58–66.

106. Ooteman, W., Naassila, M., Koeter, M. W., et al. (2009). Predicting the effect of naltrexone and acamprosate in alcohol-dependent patients using genetic indicators. *Addiction Biology, 14,* 328–337.

107. Boettiger, C. A., Kelley, E. A., Mitchell, J. M., D'Esposito, M., & Fields, H. L. (2009). Now or later? An fMRI study of the effects of endogenous opioid blockade on a decision-making network. *Pharmacology, Biochemistry, and Behavior, 93,* 291–299.

108. Ray, L. A., Chin, P. F., Heydari, A., & Miotto, K. (2011). A human laboratory study of the effects of quetiapine on subjective intoxication and alcohol craving. *Psychopharmacology, 217,* 341–351.

109. Childs, E., Roche, D. J., King, A. C., & de Wit, H. (2012). Varenicline potentiates alcohol-induced negative subjective responses and offsets impaired eye movements. *Alcoholism, Clinical and Experimental Research, 36,* 906–914.

110. Heilig, M., Goldman, D., Berrettini, W., & O'Brien, C. P. (2011). Pharmacogenetic approaches to the treatment of alcohol addiction. *Nature Reviews. Neuroscience, 12,* 670–684.

111. Morean, M. E., & Corbin, W. R. (2011). Subjective response to alcohol: a critical review of the literature. *Alcoholism, Clinical and Experimental Research, 34,* 385–395.

112. Schuckit, M. A. (2011). Comment on the paper by Quinn and Fromme entitled subjective response to alcohol challenge: a quantitative review. *Alcoholism, Clinical and Experimental Research, 35,* 1723–1725.

113. Ray, L. A., Courtney, K. E., Bujarski, S., & Squeglia, L. M. (2012). Pharmacogenetics of alcoholism: a clinical neuroscience perspective. *Pharmacogenomics, 13,* 129–132.

114. Bart, G., Kreek, M. J., Ott, J., et al. (2005). Increased attributable risk related to a functional mu-opioid receptor gene polymorphism in association with alcohol dependence in central Sweden. *Neuropsychopharmacology, 30,* 417–422.

115. Kalivas, P. W., & Volkow, N. D. (2005). The neural basis of addiction: a pathology of motivation and choice. *American Journal of Psychiatry, 162,* 1403–1413.

116. Volkow, N. D., & Li, T. K. (2005). The neuroscience of addiction. *Nature Neuroscience, 8,* 1429–1430.

117. Kendler, K. S., Myers, J., & Prescott, C. A. (2007). Specificity of genetic and environmental risk factors for symptoms of cannabis, cocaine, alcohol, caffeine, and nicotine dependence. *Archives of General Psychiatry, 64,* 1313–1320.

118. Kendler, K. S., Jacobson, K. C., Prescott, C. A., & Neale, M. C. (2003). Specificity of genetic and environmental risk factors for use and abuse/dependence of cannabis, cocaine, hallucinogens, sedatives, stimulants, and opiates in male twins. *American Journal of Psychiatry, 160,* 687–695.

6 Subjective Drug Effects as Intermediate Phenotypes for Substance Abuse

Leah M. Mayo, Abraham A. Palmer, and Harriet de Wit

This chapter will discuss the use of drug-induced subjective effects in humans as an intermediate phenotype for substance abuse. We will first discuss the advantages to this approach, as well as the common methodology used for studies involving subjective drug effects. Next, we will review the current literature involving the use of subjective effects of caffeine, amphetamine, and nicotine as intermediate phenotypes. Finally, we will discuss the limitations of this methodology and the future directions this research may take.

Drugs from different pharmacological classes produce prototypic effects on how people "feel" (e.g., "stimulants," "sedatives," "hallucinogens"). These "subjective" effects are quantifiable, time dependent, and dose dependent, and they are consistent with the drugs' known effects in nonhumans. However, there are also marked individual differences in the subjective effects of drugs. For example, it is known that some people experience stimulant effects from alcohol whereas others experience sedation [1, 2]. Nonhumans exhibit comparable heterogeneity [3]. Similarly, some people feel energized and confident after taking amphetamine whereas others feel anxious [4]. There is also evidence that some of these individual differences are genetically based. Twin studies [5–7] demonstrated that the subjective effects of these drugs are more similar in monozygotic twins than dizygotic twins, and epidemiological studies also support the heritability of both stimulant use and abuse (e.g., [8]). In addition, several association studies discussed below indicate a genetic basis in the acute subjective response to certain drugs. Therefore, there may be a class of alleles of genes that influence both the sensitivity to certain effects of drugs and the risk for development of drug abuse.

Although we focus here on subjective drug effects as intermediate phenotypes in the development of drug dependence, acute responses to drugs may also predict other end points related to nonproblem drug use or other psychiatric conditions. For example, subjective responses to drugs of abuse may predict future use of the drug without predicting either abuse or dependence. Also, subjective responses to drugs may serve as intermediate phenotypes for psychiatric or neurobiological conditions such as panic disorder, attention-deficit/hyperactivity disorder (ADHD), or depression. Alternatively, subjective drug effects may predict therapeutic responses to drugs used in the treatment of psychiatric disorders (e.g., stimulants for ADHD). Finally, to the extent that drugs of abuse act on specific endogenous neurotransmitter systems, they may shed light on individual differences in basic underlying neurobiological mechanisms. We will discuss some of these possibilities, but the thrust of this chapter is to examine the use of subjective drug effects as an intermediate phenotype for substance dependence and related disorders.

6.1   Advantages

There are several benefits of assessing subjective responses to acute drug administration as intermediate phenotypes. These advantages, described in detail below, include their face validity, their biological relevance, their ability to provide qualitatively rich data in a noninvasive manner, and their translational potential in relation to animal models. While studies with nonhumans provide essential information about the underlying neurobiology of drug effects, studies with humans provide unique insights into the subjective states induced by drugs, which can only be indirectly inferred from nonhuman behavior.

6.1.1   Face Validity

Subjective drug effects as intermediate phenotypes for drug abuse have substantial face validity. Drug-induced euphoria is considered to be a primary indicator of abuse potential, and pleasurable drug effects are widely thought to contribute to repeated drug use and possible abuse [9, 10]. Most abused drugs produce pleasurable mood states in a large proportion of users, and these pleasurable states are thought to facilitate repeated drug use. However, individuals differ in the extent to which the drugs produce euphoria (see below), and variations in this early drug response may influence repeated drug use, misuse, or abuse. In addition, it is possible that individual differences in sensitivity to negative drug-induced effects may also influence repeated drug use. For example, males with a family history of alcoholism are relatively insensitive to ethanol-induced static ataxia, or body sway (e.g., [11]; for a detailed review of subjective responses to alcohol, see chapter 5). Therefore, insensitivity to unfavorable drug responses could also facilitate repeated use.

The predictive relationship of drug-induced euphoria to subsequent drug use has been studied using both retrospective and prospective designs. In retrospective studies [12, 13], when drug abusers were asked to recall their early drug use experiences they reported that their initial responses to drugs were pleasant. These studies are limited by the fact that they rely on the memory of the drug users who might be biased in their recollections. Stronger evidence comes from longitudinal prospective studies, which support the predictive validity of drug-induced euphoria or other subjective responses as an indicator of subsequent substance abuse. In these studies, initial responses in individuals with little or no prior drug use are examined, and the participants' recreational drug use is then followed over months or years. Many studies support the idea that individuals who report pleasant drug effects are more likely to use again although there are also examples of possibly unpleasant subjective feelings such as dizziness as positive predictors of future use [10]. Prospective studies require significant expense and effort and typically focus on legal substances (i.e., cigarettes) because a greater fraction of the study population is expected to become dependent. Although both retrospective and prospective studies have limitations, when taken together, they support the connection between initial drug responses and future drug use.

One limitation of both retrospective and prospective studies is that they fail to control expectancies that may influence how a drug affects the user. This problem is minimized in controlled laboratory studies using double-blind, placebo-controlled designs, in which neither the subject nor the experimenter knows the identity or dose of a drug at the time it is administered. Data from controlled double-blind studies show that positive subjective effects predict whether a subject will subsequently choose to take that drug again in a laboratory setting [14, 15], supporting the notion that initial drug responses predict subsequent drug taking. In addition, one recent study [2] found that positive, stimulant-like acute responses to alcohol in the laboratory were predictive of habitual use over the subsequent two-year period, further supporting the value of acute responses as predictors of future use. Thus, these findings, from retrospective, longitudinal, and laboratory-based studies, lend support to the idea that drug-induced euphoria is a useful intermediate phenotype for future drug use or abuse. They do not, however, clarify whether this correlation is due to genetic or nongenetic influences, which is a critical question if they are to be employed as intermediate phenotypes.

6.1.2 Biological Relevance

The second advantage to using subjective drug effects as intermediate phenotypes is their biological relevance. With a few notable exceptions, such as ethanol and inhalants which have multiple sites of action, most drugs of abuse act on well-characterized, discrete biological targets within the brain (receptors, neurotransmitter transporters) and have unique pharmacodynamic and pharmacokinetic properties that determine the amount of active drug at the site of action, as well as the drug's efficacy once it reaches its target. Once in the body, the drug is metabolized by enzymes, and differences in the amount or efficacy of these enzymes control the amount of drug in the body. Genetic or environmentally induced differences in either receptor function or enzymatic activity may alter the subjective effects of drugs. Heritable individual differences in drug metabolizing enzymes are known to influence use of certain drugs, most notably cigarette smoking [16], alcohol consumption [17], and caffeine consumption [18, 19]. Usually, individuals who are slow or poor metabolizers of these drugs use less of them. A large body of research has investigated heritable individual differences in the pharmacokinetic profiles of drugs, and a smaller body of work has examined genetic variations in the pharmacodynamic effects of the drugs that lead to their subjective effects.

The main target for studying heterogeneity in the pharmacodynamic properties of drugs of abuse is in the neurotransmitter receptor or transporter systems where the drugs act. These targets are typically located in brain systems thought to be involved in motivation and reinforcement value of drug taking, such as the dopaminergic and opioid systems [20], but also include other targets such as the acetylcholine, norepinephrine, serotonin, GABA, glutamate, adenosine, and cannabinoid systems. Some drugs have one primary site of action (e.g., nicotine as an agonist at nicotinic cholinergic receptors), but many act on several targets within the central nervous system (e.g., alcohol acts through multiple neurotransmitter systems). In addition to acting on more than one site, drugs can either inhibit or activate transmission. Indeed, some drugs can have different effects depending on brain region. For example amphetamine increases cellular activity in prefrontal areas but decreases activity in motor and visual cortices [21]. It is also possible for the effect of a drug on a transmitter system to change over time. Nicotine is believed to activate nicotinic acetylcholine receptors during initial use [22], but chronic use leads to inhibition of these receptors through desensitization [23]. Finally, specific receptor subtypes mediate different effects of drugs, as in the case of benzodiazepine-like drugs, in which specific subtypes of the GABA-A receptor mediate the anxiolytic, abuse-related, and motor effects [24]. Because drugs produce their subjective effects through these endogenous neurotransmitter systems, the subjective responses to the drugs in humans may serve as a biological assay for pharmacodynamic differences at these sites of action. In general, interindividual variations in neurotransmitter systems are difficult to identify, but acute responses to challenge doses of known drugs may provide a means to characterize these variations. Thus, a better understanding of specific actions of drugs and resulting effects on the organism will help to elucidate differences in neurotransmitter receptor or transporter function.

Subjective drug responses may also provide insight into psychiatric conditions or potential therapeutic responses to psychiatric medications. Because many psychiatric conditions are sensitive to the manipulation of specific neurotransmitter systems, acute responses to drugs with known mechanisms of action can bring such dysfunction to light. For example, drug challenge studies have been used to study vulnerability to anxiety-related disorders. Specifically, acute administration of moderate doses of caffeine induces anxiety in susceptible individuals. One component of this susceptibility may include a specific allele of the gene for the adenosine 2A receptor [25]. This same polymorphism is also implicated in susceptibility to panic disorder [26], and, interestingly, individuals with panic disorder are more likely to experience anxiety after caffeine administration [27]. Thus, assessment of subjective response to caffeine may help to identify individuals vulnerable to developing anxiety disorders or to characterize a specific subtype of anxiety. In a similar manner, subjective drug effects may be useful in determining efficacy of therapeutic response for psychiatric disorders. For example, polymorphisms within the dopamine transporter (DAT) and dopamine receptor D4 have been reported to predict both subjective response to stimulant drugs and therapeutic response to stimulant treatment in ADHD [28]. Therefore, knowledge gained from studies of acute subjective effects of drugs is relevant not only to drug abuse but to other psychiatric conditions and their treatment as well.

6.1.3 Qualitative Richness
Subjective drug effects provide qualitatively rich information in a relatively noninvasive manner. Based on responses to a single dose of a drug, researchers can readily assess the quality of the effect (e.g., stimulant-like, sedative-like), the sensitivity to the drug (e.g., magnitude of response), the affective valence of the effect (e.g., liking, disliking), and the time course of effects (e.g., onset or duration of the effect) using brief, repeated self-report or mood questionnaires (discussed further below). With relatively little additional effort, these responses can be assessed after different drug doses, in different subpopulations, or during different times of the day. Responses to questionnaires may be obtained confidentially, minimizing bias that might occur in verbal descriptions of effects. The self-report measures are less expensive and less invasive than some other technologies used to study brain function (e.g., imaging techniques), and allow researchers to characterize the magnitude, quality, and time course of drug administrations.

6.1.4 Translational Appeal
Subjective drug effects provide a point of entry for human to nonhuman translational genetic studies. Subjective effects of drugs in humans may correspond to parallel behavioral outcomes in nonhumans, including discriminative stimulus effects, and behavioral measures of anxiety, depression, arousal, and euphoria. As previously mentioned, it is not possible to know the subjective experiences of nonhumans. However, indirect inferences may be made by observing their behavior. First, there are animal behavioral models of anxiety (e.g., elevated plus maze) and depression (e.g., forced swim test), which are thought to model corresponding mood states in humans. Second, in some cases it is possible to measure the same behaviors in humans and nonhumans and, in the human version, to also measure subjective states. One example

of this is drug discrimination studies, which can be conducted in the same manner in humans and nonhumans. In animals, drug discrimination studies are thought to model human subjective effects. In the human studies, subjective reports can also be obtained. Thus, the combined findings from human and nonhuman drug discrimination studies can provide indirect evidence that nonhumans experience similar subjective drug effects. In the drug discrimination procedure [29, 30] animals are trained to perform one response in the presence of a certain drug and another response in its absence. Then, generalization tests are performed to determine whether other drugs produce effects that are similar to the training drug (i.e., also produce the drug-appropriate response). Interestingly, the same procedure has been used with humans [31], and the findings are strikingly similar. Moreover, the human studies also provide valuable information about the subjective effects that form the basis of a discriminative stimulus [32].

The conditioned place preference (CPP) procedure [33, 34] is another example of a procedure that can be used in both humans and nonhumans. In nonhumans it is attractive because it can be used to infer subjective drug effects. In this procedure, animals are given experience with one environment paired with a drug and another paired with vehicle and then are given a choice to spend time in one or the other environment. The extent to which they prefer the drug-paired environment on a subsequent challenge day is an index of the positive or motivational effects of the drug. Recently, this CPP procedure was extended to humans [35, 36] and included a simultaneous assessment of the participants' subjective drug liking. It was shown that humans, like nonhumans, preferred a place paired with amphetamine administration. In addition, this work established that the extent of drug liking was positively correlated with the preference for the drug-paired environment.

Drug self-administration studies have also been conducted with both nonhumans [37, 38] and humans [14]. In these studies, the subjects are permitted to regulate their own level of drug intake during sessions, and nonhumans readily self-administer the same drugs that are used and abused by humans [39, 40]. As with CPP, the human versions of these studies explicitly link drug self-administration in humans with subjective measures of drug liking [15], providing a direct validation that drug taking is related to subjective effects. Although there are some exceptions to this concordance (e.g., [41]), in general drug self-administration in humans is directly related to positive subjective drug effects. The concordance in self-administration of drugs in humans and nonhumans, combined with the concordance between drug-induced euphoria and drug self-administration in humans, suggests that drugs may also produce unmeasurable positive interoceptive effects in nonhumans, as well. Thus, self-administration by nonhumans may indicate "liking" of the drug.

Thus, there are several examples of translational research that provide indirect information about the subjective effects of drugs in nonhumans. Improvements in these translational approaches will make it possible to use animals to model states that are difficult to study in human laboratory studies, such as full-fledged drug abuse and the transition into such a state. However, such procedures must be validated by demonstrating a correlation between the behavioral and self-reported drug effects in human subjects.

6.2 Methodological Considerations

The assessment of subjective drug effects requires careful experimental control. Many factors can influence subjective response to drugs, including the participants' psychiatric condition [42], expectancies [43–45], recent drug use, drug use history [14, 46], sex [47, 48], age [49], setting [50], weight, menstrual cycle stage [51], circadian factors [52], and amount and quality of sleep [53, 54]. Therefore, considerable care must be taken to control for these other factors that could influence subjective drug effects.

Essential features of drug challenge studies are the use of double-blind procedures, inclusion of a placebo condition, control over expectancies, and the inclusion of several doses of the drug. Double-blind procedures and inclusion of placebo are essential to reduce the influence of expectancies or experimenter influence. Expectancies are further controlled by instructions to participants: expectancies can be reduced by informing subjects that they may receive one of several drugs types (e.g., sedative, stimulant, alcohol, or a placebo), even if the study involves only a single drug. Although this introduces a minor element of deception, the benefits of measuring the pure pharmacological effects of the drug outweigh the risks. Use of several doses of a drug provides full information about the dose effect, it sometimes provides some data regarding test–retest reliability, and it further ensures blinding of the participants and experimenters. If more than one session is used, the same order can be used for all subjects or a randomized order can be used; the first makes it impossible to keep the experimenters blind over the course of the study, and so randomization is often used.

6.2.1 Standardized Self-Report Measures
Self-reported subjective states and mood are typically assessed using validated, standardized questionnaires appropriate for the drug and the subject sample. These questions focus on states which are expected to be labile over the course of the study. Table 6.1 summarizes some of the most commonly used scales. These questionnaires are typically administered at multiple time points, including a measure before drug administration, to provide a baseline assessment, and several measures after drug administration, to capture the time course of the effects. Concurrent assessment of physiological measures, such as heart rate and blood pressure, is often valuable to complement the subjective reports, as it provides a means for ensuring the safety of

Table 6.1
Commonly used rating questionnaires used in drug abuse research

Name	Reference	Use	Measurement	Subscales	Features
modified Profile of Mood States (POMS)	Johanson & Uhlenhuth, 1980 (adapted from McNair & Droppleman, 1971)	Subjective states and mood	5 point scale (1-"Not at all" to 5-"very much") 72 adjectives (modified from 90)	Anxiety, Anger, Confusion, Depression, Elation, Fatigue, Friendliness, Vigor, Arousal (derived scale) Positive Mood (derived scale)	Measures momentary mood states
Addiction Research Center Inventory (ARCI)	Martin et al., 1971	Acute subjective responses to alcohol or drugs	"True" or "False" 49 statements	Amphetamine (A), Benzedrine (BG), Lysergic Acid (LSD), Morphine-Benzedine (MBG), Pentobarbital, Chlorpromazine, and Alcohol (PCAG); Marijuana (modified in Chait et al., 1994)	Measures drug-specific subjective experiences
Drug Effects Questionnaire (DEQ)	Fraser et al., 1961	Acute subjective responses to alcohol or drugs	Visual analog scale ("not at all" to "extremely")	Feel Drug, Like Effects, Dislike Effects, High, Want More	Assesses global ratings of drug effects
Visual Analog Scales (VAS)	Folstein & Luria, 1973	Subjective states and mood	Visual analog scale ("not at all" to "extremely")	Varies (chosen by experimenter)	Assesses specific moods or states Items selected to suit study goal
modified Cigarette Evaluation Questionnaire (mCEQ)	Cappelleri et al., 2007	Subjective responses to cigarettes	7 point scale (1-"not at all" to 7-"extremely")	Smoking Satisfaction, Psychological Reward, Aversion, Enjoyment of Respiratory Tract Sensations, Craving Reduction	Assesses responses to cigarettes

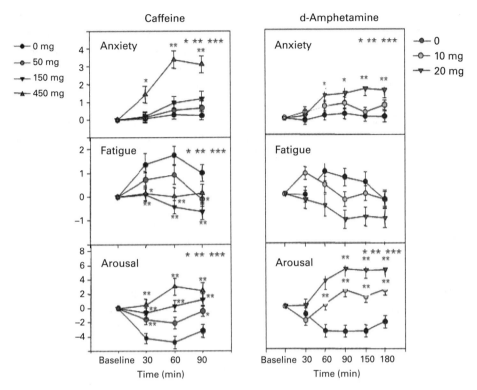

Figure 6.1
Typical subjective effects at varying time points following oral administration of caffeine (0, 50, 150, and 450 mg; left) and *d*-amphetamine (0, 10, and 20 mg; right) in healthy young adults (N = 100 each). Values are mean (and standard error of the mean) ratings on Profile of Mood States subscales Anxiety, Fatigue, and Arousal, expressed as a change in subjective ratings following capsule ingestion. The x-axis measures time from drug administration.

the participants and provides objective confirmation of the pharmacological effect of the drug in many cases.

Figure 6.1 illustrates typical subjective effects of two drugs, caffeine and amphetamine, showing their dose dependence and time course on several rating scales. The figure shows that the two drugs share certain effects (increased anxiety and arousal, decreased fatigue, as measured by the Profile of Mood States [POMS] [55]), but also that the drugs differ: doses of caffeine appear to produce greater increases in anxiety compared to amphetamine. This figure exemplifies the rich array of subjective responses that the drugs produce.

The effects shown in figure 6.1 illustrate the average responses to the drug in a relatively large sample of individuals (N = 100 for each drug). However, individuals vary in their responses to drugs, and subjective ratings provide a sensitive indicator of responses *across* individuals. Figure 6.2 depicts individual differences in the subjective responses to amphetamine (top: 20 mg, N = 195) and caffeine (bottom: 400 mg, N = 95), as measured by the peak change scores of POMS arousal ratings. The figures show that there is wide variability in the degree of subjective arousal they induce in subjects, which allows for further exploration not only as to what causes this variability but also the relationship of these subjective effects with future drug use and potential abuse.

6.3 Illustrative Cases

In the following sections, we provide examples of studies in which acute subjective drug effects have provided informative intermediate phenotypes.

6.3.1 Caffeine
Caffeine is the most widely used psychoactive substance in the world. Most often, caffeine produces mild stimulant effects, such as increased arousal and alertness. However, at high doses or in susceptible individuals, caffeine can also produce anxiety and dysphoria [56]. Neurobiologically, caffeine acts as a competitive antagonist at adenosine A1 and A2a receptors [57], preventing the effects of endogenous adenosine. These adenosine receptors also colocalize, and functionally interact with, dopamine receptors [58]. Therefore, caffeine not only acts on the adenosine neurotransmitter system but also has effects on the dopaminergic system, which is implicated in drug-taking and drug-seeking behaviors.

Individual differences in the acute subjective response to caffeine have been studied for decades (e.g., [59]), but the reasons for the individual variability in responses is only now becoming clear. Twin studies indicated that behaviors relating to caffeine use, such as consumption, intoxication, tolerance, and withdrawal, are heritable [60]. A controlled study with acute administration of caffeine in nonusers of the drug showed that two linked single nucleotide polymorphism (SNPs) on the A2a receptor gene (*ADORA2A*) were significantly associated with caffeine-induced anxiety [25]. This was one of the first candidate gene studies using an acute drug challenge to demonstrate a genetic basis for the subjective responses to a psychoactive drug. This work has since been replicated [61] and also extended to demonstrate that SNPs within the dopamine D2 receptor gene *DRD2* also have a functional role in the response to caffeine (for a detailed review of the genetics of response to caffeine, see [62]). This study provided early evidence that genetic differences contribute to the subjective response to caffeine, such that adenosine- and dopamine-related mechanisms may underlie caffeine-induced anxiety.

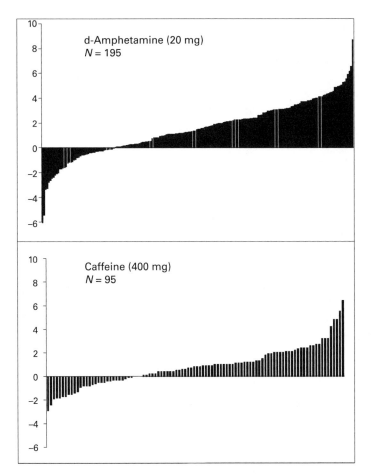

Figure 6.2
Individual subjects' ratings of arousal on the Profile of Mood States after d-amphetamine (20 mg; top) and caffeine (400 mg; bottom) were administered, calculated as the peak change score on the drug session minus the peak change score on the placebo session. The x-axis represents response of individual subjects, ranked by magnitude of drug response.

More recently, two genome-wide association studies (GWASs) of caffeine consumption [18, 19] have identified polymorphisms in two genes involved in caffeine metabolism. Neither GWAS identified *ADORA2A* or *DRD2*. This could reflect the limited importance of differences in subjective response for caffeine consumption. Such an explanation contradicts the hypothesis that subjective response is a predictor of subsequent drug use. Alternatively, it could indicate that the influence of polymorphisms in these genes on the subjective effects of caffeine is much smaller than indicated by the Alsene et al. [25] and Childs et al. [61] or that these polymorphisms are only important in occasional but not chronic caffeine users.

6.3.2 Amphetamine

Amphetamine is a prototypical stimulant that induces feelings of euphoria, alertness, arousal, and well-being in humans. Although amphetamine has therapeutic value in the treatment of ADHD and narcolepsy, it also has a high abuse potential. Across several contexts, including laboratory studies, recreational use, and therapeutic settings, individuals are known to vary in the quality and magnitude of subjective effects and in clinical response to amphetamine [1, 5, 28], and at least some of this variability appears to be due to genetic variability in the integral proteins that amphetamine acts upon.

Amphetamine exerts its effects on multiple neurotransmitter systems, including dopamine, norepinephrine, and serotonin. Once it reaches the brain, it acts presynaptically to release neurotransmitters from vesicular stores by acting on the vesicular monoamine transporter. In addition, it also causes a reversal of the reuptake transporters, the DAT, norepinephrine transporter (NET), and serotonin transporter, resulting in these proteins transporting neurotransmitter *out* of the neuron. This effectively increases levels of active transmitters in the synaptic cleft, which in turn causes activation of pre- and postsynaptic receptors. With so many proteins involved in the neurobiological mechanism of amphetamine action, it is not surprising that variations in several different genes have been found to play some role in this response.

Several studies have examined variations in the many targets of amphetamine on differences in the subjective effects of the drug. Genes that code for the norepinepherine transporter (*NET*) [63], dopamine transporter (*DAT*) [64], and serotonin transporter (*5HTT*) [65] all seem to play a role in the subjective response to amphetamine. Similar to caffeine, *ADORA2A* contains SNPs that influence amphetamine-induced anxiety [4]. In addition, genes encoding less obvious molecules, such as fatty acid amide hydroxylase (*FAAH*) [66], the mu-opioid receptor 1 (*OPRM1*) [67], casein kinase 1 epsilon (*CSNK1E*) [68], and brain-derived neurotrophic factor (*BDNF*) [69], were also found to have relevant SNPs. Conclusions from these studies must be interpreted with caution until they can be replicated in independent data sets. For a complete review of the genetic factors that modulate amphetamine-induced subjective effects, see Hart et al. [70].

6.3.3   Nicotine

Nicotine is commonly administered through inhalation of cigarette smoke, which allows for quick absorption in the brain. The subjective effects of acute nicotine vary greatly among individuals [71]. Although many established smokers claim to use cigarettes to reduce anxiety, enhance pleasure, or alleviate negative affect [72], most inexperienced users report negative effects upon initial nicotine consumption [73]. Once nicotine enters the brain, it binds to the nicotinic acetylcholine receptors (nAChRs), comprised of a combinations of nine α subunit isoforms and three isoforms of the β subunit [74]. Within the mammalian brain, most of the nAChRs consists of complexes of α4 and β2 subunits, which are also the sites of highest nicotine binding affinity [75]. In addition to its direct effects on nAChRs, nicotine has indirect effects dopamine and opioid systems, which have been implicated in the motivational and reinforcing effects of nicotine [76, 77].

GWASs involving nicotine or tobacco smoking have implicated several genes in nicotine dependence [78, 79]. The phenotypes assessed in these studies include smoking onset, smoking status, smoking quantity, or development of smoking-related disease, rather than the subjective effects of smoking. The strongest finding to come from these studies is a risk for smoking dependence associated with a cluster of three nicotinic subunit genes: *CHRNA5, CHRNA3, and CHRNB4*. It remains to be determined whether this locus influences smoking dependence by influencing the subjective effects of nicotine; however, some evidence supports this view. For example, it was reported that, in addition to its relationship with smoking status, this locus was also associated with pleasurable buzz during initial smoking experiences [80]. Another nicotinic receptor gene cluster, *CHRNA6* and *CHRNB3*, has also been implicated in development of nicotine dependence [79, 81], although to a lesser degree than the previously mentioned group of genes. Regardless, one study [82] found that SNPs within both of these genes were implicated in initial subjective responses to smoking. GWASs have also implicated a number of other genes associated with smoking dependence. However, since the phenotype assessed in most of these studies is related to smoking dependence or associated disease development (for a review, see [83]), it is unclear as to whether or not these genetic findings play a role in the subjective response to nicotine.

In addition to the genes implicated by GWAS, candidate-gene studies have implicated other polymorphisms in the subjective response to nicotine. In a laboratory setting, the level of nicotine-induced effects, including dizziness, buzzing, heart pounding, nausea, and sweat, was associated with polymorphisms within *CHRNB2* [84]. In addition, SNPs within *CHRNA4* have been associated both with subjective effects of cigarette smoking, such as sensitivity to cigarette-induced rush or high or a greater feeling of reward after smoking, and treatment outcome in a smoking cessation program in adult smokers [85]. Therefore, not only do nicotine-induced subjective responses appear to be associated with genetic variations but they also suggest a correlation with treatment outcome. However, the above-mentioned findings have not been detected by GWAS for nicotine dependence, suggesting either that these results are incorrect or that subjective effects as measured by these studies have only a modest impact on smoking dependence and related phenotypes.

6.4   Limitations

As with most methodologies, subjective drug effects as intermediate phenotypes also have limitations. There are ethical and safety limitations to the drug dosage and route of administration possible in humans. Ethical constraints limit the ability of researchers to study the high doses or riskier routes of administration commonly used by recreational drug users (e.g., intravenous) or to model the protracted binge use in order to ensure the safety of the participant. Some studies select individuals with little or no history of drug use, to avoid exposing at-risk individuals (i.e., those with a history of use) to the drugs. Other studies select individuals with established drug use histories, to avoid exposing drug-naive individuals to an abused drug. The most appropriate population depends on the details of the study and on the goal of the study (e.g., modeling initiation of use or persistence of use). Inclusion of individuals with extensive drug use history may introduce a bias in self-reported effects because of expectations and prior learning about the drug. In addition to these ethical constraints, it is also possible that the standardized self-report questionnaires do not capture the full effects of the drug.

One potential new avenue for studying acute drug responses is to combine the self-reported, subjective effects with brain imaging or electrophysiological methods to ascertain specific brain regions involved in the responses. Further, subjective drug effects may also be assessed along with cognitive performance tests. Taken together, these approaches provide powerful tools to study vulnerability to substance abuse.

6.5   Conclusions

Subjective effects induced by drugs are believed to influence future drug use, which can lead to drug abuse or dependence. Therefore, these subjective effects may be valuable intermediate phenotypes to study vulnerability to drug abuse, and they may also provide insights into the neurobiology underlying drug use and facilitate cross-species translational research. We have reviewed evidence that these intermediate phenotypes may be genetically determined and are useful targets for study. Although many factors influence drug use and abuse, acute subjective responses to drugs remain a key element in the phenomenology of human drug use.

Acknowledgments

This work was supported by National Institutes of Health grants DA021336, GM097737, and MH079103 (AAP); DA02812 (HdW); as well as National Institute on Drug Abuse grant T32 DA07255 and Howard Hughes Medical Institute grant 56006772 (LMM).

References

1. Holdstock, L., & de Wit, H. (2001). Individual differences in responses to ethanol and d-amphetamine: a within-subject study. *Alcoholism, Clinical and Experimental Research, 25,* 540–548.

2. King, A. C., de Wit, H., McNamara, P. J., & Cao, D. (2011). Rewarding, stimulant, and sedative alcohol responses and relationship to future binge drinking. *Archives of General Psychiatry, 68,* 389–399.

3. McKinzie, D. L., Nowak, K. L., Murphy, J. M., et al. (1998). Development of alcohol drinking behavior in rat lines selectively bred for divergent alcohol preference. *Alcoholism, Clinical and Experimental Research, 22,* 1584–1590.

4. Hohoff, C., McDonald, J. M., Baune, B. T., et al. (2005). Interindividual variation in anxiety response to amphetamine: possible role for adenosine A2A receptor gene variants. *American Journal of Medical Genetics. Part B, Neuropsychiatric Genetics, 139B,* 42–44.

5. Nurnberger, J. I., Jr., Gershon, E. S., Simmons, S., et al. (1982). Behavioral, biochemical and neuroendocrine responses to amphetamine in normal twins and "well-state" bipolar patients. *Psychoneuroendocrinology, 7,* 163–176.

6. Crabbe, J. C., Jarvik, L. F., Liston, E. H., & Jenden, D. J. (1983). Behavioral responses to amphetamines in identical twins. *Acta Geneticae Medicae et Gemellologiae, 32,* 139–149.

7. Haberstick, B. C., Zeiger, J. S., Corley, R. P., et al. (2011). Common and drug-specific genetic influences on subjective effects to alcohol, tobacco and marijuana use. *Addiction, 106,* 215–224.

8. Kendler, K. S., Karkowski, L. M., Neale, M. C., & Prescott, C. A. (2000). Illicit psychoactive substance use, heavy use, abuse, and dependence in a US population-based sample of male twins. *Archives of General Psychiatry, 57,* 261–269.

9. Fergusson, D. M., Horwood, L. J., Lynskey, M. T., & Madden, P. A. (2003). Early reactions to cannabis predict later dependence. *Archives of General Psychiatry, 60,* 1033–1039.

10. Chen, X., Stacy, A., Zheng, H., et al. (2003). Sensations from initial exposure to nicotine predicting adolescent smoking in China: a potential measure of vulnerability to nicotine. *Nicotine & Tobacco Research, 5,* 455–463.

11. Schuckit, M. A. (1985). Ethanol-induced changes in body sway in men at high alcoholism risk. *Archives of General Psychiatry, 42,* 375–379.

12. Davidson, E. S., & Schenk, S. (1994). Variability in subjective responses to marijuana: initial experiences of college students. *Addictive Behaviors*, *19*, 531–538.

13. Haertzen, C. A., Kocher, T. R., & Miyasato, K. (1983). Reinforcements from the first drug experience can predict later drug habits and/or addiction: results with coffee, cigarettes, alcohol, barbiturates, minor and major tranquilizers, stimulants, marijuana, hallucinogens, heroin, opiates and cocaine. *Drug and Alcohol Dependence*, *11*, 147–165.

14. de Wit, H., & Griffiths, R. R. (1991). Testing the abuse liability of anxiolytic and hypnotic drugs in humans. *Drug and Alcohol Dependence*, *28*, 83–111.

15. Johanson, C. E., & Uhlenhuth, E. H. (1980). Drug preference and mood in humans: d-amphetamine. *Psychopharmacology*, *71*, 275–279.

16. Mwenifumbo, J. C. & Tyndale, R. F. (2009). Molecular genetics of nicotine metabolism. In J. E. Henningfield, E. D. London, & S. Pogun (Eds.), Handbook of Experimental Pharmacology: Vol. 192. Nicotine Psychopharmacology (pp. 235–259). Berlin: Springer-Verlag.

17. Agarwal, D. P., & Goedde, H. W. (1992). Pharmacogenetics of alcohol metabolism and alcoholism. *Pharmacogenetics*, *2*, 48–62.

18. Sulem, P., Gudbjartsson, D. F., Geller, F., et al. (2011). Sequence variants at CYP1A1–CYP1A2 and AHR associate with coffee consumption. *Human Molecular Genetics*, *20*, 2071–2077.

19. Cornelis, M. C., Monda, K. L., Yu, K., et al. (2011). Genome-wide meta-analysis identifies regions on 7p21 (AHR) and 15q24 (CYP1A2) as determinants of habitual caffeine consumption. *PLOS Genetics*, *7*, e1002033.

20. Mackillop, J., Obasi, E., Amlung, M. T., McGeary, J. E., & Knopik, V. S. (2010). The role of genetics in nicotine dependence: mapping the pathways from genome to syndrome. *Current Cardiovascular Risk Reports*, *4*, 446–453.

21. Devous, M. D., Sr., Trivedi, M. H., & Rush, A. J. (2001). Regional cerebral blood flow response to oral amphetamine challenge in healthy volunteers. *Journal of Nuclear Medicine*, *42*, 535–542.

22. Galzi, J. L., & Changeux, J. P. (1995). Neuronal nicotinic receptors: molecular organization and regulations. *Neuropharmacology*, *34*, 563–582.

23. Giniatullin, R., Nistri, A., & Yakel, J. L. (2005). Desensitization of nicotinic ACh receptors: shaping cholinergic signaling. *Trends in Neurosciences*, *28*, 371–378.

24. Rowlett, J. K., Platt, D. M., Lelas, S., Atack, J. R., & Dawson, G. R. (2005). Different GABAA receptor subtypes mediate the anxiolytic, abuse-related, and motor effects of benzodiazepine-like drugs in primates. *Proceedings of the National Academy of Sciences of the United States of America*, *102*, 915–920.

25. Alsene, K., Deckert, J., Sand, P., & de Wit, H. (2003). Association between A2a receptor gene polymorphisms and caffeine-induced anxiety. *Neuropsychopharmacology*, *28*, 1694–1702.

26. Hamilton, S. P., Slager, S. L., De Leon, A. B., et al. (2004). Evidence for genetic linkage between a polymorphism in the adenosine 2A receptor and panic disorder. *Neuropsychopharmacology, 29,* 558–565.

27. Nardi, A. E., Lopes, F. L., Valenca, A. M., Freire, R. C., Veras, A. B., de-Melo-Deto, V. L., et al. (2007). Caffeine challenge test in panic disorder and depression with panic attacks. *Comprehensive Psychiatry, 48,* 257–263.

28. Froehlich, T. E., Epstein, J. N., Nick, T. G. et al. (2011). Pharmacogenetic predictors of methylphenidate dose-response in attention-deficit/hyperactivity disorder. *Journal of the American Academy of Child and Adolescent Psychiatry, 50,* 1129–1139 e2.

29. Overton, D. A. (1961). Discriminative behavior based on the presence or absence of drug effects. *American Psychologist, 16,* 453–454.

30. Stolerman, I. (1992). Drugs of abuse: behavioural principles, methods and terms. *Trends in Pharmacological Sciences, 13,* 170–176.

31. Schuster, C. R., & Johanson, C. E. (1988). Relationship between the discriminative stimulus properties and subjective effects of drugs. *Psychopharmacology Series, 4,* 161–175.

32. Preston, K. L., & Jasinski, D. R. (1991). Abuse liability studies of opioid agonist-antagonists in humans. *Drug and Alcohol Dependence, 28,* 49–82.

33. Beach, H. D. (1957). Morphine addiction in rats. *Canadian Journal of Psychology, 11,* 104–112.

34. Tzschentke, T. M. (2007). Measuring reward with the conditioned place preference (CPP) paradigm: update of the last decade. *Addiction Biology, 12,* 227–462.

35. Childs, E., & de Wit, H. (2011). Contextual conditioning enhances the psychostimulant and incentive properties of d-amphetamine in humans. *Addiction Biology.* doi:10.1111/j.1369-1600.2011.00416.x.

36. Childs, E., & de Wit, H. (2009). Amphetamine-induced place preference in humans. *Biological Psychiatry, 65,* 900–904.

37. Spragg, S. D. S. (1940). Morphine addiction in chimpanzees. *Comparative Psychology Monographs, 15,* 1–132.

38. Weeks, J. R. (1962). Experimental morphine addiction: method for automatic intravenous injections in unrestrained rats. *Science, 138,* 143–144.

39. Johanson, C. E., & Balster, R. L. (1978). A summary of the results of a drug self-administration study using substitution procedures in rhesus monkeys. *Bulletin on Narcotics, 30,* 43–54.

40. Woolverton, W. L., & Nader, M. A. (1990). Experimental evaluation of the reinforcing effects of drugs. In Adler, M. W., and Cowan, A., eds., Testing and Evaluation of Drugs of Abuse (pp. 165–192). New York: Wiley-Liss.

41. Lamb, R. J., Preston, K. L., Schindler, C. W., et al. (1991). The reinforcing and subjective effects of morphine in post-addicts: a dose-response study. *Journal of Pharmacology and Experimental Therapeutics, 259*, 1165–1173.

42. Kollins, S. H., English, J., Robinson, R., Hallyburton, M., & Chrisman, A. K. (2009). Reinforcing and subjective effects of methylphenidate in adults with and without attention deficit hyperactivity disorder (ADHD). *Psychopharmacology, 204*, 73–83.

43. Kirk, J. M., Doty, P., & de Wit, H. (1998). Effects of expectancies on subjective response to oral delta-9-tetrahydrocannabinol. *Pharmacology, Biochemistry, and Behavior, 59*, 287–293.

44. Mitchell, S. H., Laurent, C. L., & de Wit, H. (1996). Interaction of expectancy and the pharmacological effects of d-amphetamine: subjective effects and self-administration. *Psychopharmacology, 125*, 371–378.

45. Southwick, L., Steele, C., Marlatt, A., & Lindell, M. (1981). Alcohol-related expectancies: defined by phase of intoxication and drinking experience. *Journal of Consulting and Clinical Psychology, 49*, 713–721.

46. Kirk, J. M., & de Wit, H. (1999). Responses to oral delta9-tetrahydrocannabinol in frequent and infrequent marijuana users. *Pharmacology, Biochemistry, and Behavior, 63*, 137–142.

47. Carroll, M. E., Lynch, W. J., Roth, M. E., Morgan, A. D., & Cosgrove, K. P. (2004). Sex and estrogen influence drug abuse. *Trends in Pharmacological Sciences, 25*, 273–279.

48. Chi, H., & de Wit, H. (2003). Mecamylamine attenuates the subjective stimulant-like effects of alcohol in social drinkers. *Alcoholism, Clinical and Experimental Research, 27*, 780–786.

49. de Wit, H., Uhlenhuth, E. H., & Johanson, C. E. (1985). Drug preference in normal volunteers: effects of age and time of day. *Psychopharmacology, 87*, 186–193.

50. Doty, P., & de Wit, H. (1995). Effect of setting on the reinforcing and subjective effects of ethanol in social drinkers. *Psychopharmacology, 118*, 19–27.

51. Terner, J. M., & de Wit, H. (2006). Menstrual cycle phase and responses to drugs of abuse in humans. *Drug and Alcohol Dependence, 84*, 1–13.

52. Shappell, S. A., Kearns, G. L., Valentine, J. L., Neri, D. F., & DeJohn, C. A. (1996). Chronopharmacokinetics and chronopharmacodynamics of dextromethamphetamine in man. *Journal of Clinical Pharmacology, 36*, 1051–1063.

53. Pigeau, R., Naitoh, P., Buguet, A., et al. (1995). Modafinil, d-amphetamine and placebo during 64 hours of sustained mental work: I. effects on mood, fatigue, cognitive performance and body temperature. *Journal of Sleep Research, 4*, 212–228.

54. Childs, E., & de Wit, H. (2008). Enhanced mood and psychomotor performance by a caffeine-containing energy capsule in fatigued individuals. *Experimental and Clinical Psychopharmacology, 16*, 13–21.

55. McNair, D. M., Lorr, M., & Droppleman, L. F. (1971). Manual for the Profile of Mood States. San Diego, CA: Educational and Industrial Testing Service.

56. Childs, E., & de Wit, H. (2006). Subjective, behavioral, and physiological effects of acute caffeine in light, nondependent caffeine users. *Psychopharmacology, 185,* 514–523.

57. Fredholm, B. B. (2010). Adenosine receptors as drug targets. *Experimental Cell Research, 316,* 1284–1288.

58. Ferre, S., Fuxe, K., von Euler, G., Johansson, B., & Fredholm, B. B. (1992). Adenosine–dopamine interactions in the brain. *Neuroscience, 51,* 501–512.

59. Loke, W. H. (1988). Effects of caffeine on mood and memory. *Physiology & Behavior, 44,* 367–372.

60. Kendler, K. S., & Prescott, C. A. (1999). Caffeine intake, tolerance, and withdrawal in women: a population-based twin study. *American Journal of Psychiatry, 156,* 223–228.

61. Childs, E., Hohoff, C., Deckert, J., et al. (2008). Association between ADORA2A and DRD2 polymorphisms and caffeine-induced anxiety. *Neuropsychopharmacology, 33,* 2791–2800.

62. Yang, A., Palmer, A. A., & de Wit, H. (2010). Genetics of caffeine consumption and responses to caffeine. *Psychopharmacology, 211,* 245–257.

63. Dlugos, A., Freitag, C., Hohoff, C., et al. (2007). Norepinephrine transporter gene variation modulates acute response to D-amphetamine. *Biological Psychiatry, 61,* 1296–1305.

64. Lott, D. C., Kim, S. J., Cook, E. H., Jr., & de Wit, H. (2005). Dopamine transporter gene associated with diminished subjective response to amphetamine. *Neuropsychopharmacology, 30,* 602–609.

65. Lott, D. C., Kim, S. J., Cook, E. H., & de Wit, H. (2006). Serotonin transporter genotype and acute subjective response to amphetamine. *American Journal on Addictions, 15,* 327–335.

66. Dlugos, A. M., Hamidovic, A., Hodgkinson, C. A., et al. (2010). More aroused, less fatigued: fatty acid amide hydrolase gene polymorphisms influence acute response to amphetamine. *Neuropsychopharmacology, 35,* 613–622.

67. Dlugos, A. M., Hamidovic, A., Hodgkinson, C., et al. (2011). OPRM1 gene variants modulate amphetamine-induced euphoria in humans. *Genes, Brain, and Behavior, 10,* 199–209.

68. Veenstra-VanderWeele, J., Qaadir, A., Palmer, A. A., Cook, E. H., Jr., & de Wit, H. (2006). Association between the casein kinase 1 epsilon gene region and subjective response to D-amphetamine. *Neuropsychopharmacology, 31,* 1056–1063.

69. Flanagin, B. A., Cook, E. H., Jr., & de Wit, H. (2006). An association study of the brain-derived neurotrophic factor Val66Met polymorphism and amphetamine response. *American Journal of Medical Genetics. Part B, Neuropsychiatric Genetics, 141B,* 576–583.

70. Hart, A. B., de Wit, H., & Palmer, A. A. (2012). Genetic factors modulating the response to stimulant drugs in humans. *Current Topics in Behavioral Neurosciences.* [Epub ahead of print]

71. Ashare, R. L., Baschnagel, J. S., & Hawk, L. W. (2010). Subjective effects of transdermal nicotine among nonsmokers. *Experimental and Clinical Psychopharmacology*, *18*, 167–174.

72. Kalman, D., & Smith, S. S. (2005). Does nicotine do what we think it does? A meta-analytic review of the subjective effects of nicotine in nasal spray and intravenous studies with smokers and nonsmokers. *Nicotine & Tobacco Research*, *7*, 317–333.

73. Perkins, K. A., Lerman, C., Coddington, S., & Karelitz, J. L. (2008). Association of retrospective early smoking experiences with prospective sensitivity to nicotine via nasal spray in nonsmokers. *Nicotine & Tobacco Research*, *10*, 1335–1345.

74. Role, L. W. (1992). Diversity in primary structure and function of neuronal nicotinic acetylecholine receptor channels. *Current Opinion in Neurobiology*, *2*, 254–262.

75. Colquhoun, L. M., & Patrick, J. W. (1997). Pharmacology of neuronal nicotinic acetylcholine receptor subtypes. *Advances in Pharmacology (San Diego, Calif.)*, *39*, 191–220.

76. Hadjiconstantinou, M., & Neff, N. H. (2011). Nicotine and endogenous opioids: neurochemical and pharmacological evidence. *Neuropharmacology*, *60*, 1209–1220.

77. De Biasi, M., & Dani, J. A. (2011). Reward, addiction, withdrawal to nicotine. *Annual Review of Neuroscience*, *34*, 105–130.

78. Liu, J. Z., Tozzi, F., Waterworth, D. M., et al. (2010). Meta-analysis and imputation refines the association of 15q25 with smoking quantity. *Nature Genetics*, *42*, 436–440.

79. Thorgeirsson, T. E., Gudbjartsson, D. F., Surakka, I., et al. (2010). Sequence variants at CHRNB3-CHRNA6 and CYP2A6 affect smoking behavior. *Nature Genetics*, *42*, 448–453.

80. Sherva, R., Wilhelmsen, K., Pomerleau, C. S., et al. (2008). Association of a single nucleotide polymorphism in neuronal acetylcholine receptor subunit alpha 5 (CHRNA5) with smoking status and with "pleasurable buzz" during early experimentation with smoking. *Addiction*, *103*, 1544–1552.

81. Bierut, L. J., Madden, P. A., Breslau, N., et al. (2007). Novel genes identified in a high-density genome wide association study for nicotine dependence. *Human Molecular Genetics*, *16*, 24–35.

82. Zeiger, J. S., Haberstick, B. C., Schlaepfer, I., et al. (2008). The neuronal nicotinic receptor subunit genes (CHRNA6 and CHRNB3) are associated with subjective responses to tobacco. *Human Molecular Genetics*, *17*, 724–734.

83. Bierut, L. J. (2011). Genetic vulnerability and susceptibility to substance dependence. *Neuron*, *69*, 618–627.

84. Hoft, N. R., Stitzel, J. A., Hutchison, K. E., & Ehringer, M. A. (2011). CHRNB2 promoter region: association with subjective effects to nicotine and gene expression differences. *Genes, Brain, and Behavior*, *10*, 176–185.

85. Hutchison, K. E., Allen, D. L., Filbey, F. M., et al. (2007). CHRNA4 and tobacco dependence: from gene regulation to treatment outcome. *Archives of General Psychiatry*, *64*, 1078–1086.

7 Developmental Considerations in Gene Identification Efforts

Danielle M. Dick

7.1 The Endophenotype Concept: History and Legacy

As an undergraduate working with Irving Gottesman at the University of Virginia in the mid to late 1990s, I was introduced early to the endophenotype concept. At that time, Irv talked a lot about endophenotypes, but no one else did. This was despite the fact that the endophenotype concept had been introduced to the field of psychiatry in 1972 by Gottesman and Shields [1], when they characterized endophenotypes as internal phenotypes discoverable by a "biochemical test of microscopic examination," adapting the term from a 1966 paper addressing evolution and insect biology [2]. After decades of being ignored, the idea of using endophenotypes to aid in gene identification has gained widespread popularity as evidenced by the exponential growth of PubMed citations for the keyword "endophenotype" in recent years. The concept began to gain widespread attention when it was reintroduced in a 2003 review paper by Gottesman and Gould entitled "The Endophenotype Concept in Psychiatry" [3]. This paper entered the field at a time of growing frustration surrounding failure to identify genes reliably associated with psychiatric disorders and offered a potential explanation for this effect and means to deal with it. The argument was that psychiatric classification systems by their very nature create heterogeneous groups of affected individuals, and that this heterogeneity hampered our ability to detect susceptibility genes. Further, these broad binary classifications likely represented end points that were quite distal from the level of gene action; surely there was no gene "for" schizophrenia, alcohol dependence, and so forth; rather, genes affected certain biochemical processes and pathways that resulted in a downstream alteration in susceptibility to psychiatric disorder(s). Gottesman and Gould defined endophenotypes as "measurable components unseen by the unaided eye along the pathways between disease and distal genotype" (p. 636) and argued that these endophenotypes should be "simpler clues to genetic underpinnings than the disease syndrome itself" (p. 636) [3].

Gottesman and Gould delineated a series of criteria for endophenotypes [3], which has been modified and expanded upon in a number of subsequent papers [4, 5]. Still others have attempted to organize the psychiatric genetics literature on endopheno-types, delineating categories of endophenotypes that have been studied: "anatomical, developmental, electrophysiological, metabolic, sensory, and psychological/cognitive" (p. 164) [6]. Further, Flint and Munafo challenged the idea that endophenotypes represent a simpler genetic architecture, suggesting that a review of the literature provided no support for the idea that the effect sizes of loci contributing to endophenotypes are any larger than those for genes contributing to psychiatric diseases themselves [6]. Perhaps this is unsurprising, as with the rapid growth in popularity of the term "endophenotype," the kind of phenotypes characterized as "endophenotypes" has strayed far from the original conceptualization of endophenotypes as phenotypes discoverable by a "biochemical test or a microscopic examination" (p. 319) [1] or "unseen by the unaided eye" (p. 636) [3]. For example, personality traits have been suggested as endophenotypes [7, 8]. The study of personality is incredibly complex, with entire journals, textbooks, and courses devoted to the topic; the idea that personality will present a "simpler clue" to the genetic underpinnings of psychiatric disease may be wishful thinking. In some cases, the rationale for studying a particular trait (e.g., neuroticism as opposed to depression or anxiety diagnoses) may simply be that quantitative traits are more powerful for genetic studies than a binary disease outcome. And since neuroticism and depression/anxiety disorders are known to share a genetic basis [9], genes identified as influencing neuroticism are also likely to influence susceptibility for depression/anxiety disorders. This may be a strong rationale for using neuroticism in gene identification efforts, but it doesn't necessarily make neuroticism an endophenotype—at least not in the traditional sense.

This book has omitted the word endophenotype altogether from its title in favor of "intermediate phenotype" in recognition of the fact that many of the phenotypes reviewed in this volume stray from the original endophenotype concept. Even the term "intermediate phenotype" may not always be accurate, as delineated in a 2009 paper in which Kendler and Neale usefully differentiate between mediational and liability index models of endophenotypes [10]. Both involve a correlation of genetic effects on the endophenotype and psychiatric outcome, but only in the mediational model does the genetic risk for psychiatric outcome actually pass through the endophenotype and represent an "intermediate" phenotype between genetic susceptibility and disease outcome.

So where do all these complications leave the endophenotype concept? With the rising popularity and excitement surrounding endophenotypes, we have blurred the conceptualization of endophenotypes and strayed from their original meaning. However, the endophenotype concept has made a lasting and critical contribution by challenging the field of psychiatric genetics to move beyond binary clinical diagnostic affection status outcomes. A field that was once dominated by case–control studies has now expanded to appreciate the necessity of utilizing additional phenotypes in gene identification. These phenotypes can help us understand something about the component processes involved in complex disease outcomes. They can be useful in delineating underlying mechanisms and pathways of risk in ways that clinical outcomes do not. They can help conceptually organize the different pathways involved in genetic vulnerability and the routes by which a particular gene may influence psychiatric outcome. Whether or not psychiatric endophenotypes represent "simpler genetic clues" and advance gene identification remains to be seen. However, even absent that, the endophenotype concept will have a lasting legacy in psychiatry by challenging us to think more deeply about our phenotypes in gene identification efforts.

7.2 Developmental Considerations in Gene Identification

It may seem strange to have a chapter on development in a book on intermediate phenotypes; after all, development is not a phenotype (or endophenotype or intermediate phenotype) per se. However, it is in the spirit of the endophenotype legacy's creating a heightened awareness of the importance of phenotype in gene identification that a chapter on developmental considerations is well-placed in this volume. Interestingly, the importance of developmental stage on defining phenotypes for gene identification still has not received widespread attention. This is particularly problematic in the field of addiction, where alcohol use or problems in a 13-year-old likely represents a very different thing (and associated etiological process) than alcohol use/problems in a 23-year-old individual, which may represent a different thing than alcohol use/problems in a 53-year-old individual.

Twin studies have provided evidence of dramatic shifts in the importance of genetic influences on alcohol-related outcomes across time. In the era of rapid advances in statistical and molecular genetics, some have suggested that twin studies are no longer relevant. However, until we have identified all of the specific genes involved in the predispositions to psychiatric outcomes (a goal which we are still far from achieving), this study design remains the only way to study the overall genetic contribution to outcome. Further, the more sophisticated kinds of questions addressed by twin studies today, such as changes in the importance of genetic effects across time, moderation of genetic effects by specific environmental factors, and the degree of genetic overlap between disorders, can provide clues that can be useful in gene identification efforts. Accordingly, rather than ignoring twin studies as a thing of the past, it would behoove the fields of statistical and molecular genetics to take advantage of the twin literature as another method of directing and informing gene identification efforts (we need all the help we can get!). Here I review the developmental twin literature on substance use outcomes and provide evidence that these developmental considerations have implications for gene identification efforts.

7.2.1 Changing Importance of Genetic Effects across Time

A complete understanding of the etiology of substance use disorders requires viewing these disorders as end states of developmental processes that begin early in life. Although there is substantial evidence for a significant genetic component to adult alcohol dependence, with heritability in the range of 50% to 60% [11], this figure fails to capture the dynamic nature of genetic and environmental influences across development. There are a number of longitudinal studies of alcohol use, most of which focus on the period from early adolescence to young adulthood. This period represents a critical time frame in the developmental course for alcohol use disorders as it is during this time that most adolescents initiate alcohol use and, subsequently, move from early experimentation to more established patterns of use. Twin studies demonstrate that the importance of genetic and environmental influences changes dramatically over this developmental period. Data from two population-based longitudinal Finnish twin studies illustrate the striking shift in the relative importance of genetic and environmental influences that occurs from early adolescence to young adulthood: there is a steady increase in the relevance of genetic factors across adolescence and a corresponding and sharp decrease in the relevance of common environmental influences [12] (see figure 7.1).

These data demonstrate that while alcohol initiation is largely environmentally influenced [13], as drinking patterns become more regular and established across adolescence, genetic factors assume increasing importance; however, alcohol use early in adolescence is influenced largely by environmental factors found in the family, school, and neighborhood [14, 15]. A very similar pattern of results for alcohol use was obtained by a life history method in male twin pairs from the Virginia Adult Twin Study of Psychiatric and Substance Use Disorders [16]. At age 14, all twin resemblance

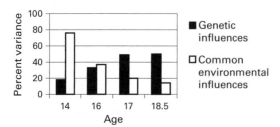

Figure 7.1
Alcohol heritability by age.
Source: Rose et al., 2001 [12].

resulted from shared environmental factors. From ages 14 to 23, shared environment became progressively less important and genetic factors more important. These changes that occur across adolescence and into adulthood likely reflect the increasing relevance of genetic variation in alcohol response as individuals move from early experimentation to more regular patterns of alcohol use, as well as the ability of individuals to increasingly select and shape their social worlds as they gain autonomy across this age range. A parallel pattern has been observed for nicotine and cannabis use [16].

Not only does the twin literature indicate that there is a shift in the importance of genetic influences on alcohol *use* from adolescence to adulthood, there is also evidence of differing genetic influence on alcohol *dependence* between young adolescents and adults. Although adult alcohol dependence shows consistent evidence of heritability across many independent studies in the range of 50% to 60% [11], the more limited number of studies investigating alcohol *dependence* symptoms in early adolescence suggest a very different underlying etiology than studies of alcohol dependence in adults. Analyses of alcohol dependence symptoms at age 14 in a large sample of Finnish twins found no evidence of genetic effects in either girls or boys at this age [17]. Data from the Missouri Adolescent Female Twin Study showed a similar pattern of results, with alcohol dependence symptoms in adolescence largely influenced by environmental factors [18]. A recent set of longitudinal analyses of symptoms of alcohol abuse and dependence help to clarify these results. Data from the Netherlands Twin Register were used to model symptoms of alcohol abuse and dependence between 15 and 32 years of age. Shared environmental influences were most important early in adolescence and decreased across time, accounting for 57% of the variance at age 15 to 17 but only 2% by age 18 to 20. Genetic influences increased from 28% at age 15 to17 to 58% at age 21 to 23 and stayed high in magnitude until age 30 to 32 [19]. These studies suggest that the factors influencing alcohol dependence symptoms that manifest very early in adolescence may differ from the etiological causes of adult alcohol dependence. Alcohol-related problems early in adolescence appear to be more heavily influenced by shared environmental factors and less so by genetic influences whereas alcohol-related problems in adulthood show greater evidence of genetic effects.

This has implications for gene identification studies. The twin findings suggest that we need to be keenly aware of the age of our participants when conducting genetic association studies. Simply using age as a covariate will not capture the important developmental shifts described above. The twin literature would suggest that specific genes that are associated with adult alcohol dependence may not be strongly associated with alcohol dependence problems in early adolescence, which appear to represent a different, more environmentally influenced etiology. And in fact, this is exactly what has been found with respect to some of the specific genes that have been identified as relevant for alcohol-related outcomes.

The gene *GABRA2* was first identified by the Collaborative Study on the Genetics of Alcoholism (COGA) as associated with adult alcohol dependence [20] in the high-risk COGA sample selected through probands in treatment facilities. The association between *GABRA2* and adult alcohol dependence was subsequently replicated in several independent samples from around the world [21]. In order to understand the effect of this gene earlier in development, we genotyped a subset of COGA participants between the ages of 7 and 17 years who were children from the families in which association with alcohol dependence in the adult generation was detected. As would be predicted by the twin studies, there was no evidence of association with alcohol dependence in the child/adolescent sample [22]. Similarly, in an independent study of the association between *GABRA2* and adolescent alcohol use disorders using a treatment-ascertained sample of individuals between the ages of 13 and 18 years, no association with alcohol problems was found [23]. Further, in COGA when we combined the child and adult samples and fit survival curves for alcohol dependence for participants ranging in age from 7–91, we found that differences in rates of alcohol dependence by genotype were only evident after age 20 (see figure 7.2) [22]. This same

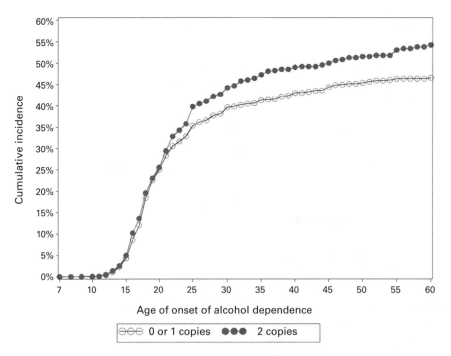

Figure 7.2
Incidence of the onset of alcohol dependence by *GABRA2* genotype.
Source: Sakai et al., 2010 [23].

pattern has been found across independent studies that have examined other genes thought to be associated with alcohol dependence. In a study using AddHealth data from nearly 2,500 individuals, associations between five candidate genes involved in the monoamine system (*DRD4*, *DRD2*, *5HTT*, *DAT1*, and *MAOA*) and alcohol use were observed only among individuals over the age of 19 years, not in adolescents [24]. Data from a Minnesota study demonstrated that the effect of the *ALDH2* genotype on drinking patterns was evident only in late adolescence/early adulthood, and not in early and mid-adolescence [25]. These studies illustrate a pattern across multiple different genes and independent samples whereby the effect of specific genes on alcohol use and problems does not become detectable until late adolescence or emerging adulthood. This maps onto the twin literature which has previously indicated that alcohol use and problems earlier in adolescence are more environmentally influenced.

One limitation of these studies is that they have used cross-sectional data from individuals of different ages. Recently, we have begun to study trajectories of drunkenness from ages 12 to 25 in a sample of COGA participants followed longitudinally. We found that *GABRA2* is not associated with differences in rates of drunkenness before age 18; rather the genotype previously associated with adult alcohol dependence in the COGA sample was associated with a larger "jump" in drunkenness after age 18 (Cho et al., in preparation). This suggests that the enhanced freedom associated with adulthood allows greater opportunity to express genetic predispositions.

7.2.2 Same Genes, Different Disorder

Alcohol problems, like most psychiatric and substance use disorders, are frequently comorbid with other problems. One of the most robust predictors of both concurrent and future alcohol problems is childhood conduct disorder [26–28]. Numerous twin studies indicate that the overlap between childhood conduct disorder and adult alcohol dependence is due largely to shared genetic factors [29–32]. This common genetic liability is thought to be a predisposition toward behavioral undercontrol/disinhibition, which can manifest as conduct disorder in childhood and alcohol dependence later in life [33]. This suggests that while genes associated with adult alcohol dependence may not be associated with early adolescent alcohol dependence (as reviewed above), they may be associated with conduct disorder at an earlier developmental stage. This is what we found in the age 7–17 COGA sample. While there was no association with alcohol dependence symptoms, there was an association with conduct disorder in the children: children carrying one or more copies of the *GABRA2* allele associated with alcohol dependence in the adult sample had nearly twice the rate of conduct problems [22]. We have subsequently replicated these findings in an independent, community-based cohort of more than 500 children, followed annually from kindergarten through age 25. Using longitudinal data on externalizing behavior from ages 12 to 25, we found that *GABRA2* was associated with trajectories of externalizing behavior across adolescence, with the genotype previously associated with adult alcohol dependence in COGA associated with persistent, elevated levels of behavior problems across adolescence [34]. In fact, two of the genes originally associated with adult alcohol dependence in COGA (both *GABRA2* and *CHRM2*) have been related to behavioral problems earlier in development, both in COGA and independent samples [22, 34, 35]. These findings remind us that the same gene may be associated with different phenotypes across different developmental stages. This provides another layer of complication for gene identification efforts.

7.2.3   Moving beyond Cross-Sectional Phenotypes

Most of the early gene identification work grew out of the tradition of psychiatric genetics, with lifetime clinical diagnoses being the outcome of interest. Accordingly, there was little attention paid to the phenotype over time; once an individual was affected, that was considered the end point. When multiple data points were collected, often these were collated to create affection status based on affection at any given time point (or unaffected if no diagnosis was met at any time point). However, there are many potential advantages associated with incorporating longitudinal data to create more sophisticated phenotypes for gene identification efforts.

First, incorporating data from multiple time points allows us to create more reliable phenotypes and remove measurement error variance, potentially reducing noise in gene identification. In addition, studying trajectories of behavior, rather than behavior at any one time point, allows us to study not only mean levels of a behavior across time but also changes in the behavior across time. Techniques such as latent growth curve analysis allow us to test whether genetic factors impact rate of change (e.g., through the testing of a slope parameter) in addition to mean levels of behavior. This may be particularly relevant for addiction phenotypes where growth in a behavior (e.g., how rapidly one escalates in use) may reflect an important genetic loading. In a related example from another psychiatric area, Petersen and colleagues [36] fitted individual growth models to symptoms of anxiety and depression from ages 12 to 17 and tested the effect of the serotonin transporter polymorphism gene (*5-HTTLPR*) and stressful life events on trajectories of anxious/depressed symptoms. They found that adolescents with lower serotonin transcriptional efficiency genotypes whose mothers reported more stressful events showed more anxious/depressed symptoms (differences in mean) and greater increases in the development of symptoms of anxiety and depression (differences in acceleration) than did higher transcriptional efficiency adolescents, particularly at ages 16 and 17.

Longitudinal data also allow for the application of more sophisticated analytic methods to test for measurement invariance across developmental stage. Measurement invariance is generally assumed in longitudinal analyses, for example, that variables assessed at multiple time points are indexing the same underlying construct. Item response theory (IRT) provides a method to explicitly test this. For example, IRT can be applied to the testing of potential interindividual changes in the construct-related meaning of items over time. That is, within the same measure, the importance of individual items may vary in magnitude across development. Data from the Child Development Project on externalizing behavior, as measured using maternal reports on the Child Behavior Checklist [37] at ages 5 and 16, illustrate this (S. Latendresse, unpublished data). Figure 7.3 shows the item characteristic curves (ICCs) for four separate items at each age. Each ICC is positioned along an x-axis reflecting the latent externalizing behavior construct and a y-axis reflecting the probability of item endorsement,

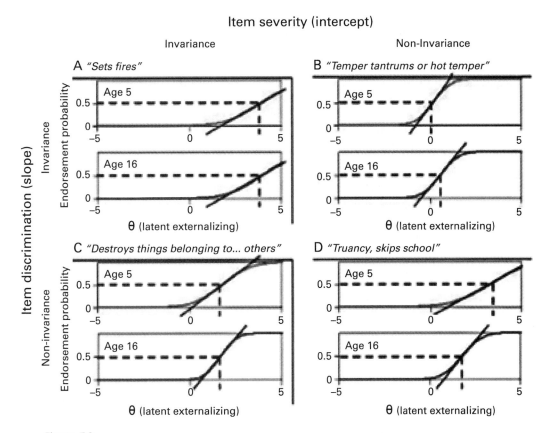

Figure 7.3

Item characteristic curves (ICCs) for four separate items on the Child Behavior Checklist at ages 5 and 16. In each panel, the x-axis reflects the latent externalizing behavior construct and the y-axis reflects the probability of item endorsement. Discrimination (slope) indicates the extent to which the item discriminates between individuals at high and low levels on the latent continuum.

and reflects two distinct parameters: severity (intercept), an indicator of the level of latent externalizing behavior at which the likelihood of endorsement equals 0.5; and discrimination (slope), an indicator of the extent to which the item discriminates between individuals at high and low levels on the latent continuum. Panel (a) shows the ICCs for "Sets fires," for which both severity and discrimination are invariant across age. That is, mothers' reports suggested that this behavior was always extremely severe and that it discriminated between levels of externalizing similarly, regardless of age. Panel (b) shows that "Temper tantrums or hot temper" discriminated similarly across age but indicated more severe behavior at age 16. In contrast, panel (c) suggests that "Destroys things belonging to ... others" was more discriminating of externalizing behavior at age 16 but was equally severe at ages 5 and 16. Finally, panel (d) indicates that "Truancy, skips school" is more discriminating of externalizing behavior at age 16, but the item indicates more severe behavior at age 5.

These analyses are important because they indicate that the same item, as measured at different developmental periods, may be differentially indexing the underlying construct of interest. If this is the case, then it will introduce noise into our gene finding efforts and potentially hamper our ability to find genes involved in the behavior under study. A related application of IRT is to evaluate different measures of the same construct by mapping items from different instruments onto the same underlying metric in order to derive individual-specific construct scores. This is relevant because often different studies will use different measures that purportedly tap into the same underlying construct; the measures are assumed to be equivalent in efforts to replicate genes across different studies, but this is rarely explicitly tested.

7.2.4 Incorporating Developmental Heterogeneity

Finally, we know that patterns of alcohol use are characterized by marked heterogeneity. There is a considerable literature aimed at identifying subtypes of alcohol-dependent individuals [38–40]. Although the exact characteristics of the subtypes varies between studies, a consistent finding is the differentiation of alcohol dependence characterized by antisocial behavior and other drug dependence (a so-called externalizing subtype) and alcohol dependence associated with more depressive and anxious features (a so-called internalizing subtype). Further, there is evidence that these different pathways start early in development [41]. The concern is that failing to differentiate on comorbid and clinical features by treating all alcohol-dependent individuals as equivalent may introduce noise into efforts at gene identification if heterogeneity in the alcohol dependence phenotype is associated with heterogeneity in etiological factors. Data from the COGA project illustrate this. The association between *GABRA2* and adult alcohol dependence was subsequently found to be limited to alcohol-dependent individuals with comorbid drug dependence [42]. In other identified genes, the association has been limited to early onset adult alcohol dependence (onset between 18–25 years, not to be confused with the environmentally influenced early adolescent alcohol dependence described above) [43, 44].

Developmental data also suggest that another complication may exist, that is, individuals may arrive at the same phenotype via different pathways. These may represent different etiological underpinnings which would be masked by simply assessing an outcome such as alcohol problems at any one given time point. Data from the Avon Longitudinal Study of Parents and Children (ALSPAC) illustrate this. ALSPAC is an epidemiological study that has followed a large cohort of more than 13,000 children and their parents from early in the mother's pregnancy through childhood and adolescence, with assessments ongoing as the participants currently enter young adulthood. The study enrolled all pregnant mothers resident in a defined geographical area with expected dates of delivery between April 1, 1991, and December 31, 1992, and collected comprehensive health-related information from both the mothers and the children. With this rich existent data set we sought to assess longitudinal pathways of risk for alcohol problems, starting very early in childhood. Using measures of temperament collected from the mother at multiple assessments from 4 months through 5 years of age, we attempted to predict self-reports of alcohol related problems at age 15.5. This project was unusual in that most of the research on risk factors for alcohol use has focused on adolescence, due to the importance of this period for initiation and escalation of alcohol use [45]. However, most risk and protective factors for alcohol use do not arise de novo in early adolescence but rather have roots in developmental processes stretching back into early childhood [46]. Accordingly, clarifying the roots of these developmental patterns is likely to be vital in acquiring a more complete understanding of the origins of risk factors for the acquisition and maintenance of patterns of alcohol use. In addition to testing for connections between early childhood temperament and adolescent alcohol use, we sought to understand factors that mediate these associations. We focused on personality and peers, two sets of factors known to play a role in patterns of adolescent alcohol use.

We found that childhood temperament prior to age 5 predicted adolescent alcohol use and problems at age 15.5 years. One of the most interesting findings to emerge from this study was that two uncorrelated and distinct temperament styles—children who were rated as consistently *sociable* through age 5 and children who were rated as having consistent *emotional and conduct difficulties* through age 5—both showed elevated rates of alcohol problems at age 15.5. Each of these early temperament styles was associated with different personality factors in midadolescence: consistently *sociable* children showed higher levels of extraversion and emotional stability in midchildhood whereas children who showed early *emotional and conduct difficulties* reported lower agreeableness, conscientiousness, and emotional stability. These two temperament factors were uncorrelated. However, both of these very different temperamental styles were associated with elevated alcohol problems at age 15.5 years. The effect of analyzing alcohol problems in adolescence in a gene identification effort without taking into account this developmental heterogeneity is unknown. One common mediating variable was associated with both temperaments: sensation seeking; both *sociable* children and children with *emotional and conduct difficulties* had higher levels of sensation seeking in late childhood/early adolescence. Accordingly, genes involved in elevated risk for alcohol problems via influence on sensation seeking may be shared across the children and detected in an analysis. However, if part of the predisposition to alcohol problems is via genes associated with the other mediating variables (e.g., emotional stability) on which the children with varying early temperamental styles differed, the ability to detect these genes may be obscured in an analysis of adolescent alcohol problems that fails to differentiate across these very different pathways of arriving at that outcome. Again, this speaks to the importance of utilizing intermediate phenotypes in gene identification efforts for alcohol-related outcomes and underscores the potential utility of longitudinal data in mapping developmental heterogeneity.

7.3 Conclusions

A legacy of the endophenotype concept in psychiatry has been a heightened awareness of the importance of phenotype in gene identification efforts. This has inspired the field to move beyond simply focusing on clinical affection status as an outcome and to think more deeply about the potential intermediate variables by which risk for clinical outcomes may eventually be conferred. One area that has yet to receive widespread attention is the importance of developmental stage in defining phenotypes. There is evidence that this is important for a number of reasons. The importance of genetic influences on an outcome may vary across developmental stage. Further, the same genetic predisposition may manifest as different disorders at different developmental stages. Twin studies provide evidence that both of these are the case in the area of alcohol dependence, and there are already emerging examples demonstrating that taking this into account can aid in gene identification. Further, collecting longitudinal information on participants can be beneficial for a number of additional reasons, including the ability to reduce phenotypic noise by creating more reliable phenotypes and taking into account potential measurement invariance. Longitudinal data allow one to test for genetic influences on *changes* in a phenotype across time (e.g., trajectories or acceleration/deceleration), which may differ from genetic influences on mean levels of an outcome. Further, without developmental data, it is possible that individuals may reach the same clinical end point by very different pathways. Failure to take this into account may thwart efforts to identify genes involved in susceptibility. Identifying genes involved in complex psychiatric outcomes is a challenging task. Accordingly, it would be prudent to harness information about these outcomes from multiple literatures to aid in gene identification. The developmental literature is an area that likely has much to contribute to our ability to identify susceptibility genes and understand the pathways by which they eventually confer risk for clinical outcomes.

References

1. Gottesman, I. I., & Shields, J. (1972). *Schizophrenia and Genetics: A Twin Study Vantage Point.* New York: Academic.

2. John, B., & Lewis, K. R. (1966). Chromosome variability and geographical distribution in insects: chromosome rather than gene variations provide the key to differences among populations. *Science, 152,* 711–721. doi:10.1126/science.152.3723.711

3. Gottesman, I. I., & Gould, T. D. (2003). The endophenotype concept in psychiatry: etymology and strategic intentions. *American Journal of Psychiatry, 160,* 636–645. doi:10.1176/appi.ajp.160.4.636

4. Cannon, T. D., & Keller, M. C. (2006). Endophenotypes in the genetic analyses of mental disorders. *Annual Review of Clinical Psychology, 2,* 267–290. doi:10.1146/annurev.clinpsy .2.022305.095232

5. Almasy, L., & Blangero, J. (2001). Endophenotypes as quantitative risk factors for psychiatric disease: rationale and study design. *American Journal of Medical Genetics, 105,* 42–44. doi:10. 1002/1096-8628(20010108)105:1<42::AID-AJMG1055>3.0.CO;2-9

6. Flint, J., & Munafo, M. R. (2007). The endophenotype concept in psychiatric genetics. *Psychological Medicine, 37,* 163–180. doi:10.1017/S0033291706008750

7. Baker, L. A., Isen, J., Bezdjian, S., & Raine, A. (2005). Are personality traits endophenotypes for antisocial behavior? *Behavior Genetics, 35,* 831–832. doi:1007s/10519-005-7287-9

8. Singh, A. L., & Waldman, I. (2005). Genetic and environmental influences on the covariation of negative emotionality and internalizing symptoms in children. *Behavior Genetics, 35,* 832. doi:1007s/10519-005-7287-9

9. Hettema, J. M., Neale, M. C., Myers, J. M., Prescott, C., & Kendler, K. S. (2006). A population-based twin study of the relationship between neuroticism and internalizing disorders. *American Journal of Psychiatry, 163,* 857–864. doi:10.1176/appi.ajp.163.5.857

10. Kendler, K. S., & Neale, M. (2010). Endophenotype: a conceptual analysis. *Molecular Psychiatry, 15,* 789–797. doi:10.1038/mp.2010.8

11. Dick, D. M., Prescott, C., & McGue, M. (2009). The genetics of substance use and substance use disorders. In Y.-K. Kim (Ed.), *Handbook of Behavior Genetics* (pp. 433–453). New York: Springer. doi:10.1007/978-0-387-76727-7_29

12. Rose, R. J., Dick, D. M., Viken, R. J., & Kaprio, J. (2001). Gene–environment interaction in patterns of adolescent drinking: regional residency moderates longitudinal influences on alcohol

use. *Alcoholism, Clinical and Experimental Research, 25*, 637–643. doi:10.1111/j.1530-0277.2001. tb02261.x

13. Hopfer, C. J., Crowley, T. J., & Hewitt, J. K. (2003). Review of twin and adoption studies of adolescent substance use. *Journal of the American Academy of Child and Adolescent Psychiatry, 42*, 710–719. doi:10.1097/01.CHI.0000046848.56865.54

14. Rose, R. J., Dick, D. M., Viken, R. J., Pulkkinen, L., & Kaprio, J. (2001). Drinking or abstaining at age 14? A genetic epidemiological study. *Alcoholism, Clinical and Experimental Research, 25*, 1594–1604. doi:10.1111/j.1530-0277.2001.tb02166.x

15. Rose, R. J., Viken, R. J., Dick, D. M., Bates, J. E., Pulkkinen, L., & Kaprio, J. (2003). It does take a village: nonfamilial environments and children's behavior. *Psychological Science, 14*, 273–277. doi:10.1111/1529-1006.03434

16. Kendler, K. S., Schmitt, J. E., Aggen, S. H., & Prescott, C. A. (2008). Genetic and environmental influences on alcohol, caffeine, cannabis, and nicotine use from adolescence to middle adulthood. *Archives of General Psychiatry, 65*, 674–682. doi:10.1001/archpsyc.65.6.674

17. Rose, R. J., Dick, D. M., Viken, R. J., Pulkkinen, L., Nurnberger, J. I., Jr., & Kaprio, J. (2004). Genetic and environmental effects on conduct disorder, alcohol dependence symptoms, and their covariation at age 14. *Alcoholism, Clinical and Experimental Research, 28*, 1541–1548. doi:10.1097/01.ALC.0000141822.36776.55

18. Knopik, V., Heath, A. C., Bucholz, K. K., Madden, P. A., & Waldron, M. (2009). Genetic and environmental influences on externalizing behavior and alcohol problems in adolescence: A female twin study. *Pharmacology Biochemistry and Behavior, 93*, 313–321. doi:10.1016/j.pbb.2009.03.011.

19. van Beek, J. H., Kendler, K. S., de Moor, M. H., et al. (2012). Stable genetic effects on symptoms of alcohol abuse and dependence from adolescence into early adulthood. *Behavior Genetics, 42*, 40–56. doi:10.1007/s10519-001-9488-8

20. Edenberg, H. J., Dick, D. M., Xuei, X., et al. (2004). Variations in GABRA2, encoding the a2 subunit of the GABA-A receptor, are associated with alcohol dependence and with brain oscillations. *American Journal of Human Genetics, 74*, 705–714. doi:10.1086/383283

21. Enoch, M. A. (2008). The role of GABA(A) receptors in the development of alcoholism. *Pharmacology, Biochemistry, and Behavior, 90*, 95–104. doi:10.1016/j.pbb.2008.03.007

22. Dick, D. M., Bierut, L., Hinrichs, A. L., et al. (2006). The role of GABRA2 in risk for conduct disorder and alcohol and drug dependence across developmental stages. *Behavior Genetics, 36*, 577–590. doi:10.1007/s10519-005-9041-8

23. Sakai, J. T., Stallings, M. C., Crowley, T. J., Gelhorn, H. L., McQueen, M. B., & Ehringer, M. A. (2010). Test of association between GABRA2 (SNP rs279871) and adolescent conduct/alcohol use disorders utilizing a sample of clinic referred youth with serious substance and conduct problems, controls and available first degree relatives. *Drug and Alcohol Dependence, 106*, 199–203. doi:10.1016/j.drugalcdep.2009.08.015

24. Guo, G., Wilhelmsen, K., & Hamilton, N. (2007). Gene–lifecourse interaction for alcohol consumption in adolescence and young adulthood: five monoamine genes. *American Journal of Medical Genetics. Part B, Neuropsychiatric Genetics, 144B*, 417–423. doi:10.1002/ajmg.b.30340

25. Irons, D., Iacono, W. G., Oetting, W., & McGue, M. (2011). Developmental trajectory and environmental moderation of the effect of ALDH2 polymorphism on alcohol use. *Alcoholism, Clinical and Experimental Research, 36*, 1882-1891. doi:10.1111/j.1530-0277.2012.01809.x

26. Crowley, T. J., Milkulich, S. K., MacDonald, M., Young, S. E., & Zerbe, G. O. (1998). Substance-dependent, conduct-disordered adolescent males: severity of diagnosis predicts 2-year outcome. *Drug and Alcohol Dependence, 49*, 225–237. doi:10.1016/S0376-8716(98)00016-7

27. White, H. R., Zie, M., Thompson, W., Loeber, R., & Stouthamer-Loeber, M. (2001). Psychopathology as a predictor of adolescent drug use trajectories. *Psychology of Addictive Behaviors, 15*, 210–218. doi:10.1037/0893-164X.15.3.210

28. Moss, H. B., & Lynch, K. G. (2001). Comorbid disruptive behavior disorder symptoms and their relationship to adolescent alcohol use disorders. *Drug and Alcohol Dependence, 64*, 75–83. doi:10.1016/S0376-8716(00)00233-7

29. Kendler, K. S., Prescott, C., Myers, J., & Neale, M. C. (2003). The structure of genetic and environmental risk factors for common psychiatric and substance use disorders in men and women. *Archives of General Psychiatry, 60*, 929–937. doi:10.1001/archpsyc.60.9.929

30. Young, S. E., Stallings, M. C., Corley, R. P., Krauter, K. S., & Hewitt, J. K. (2000). Genetic and environmental influences on behavioral disinhibition. *American Journal of Medical Genetics, 96*, 684–695. doi:10.1002/1096-8628(20001009)96:5<684::AID-AJMG16>3.0.CO;2-G

31. Slutske, W. S., Heath, A. C., Dinwiddle, S. H., et al. (1998). Common genetic risk factors for conduct disorder and alcohol dependence. *Journal of Abnormal Psychology, 107*, 363–374. doi:10.1037/0021-843X.107.3.363

32. Krueger, R. F. (1999). The structure of common mental disorders. *Archives of General Psychiatry, 56*, 921–926. doi:10.1001/archpsyc.56.10.921

33. Slutske, W. S., Heath, A. C., Madden, P. A., Bucholz, K. K., Statham, D. J., & Martin, N. G. (2002). Personality and the genetic risk for alcohol dependence. *Journal of Abnormal Psychology, 111*, 124–133. doi:10.1037/0021-843X.111.1.124

34. Dick, D. M., Latendresse, S. J., Lansford, J. E., et al. (2009). The role of GABRA2 in trajectories of externalizing behavior across development and evidence of moderation by parental monitoring. *Archives of General Psychiatry, 66*, 649–657. doi:10.1001/archgenpsychiatry.2009.48

35. Dick, D. M., Meyers, J. L., Latendresse, S. J., Creemers, H. E., et al. (2011). CHRM2, parental monitoring, and adolescent externalizing behavior: Evidence for gene-environment interaction. Psychological Science, 22, 481–489. doi:10.1177/0956797611403318.

36. Petersen, I. T., Bates, J. E., Goodnight, J. A., et al. (2012). Interaction between serotonin transporter polymorphism (5-HTTLPR) and stressful life events in adolescents' trajectories

of anxious/depressed symptoms. Developmental Psychology, 48, 1463–1475. doi:10.1037/a0027471

37. Achenbach, T. M., & Edelbrock, C. (1983). *Manual for the Child Behavior Checklist and Revised Child Behavior Profile.* Burlington, VT: Department of Psychiatry, University of Vermont.

38. Cloninger, C. (1987). Neurogenetic adaptive mechanisms in alcoholism. *Science, 236,* 410–416. doi:10.1126/science.2882604

39. Babor, T. (1996). The classification of alcoholics: typology theories from the 19th century to the present. *Alcohol Research & Health World, 20,* 6–17.

40. Hesselbrock, M. N. (1995). Genetic determinants of alcoholic subtypes. In H. Begleiter & B. Kissen (Eds.), *The Genetics of Alcoholism* (pp. 40–69). New York: Oxford University Press.

41. Zucker, R. (2006). Alcohol use and alcohol use disorders: a developmental-biopsychosocial formulation covering the life course. In D. Cicchetti & D. J. Cohen (Eds.), *Developmental Psychopathology: Vol. 3. Risk, Disorder, and Adaptation* (2nd ed., pp. 620–656). Hoboken, NJ: Wiley.

42. Dick, D. M., Agrawal, A., Wang, J. C., et al. (2007). Alcohol dependence with comorbid drug dependence: genetic and phenotypic associations suggest a more severe form of the disorder with stronger genetic contribution to risk. *Addiction, 102,* 1131–1139. doi:10.1111/j.1360-0443.2007.01871.x

43. Edenberg, H. J., Xuei, X., Wetherill, L. F., et al. (2008). Association of NFKB1, which encodes a subunit of the transcription factor NF-kappaB, with alcohol dependence. *Human Molecular Genetics, 17,* 963–970. doi:10.1093/hmg/ddm368

44. Edenberg, H., Koller, D. L., Xuei, X., et al. (2010). Genome-wide association study of alcohol dependence implicates a region on chromosome 11. *Alcoholism, Clinical and Experimental Research, 34,* 840–852. doi:10.111/j.1530-0277.2010.01156.x

45. Windle, M., Spear, L. P., Fuligni, A. J., et al. (2008). Transitions into underage and problem drinking: developmental processes and mechanisms between 10 and 15 years of age. *Pediatrics, 121*(Suppl 4), S273–S289. doi:10.1542/peds.2007-2243C

46. Zucker, R. A., Donovan, J. E., Masten, A. S., Mattson, M. E., & Moss, H. B. (2008). Early developmental processes and the continuity of risk for underage drinking and problem drinking. *Pediatrics, 121*(Suppl 4), S252–S272. doi:10.1542/peds.2007-2243B

8 Enhancing Addiction Genetics via Behavioral Economic Intermediate Phenotypes

James MacKillop and John Acker

In the current chapter, we will review potential contributions from the field of behavioral economics to understanding genetic influences on addictive behavior. More specifically, we will discuss the potential for performance on behavioral economic decision-making tasks to serve as novel endophenotypes or intermediate phenotypes. Individual variation on these tasks could reflect an intervening mechanism that connects genetic variation with either etiological liability or other aspects of an addictive disorder. First, to provide context, we will review the behavioral economic theoretical perspective on addictive behavior, which integrates concepts from psychology and microeconomics to understand these forms of persistent overconsumption. Second, we will focus on one specific mechanism, delayed reward discounting, which is a behavioral economic index of impulsivity. Variation in discounting reflects a person's orientation toward smaller immediate rewards relative to larger delayed rewards, and impulsive discounting has been linked to diverse forms of addictive behavior. Moreover, impulsive discounting is also the behavioral economic variable that has been most extensively linked to genetic factors to date. As such, the findings in this area are important in their own right but also provide a model for how behavioral economic variables can be thought of as intermediate phenotypes. The next section of the chapter will review the findings for other behavioral economic phenotypes that have been linked to genetic factors or have promise in the future, such as the Iowa Gambling Task, a neuropsychological measure of reward decision making, and the Balloon Analogue Risk Task, a behavioral assay of risk sensitivity. These represent less well developed areas of the literature, but domains that have received some study, providing some intriguing initial findings and setting the stage for future work. The following section focuses on future candidates, including probability discounting and drug demand. These are behavioral economic indices that have received virtually no examination from a genetic perspective at this point but may nonetheless be useful as intermediate phenotypes. Last, the chapter concludes with a discussion of considerations, priorities, and future directions for applying behavioral economics to contribute to a greater understanding of addiction genetics.

8.1   The Nature of Behavioral Economic Phenotypes

The field of behavioral economics is a hybrid of psychology and economics that combines theories and methodologies from each discipline to improve the understanding of individual-level economic behavior [1–3]. Importantly, however, the definition of economic behavior is broad, referring not only to personal financial transactions but to all the transactions between individuals and their environments. Behavioral economics has been defined as "the study of the allocation of behavior within a system of constraint" (p. 258) [4], which succinctly captures the breadth of what can be defined as economic behavior, including how individuals choose to spend their money, time, efforts, and other resources. The breadth of scope is also evident in the diversity of applications of behavioral economics, including individual transactions among persons [5], decision-making preferences about the values of different rewards [2], and the neural correlates of economic decision making (i.e., neuroeconomics) [6]. Across these domains, however, the common characteristic is the analysis of the choices people make in their allocation of resources. Thus, behavioral economics can be understood as the study of the systematic factors that affect values and preferences in economic decision making.

In the context of addictive behavior, a behavioral economic approach evolved from operant learning perspectives that emphasize a reinforcement-based approach to addictive behavior [7, 8]. From this perspective, addiction reflects an acquired state in which the relative value of the drug becomes prepotent relative to other reinforcers in spite of substantial and persistent costs of continued use. This conception can be thought of as overlapping with several diagnostic symptoms of substance dependence [9], such as allocating a great deal of time to the drug and continued use despite psychosocial costs or psychological/physical problems. However, unlike the atheoretical *Diagnostic and Statistical Manual of Mental Disorders* (DSM) symptoms, a behavioral economic approach specifically emphasizes the importance of the dysregulated reinforcing value of the drug to the individual.

In the context of behavioral genetics, an important point of clarification is the definition of what comprises a behavioral economic phenotype. Fundamentally, these indicators can be defined as any assessment of individual variation in choice preferences for cost–benefit trade-offs between outcomes. As will be discussed, these could be trade-offs between immediate rewards and delayed rewards, certain rewards and risky rewards, or other cost–benefit relationships, but the common thread is that these measures collectively provide fine-grained assays of different dimensions of values and preferences, and these dimensions are the phenotypes. Of note, some of these tasks were explicitly developed from a behavioral economic perspective whereas others come from other fields but nonetheless capture the same essential cost–benefit processes. We treat this distinction as largely arbitrary. Ultimately, regardless of provenance, the phenotypes in question in this chapter are individuals' decision-making profiles on these various cost–benefit tasks.

8.2 Delayed Reward Discounting as an Endophenotype

The most well-established behavioral economic phenotype is almost certainly delayed reward discounting, or how much an individual *discounts* future *rewards* based on *delay in time*. It is a measure of how much an individual is willing to give up future benefits to receive a reward immediately and is also referred to as capacity to delay gratification or intertemporal choice preference. During a typical discounting task, subjects are presented with a series of choices between two rewards, one being a smaller but immediately available reward, and the other being a larger but delayed reward [10, 11]. The selection of the former putatively reflects impulsivity, and the selection of the latter reflects self-control. The most common task comprises permutations that vary the smaller immediate reward and the lengths of delay and permit the systematic characterization of how much time impacts the value of the larger reward. Importantly, a number of variations of delay discounting tasks exist, including the famous "marshmallow test" (one marshmallow now vs. two after a delay, e.g., [12]), permuted choice tasks or questionnaires (e.g., [13, 14]), and versions using experiential delays (e.g., [15, 16]), but the common feature is the assessment of time-based reward devaluation. An advantage of the task-based assessments and more extensive questionnaire measures is that they provide more precise characterization of the individual's temporal discounting function, the summary index of how much larger future rewards are devalued by virtue of delay. Prototypic temporal discounting functions that can be generated from delay discounting tasks are provided in figure 8.1 and have been found to be best fit by nonlinear equations that are hyperbolic or hyperboloid in nature [17–19]. These equations capture a reward's characteristically sharp drop in value with the imposition of an initial delay, which is followed by smaller and more consistent subsequent decreases over longer delays. The steepness of an individual's temporal discounting function is the index of impulsivity.

Importantly, the level of impulsive discounting varies substantially in the general population as a quantitative trait, and there is consistent evidence that highly impulsive discounting is substantially associated with addictive behavior. At a descriptive level, this is intuitive. Overvaluation of immediate rewards at the cost of larger long-term rewards maps directly onto a characteristic behavioral pattern that is evident in addictive disorders. Across alcohol, nicotine, opiate, and stimulant dependence, a common feature is persistent overvaluation of immediate drug rewards at the cost of an array of future benefits. However, there is also a persuasive empirical literature supporting a link between impulsive discounting and addictive behavior. Numerous categorical studies comparing delay discounting in individuals with a criterion level of addictive behavior to a matched control group have been completed, and a consistent

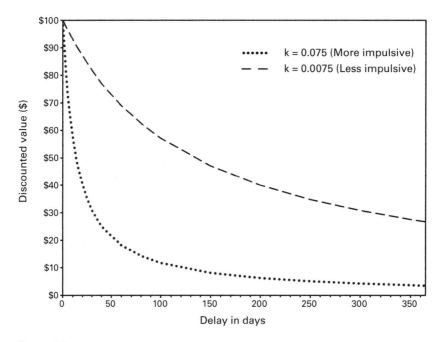

Figure 8.1
Prototypic delay discounting curves. The lines reflect two temporal discounting functions (*k*), and the values are of the present value (y-axis) of $100 at escalating levels of delay (x-axis).

finding is of more impulsive discounting in the criterion groups. This is the case for differing levels of alcohol misuse and dependence (e.g., [20, 21]), nicotine dependence (e.g., [22]), stimulant dependence (e.g., [23]), opiate dependence (e.g., [10]), and also pathological gambling (e.g., [24]). A recent meta-analysis of categorical studies revealed evidence of statistically significant differences of medium effect size between groups across studies [25]. Interestingly, the meta-analysis found larger magnitude differences between groups among studies using clinically diagnosed samples compared to those using individuals at subclinical levels of problems. This suggests that the more severely affected the sample, the larger the difference from the control group, and further implicates discounting level with severity of addictive behavior. Studies comparing different groups have also been complemented by a large number of studies using continuous analyses, which have also found a significant positive relationship between level of discounting and severity of addictive behavior (e.g., [26–29]). Taken together, cross-sectional studies reveal a very robust relationship between impulsive delay discounting and addictive behavior.

A challenge to these cross-sectional studies is disentangling whether discounting precedes addictive behavior as an etiological factor or is a consequence of persistent drug use, reflecting a symptom. Considerably fewer studies can address this, but there have been several direct investigations. One recent study by Audrain-McGovern et al. [30] found that, over a six-year period, impulsive discounting predicted smoking initiation but did not significantly itself change over time. Using the marshmallow test, impulsive discounting in preschoolers has been associated with adult drug use over 20 years later [31], albeit as an interaction with rejection sensitivity. In addition, longitudinal studies examining the temporal consistency of discounting have consistently found robust evidence of reliability in increments ranging from one week to a year or even longer [30, 32–39]. These levels of stability are generally similar to the levels observed for other putatively long-standing characteristics, such as personality traits [40]. These findings do not rule out the effects of addictive behavior on discounting but suggest it is not a characteristic that is inherently highly malleable. Finally, cross-sectional studies have also contributed to the question of causality by focusing on developmental variables. For example, a recent study using a large nationally representative sample of young adults found that tobacco, alcohol, and marijuana use was significantly associated with impulsive discounting, irrespective of sex, age, and the related personality characteristic of sensation seeking [41]. Similarly, impulsive discounting is associated with earlier onset of alcohol-related problems in retrospective studies [42, 43].

Perhaps the most convincing evidence of discounting as an etiological risk factor comes from preclinical studies using rodent models of discounting in relation to drug acquisition and self-administration escalation. Here, several studies have found that drug-naive animals with higher levels of discounting showed faster acquisition of drug self-administration behavior and greater escalation of consumption to clinically relevant levels [44–47]. As these animals have no previous exposure to addictive drugs, this clearly suggests that impulsive discounting reflects a predisposition to developing addictive behavior and the high levels observed in humans with substance use disorders are not simply artifacts of the addictive behavior itself. Importantly, however, there is also preclinical evidence that drug use itself can lead to long-standing increases in impulsive discounting [48], suggesting that discounting both may serve as an etiological risk factor and also may be significantly affected by chronic drug use.

A last source of relevant findings in this area comes from clinical studies. Early studies revealed that recovered substance-dependent individuals exhibited levels of discounting that were equivalent to controls [22] or intermediate between controls and active substance-dependent individuals [21]. This suggested that discounting may either be a negative prognostic factor or may itself be ameliorated by recovery, which would be consistent with it reflecting a symptom rather than a causative factor. At this point, several prospective studies have addressed this question. Four studies have found that delay discounting predicts negative outcomes in smoking cessation treatment [49–52]. Using a naturalistic index of discounting, Tucker et al. [53–56] have consistently found that level of discounting predicts alcohol-related problem resolution over time. In addition, several recent studies have implicated impulsive discounting as a negative prognostic factor in marijuana, cocaine, and opiate treatment [57–59]. Taken together, although most of the studies have focused on tobacco and alcohol, these findings generally suggest that impulsive discounting is a predictor of negative treatment outcome, thus serving as a maintaining factor in addictive behavior.

The preceding findings build a convincing case that impulsive discounting is an important process in the development and maintenance of addictive behavior, but they do not directly address whether this role is a function of genetic influences. There is, however, accumulating evidence that this is the case and that discounting may be a viable endophenotype for addictive behavior. The critical criteria for being an endophenotype are stability, heritability, and an association with the clinical disorder (see chapter 1), and, as reviewed above, the first and third criteria are clearly met. However, there is also increasing evidence that discounting preferences are heritable, arguably the most important endophenotype criterion.

The most extensive evidence of heritability comes from studies examining differences in discounting between inbred rodent strains, which have virtually identical levels of genetic variation and rearing environments. These comparisons between strains effectively compare differences based on systematic genetic variation between strains. In an early study, significantly higher delay discounting was found in Lewis versus Fischer 344 rats [60]. This has subsequently been replicated in one case [61], but in another, using six strains, Wilhelm and Mitchell [62] found significant differences between Fischer rats compared to Copenhagen and Noble rats (more impulsive discounting in the former), but no differences compared to Lewis rats. Complementing differences between rat strains, significant differences have also been present across different strains of mice [63, 64].

Furthermore, animal studies have linked propensity for addictive behavior with strain-based differences in discounting. For example, rats with a high preference for saccharin have been shown to have a propensity for cocaine consumption (e.g., [65]) and also exhibit significantly more impulsive discounting than low-preference rats [47]. Similarly, Roman High-Avoidance rats are more susceptible to the reinforcing effects of drugs than the Roman Low-Avoidance strain [66] and exhibit parallel differences in terms of the discounting phenotype [67]. Greater delay discounting has also been found in rats and mice from alcohol-preferring strains compared to low-alcohol-preference strains [68, 69], although not in all cases [70, 71]. Although these studies are not definitive, they nonetheless demonstrate clear evidence of general heritability and, in some cases, implicate strain-based genetic differences in vulnerabilities to addictive behavior, providing a highly suggestive link.

There are also a number of investigations that implicate genetic factors in discounting in human studies. The classic quantitative genetics methodology for inferring phenotype heritability is the twin methodology, which compares the relative pheno-typic concordance in monozygotic twins, who share almost 100% of genetic variation, to dizygotic twins, who share, on average, 50% genetic variation. This permits estimation of additive and nonadditive genetic contributions and also shared and nonshared (unique) environmental contributions. This approach has been applied extensively to addictive behavior (for a review, see [72]), but only one study has focused on discounting. In this study, a large sample of monozygotic and dizygotic adolescent twins were assessed using a brief discounting assessment at ages 12 and 14 [38]. The heritability of discounting was 30% and 51% at the respective age assessments, with evidence that the same genetic factors influenced the phenotype at both time points. With both assessments considered simultaneously, the best fitting model suggested that the same additive genetic factors influenced discounting at both ages but unique (nonshared) environmental factors significantly influenced discounting at each time point also. Converging with the findings previously reviewed, impulsive discounting was associated with alcohol, tobacco, and marijuana use at age 14, although not at age 12, potentially due to restriction of range.

Although not as robust a methodology as twin studies, a second strategy for determining heritability is examining whether a phenotype aggregates with a family history (FH) of the clinical condition with which it is putatively associated. For heritable conditions, like substance use disorders, a positive FH (FH+) is a risk factor for developing the condition, and evidence that a phenotype aggregates in FH+ individuals, even in the absence of the condition, implicates it as a possible inherited mechanism. Evidence of significantly more impulsive discounting in FH+ individuals compared to FH– individuals (assuming both groups are negative for substance use disorders) would suggest that discounting might be at least partially responsible for the increase in overall risk. Several studies have applied this approach to discounting. In an initial study, Petry et al. [73] examined alcohol dependence FH status (paternal only) in 122 adults and found significantly more impulsive discounting in FH+ women, but not men. Also examining paternal alcoholism, Crean et al. [74] found no differences between the two groups. In contrast, Herting et al. [75] assessed alcohol use disorder FH including both parents and grandparents in a relatively small sample (N = 33), finding a trend-level difference in discounting and significantly faster reaction times in FH+ adolescents compared to FH– adolescents. Thus, it appears that more inclusive definitions of FH are more sensitive to differences than paternal alcoholism only. This is further supported by another recent study, the largest to date (N = 298), which characterized the density of FH for alcohol and other drugs [76]. In this case, FH+ status was significantly associated with more impulsive discounting dichotomously and significantly linearly positively related to the FH density ($r = 0.15$). Interestingly, in the small study by Herting et al., FH density was associated with discounting at a similar magnitude (R^2 = 0.11). Considering these findings together, although there are mixed results, it appears that when FH status is carefully assessed and samples are sufficiently large, there is further support for discounting aggregating with FH of addictive behavior.

A limitation of the preceding human studies is that they suggest genetic factors play a role at an aggregate level, but do not address which specific genetic polymorphisms are involved. To date, only a small number of molecular genetic association studies have addressed this. One early study examined discounting preferences in healthy young adults based on genotype status for the *ANKK1* TaqIA single nucleotide polymorphism (SNP; rs1800497) and the variable number of tandem repeats polymorphism in exon 3 of the dopamine D receptor gene (*DRD4* VNTR) [77]. This revealed a main effect of *ANKK1* genotype, such that possession of at least A1 allele was associated with significantly greater discounting. In addition, an epistatic (gene × gene) interaction was present, such that individuals who were A1+ and possessed the long allele of *DRD4* VNTR exhibited dramatically more impulsive discounting than the other combinations. Subsequently, three studies have examined *DRD4* VNTR genotype and similarly found no main effects of genotype [78–80]. However, in a relatively large sample (N = 546), Sweitzer et al. [79] found evidence of a gene × environment interaction, such that long allele carriers with low childhood socioeconomic status exhibited highly impulsive discounting compared to short allele homozygotes but exhibited less impulsive discounting in the absence of childhood adversity.

Several studies to date have focused on discounting as a function of *COMT* Val-158Met genotype (rs4680). In an initial study of alcoholics and healthy controls (N = 19), significant differences by genotype were present [81], with homozygous Val carriers exhibiting significantly more impulsive discounting, irrespective of diagnosis. Subsequently, Paloyelis et al. [80] examined discounting in relation to *COMT* genotype in samples of attention-deficit/hyperactivity disorder (ADHD) and control participants (N = 68) but found a different pattern of associations. Whereas Val/Val homozygotes exhibited the greatest discounting in the study by Boettiger et al., Paloyelis et al. found that Met–/Met homozygotes exhibited significantly greater discounting, a finding that was independent of ADHD diagnosis. Interestingly, Paloyelis et al. also found significant differences based on the *DAT1*10/6 haplotype, finding that diagnosis interacted with genotype, such that only ADHD individuals with fewer than two copies of the *DAT1* 10/6 haplotype exhibited more impulsive discounting compared to controls. Most recently, Smith and Boettiger [82] found an interaction between age and *COMT* genotype among late adolescent and adult participants (N = 142). Among late adolescent participants, Val/Val genotype was associated with significantly greater discounting, whereas among adult participants, Met allele carriers consistently exhibited significantly greater discounting. Finally, a study of the *DRD2* C957T (rs6277) SNP by White et al. [83] found that nonclinical young adults who were C allele carriers exhibited faster discounting task responding, but not more impulsive discounting.

It is hard to make definitive conclusions in light of the small number of studies to date, particularly because the sample sizes have been modest in size, indeed quite small in some cases, and have also included both clinical and nonclinical participants. Nonetheless, these studies have demonstrated that specific genetic polymorphisms are linked to systematic variation in discounting, and it is worth noting that the variants associated with more impulsive discounting are those associated with vulnerability to addictive behavior [84–86]. This lends further credibility to these findings and further support to the superordinate hypothesis that specific genetic variants influence an individual's level of discounting and thereby confer further risk for developing an addictive disorder. Although it will be critical for future studies to establish these first findings via replication, which is no small feat, these initial findings nonetheless provide a promising start.

More generally, the preceding lines of evidence provide provisional converging evidence that impulsive delay discounting meets the criteria for being an endophenotype for addictive behavior. As a phenotype, delay discounting is stable over time and associated with both the etiology and maintenance of addictive behavior. From a genetic standpoint, the preclinical studies, twin studies, and FH studies suggest that it is a heritable characteristic, and the molecular genetic studies suggest a number of loci are specifically associated with discounting preferences. It would be premature to conclude that these data are definitive at this point, but these findings provide a strong basis for pursuing this hypothesis further.

8.3   Genetic Associations with Other Behavioral Economic Phenotypes

Delay discounting is by far the most extensively studied behavioral economic variable to have been examined in relation to genetic factors, but performance on two other validated behavioral tasks have also been examined as novel intermediate phenotypes, the Iowa Gambling Task (IGT) and the Balloon Analogue Risk Task (BART). The IGT was originally developed as a neuropsychological task to assess executive function in individuals with ventromedial prefrontal cortex damage [86] but it is conceptually a behavioral economic task. During the task, participants receive $2,000 in hypothetical money and select from four decks of cards, each with reward–punishment ratios reflecting aggregate profiles that are "advantageous" (smaller but more frequent rewards, with fewer punishments, leading to a net gain) or "disadvantageous" (larger but less frequent rewards, with more punishments, leading to a net loss). The task uses 100 trials, and, after a period of sampling, healthy participants tend to increasingly select from the advantageous decks whereas individuals with ventromedial prefrontal cortex damage [87, 88] and addictive disorders [89–93] tend not to successfully develop advantageous strategies, continuing to favor net loss decks. Importantly, performance on the IGT is typically studied in five blocks of 20 trials, with the first block considered the sampling period and an advantageous strategy being a migration to the smaller-but-more-frequent-reward decks across successive blocks. Thus, the IGT generates a phenotype that comprises reward learning under conditions of possible gain and loss. Based on this, it is often referred to as a risk-taking task, although in light of its diverse components, the best way to characterize what the IGT measures is still an active point of discussion [94].

Although no twin or other quantitative genetic studies have systematically determined the heritability of IGT performance, two studies using FH-based designs have found significantly poorer performance in individuals with family histories of alcoholism [95] and obsessive–compulsive disorder [96]. However, the study on FH of alcoholism found effects that were specific to task attention and consistency and only found differences in males [95]. Interestingly, a follow-up investigation using functional magnetic resonance imaging of IGT performance found that the two FH groups did not differ behaviorally, but the FH+ group exhibited significantly greater activation in the left anterior cingulate cortex and caudate [97], potentially suggesting genetically influenced differences in the underlying neural processing of risk.

These initial findings are complemented by a number of association studies that have directly examined IGT performance by genotype in both healthy and clinical samples. One common target has been the 5-HTTLPR polymorphism, which has been linked to the serotonin transporter gene (*SLC6A4*). Specifically, the 5-HTTLPR polymorphism refers to a 44 base pair insertion located 1 kb upstream from the transcription initiation site of *SLC6A4*; the absence of the insertion is designated the short (S) version and the presence of the insertion is referred to as the long (L) version. The S allele is associated with reduced serotonin transporter protein availability and function ([98]; for a review, see [99]).

In relation to IGT performance, the associations with 5-HTTLPR genotype have been multifarious and are summarized in table 8.1. Most straightforward, two studies have simply found no main effects of 5-HTTLPR genotype on IGT performance [100, 101]. Several others, however, have reported diverse significant associations. For example, in a study of females only, S/S individuals have been found to exhibit significantly riskier decisions in general [102]. In contrast, L/L individuals have also been found to exhibit significantly more risky decision making in a mixed sex sample, but only during the first 20 trials [103], a surprising period to be sensitive to genetic influences because it putatively reflects random initial deck sampling. Possible differences by sex have also been suggested by two other studies of healthy adults. He et al. [104] found that, compared to L/L individuals, S/S individuals exhibited poorer performance during the first 40 trials, but S/S males exhibited significantly lower performance than L/L males. Stoltenberg et al. [105] found that S/L males made more advantageous

Table 8.1

Studies using Iowa Gambling Task (IGT) performance as an intermediate phenotype

Study	Sample	Polymorphism(s)	Primary findings
Homberg et al., 2008	Healthy adults, N = 88 (100% female)	5-HTTLPR	1. S/S individuals exhibit greater riskier decision making compared to S/L and L/L carriers during trials 41–100.
Stoltenberg et al., 2011	Healthy adults, N = 391 (64.5% female)	5-HTTLPR	1. L/L genotype is associated with riskier decision making in the first block of 20 trials. 2. Childhood trauma is associated with riskier decision making during blocks 3–5 of IGT.
Must et al., 2007	Adults with unipolar major depressive disorder (MDD), N = 124 (67% female)	5-HTTLPR	1. Overall, L/L MDD patients exhibit better IGT performance relative to individuals with the S/L or S/S genotypes.
van den Bos et al., 2009	Healthy adults, N = 70 (100% female)	5-HTTLPR; *COMT* Val158Met (rs4680)	1. Met/Met individuals exhibit riskier decision making than Val/Val individuals during the IGT. 2. Interaction between *COMT* and 5-HTTLPR polymorphisms; individuals with both Met/Met and S/S genotypes perform more poorly on the IGT compared with all other individuals.
Stoltenberg et al., 2010	Healthy adults, N = 188 (62% female)	5-HTTLPR	1. No association between rs1386438 genotype and overall performance on the IGT. 2. Males with S/S and S/L 5-HTTLPR exhibit better performance during first 20 trials compared to L/L males.
Lage et al., 2011	Healthy adults, N = 127 (59% female)	5-HTTLPR	1. No significant associations.
He et al., 2010	Healthy adults, N = 572 (55% female)	5-HTTLPR	1. Males exhibited better performance on trials 1–40 compared with females. 2. S/S genotype is associated with riskier decision making on trials 1–40 compared to L+ carriers.

Table 8.1

(continued)

Study	Sample	Polymorphism(s)	Primary findings
Ha et al., 2009	Healthy adults, N = 159 (47% female)	5-HTTLPR; DRD4 VNTR	1. No main effects of the 5-HTTLPR and DRD4 genotypes 2. Significant gene × gene interaction present. Higher IGT scores for individuals with 2R+ genotype of DRD4 and S/S genotype of 5-HTTLPR relative to carriers of the 2R- genotype missing the S/S genotype. Absence of the S/S genotype in individuals with the 2R- associated with higher IGT scores relative to 2R+ individuals.
Maurex et al., 2009	Individuals with borderline personality disorder (BPD) and controls, Ns = 48 and 30, respectively (100% female)	TPH-1 haplotype	1. Threefold higher frequency of TPH-1 haplotype in BPD individuals that exhibited very risky decision making during IGT relative to BPD individuals who exhibited less risky choices.
Kang et al., 2010	Healthy adults, N = 168 (45% female)	*BDNF* Val66Met(rs6265); *COMT* Val158Met (rs4680)	1. Compared to *BDNF* Val/Val individuals, Met+ individuals exhibit both poorer performance on IGT blocks 3–5 and less improvement from block 1 to blocks 3–5. 2. No *COMT* Val158Met associations. 3. No epistatic interactions.
da Rocha et al., 2008	Individuals with obsessive–compulsive disorder, N = 49 (53% female)	5-HTTLPR	1. Low expression genotypes associated with poorer performance on the IGT from blocks 3–5.
Roussos et al., 2008	Healthy adults, N = 107 (100% male)	*COMT* (rs 4818)	1. Individuals homozygous for the *COMT*-g allelic variation exhibited significantly better performance on the IGT, compared with both heterozygous and individuals homozygous for the *COMT*-c allelic variation.

S/S = homozygotes for the short allele of the serotonin transporter linked polymorphic region (5-HTTLPR); L/L = homozygotes for the long allele of the 5-HTTLPR; S/L = heterozygotes for the short and long alleles of the 5-HTTLPR.

choices than L/L males, whereas L/L females made more advantageous choices than heterozygous females, but again during the first 20 trials of the IGT. These studies suggest possible sex moderation of 5-HTTLPR influences, however, the basis for why this would be present is not clear and, given the heterogeneity of findings and surprising associations with the sampling period, type I error cannot be ruled out.

In clinical samples, several studies have also implicated 5-HTTLPR or other variants that are part of the serotonin system, such as the tryptophan hydroxylase genes (*TPH1*, *TPH2*), which play key roles in the synthesis of serotonin. For example, Must et al. [106] found that L/L individuals performed better than S/S individuals among patients with major depressive disorder and, among individuals with a history of suicide attempts, L/L and S/L individuals performed significantly more poorly than S/S individuals. Among individuals with obsessive–compulsive disorder, individuals with low functionality genotypes exhibited significantly poorer IGT performance [107]. Finally, comparing individuals with borderline personality disorder (BPD) to controls, IGT performance was significantly worse in the clinical group and a *TPH1* haplotype was significantly more frequent in BPD patients exhibiting poor IGT performance [108]. Clearly the studies using clinical samples represent a very small and heterogeneous literature at this point.

Other loci have also been investigated in relation to the IGT, largely in healthy adult samples. For example, Kang et al. [109] found that, for the *BDNF* Val66Met (rs6265) SNP, both Met/Met and Val/Met individuals exhibited poorer performance during blocks 3–5 but also examined the *COMT* Val158Met (rs4850) SNP and found no differences by genotype. In contrast to this latter finding, van den Bos et al. [99] examined *COMT* Val158Met and found Met/Met individuals performed more poorly than Val/Val individuals. In addition, an interaction was present between *COMT* and 5-HTTLPR status, with individuals who were Met/Met homozygotes and carriers of the S/S genotype performing more poorly than any other condition. Also examining the *COMT* gene, Roussos et al. [110] examined a different SNP (C/G; rs4818), finding that GG individuals performed significantly better on the IGT compared to the other genotypes. Finally, two studies have investigated the *DRD4* VNTR polymorphism. Roussos et al. [111] found that individuals with at least one long version performed worse than individuals with short versions, but this was moderated by season of birth. Only winter-born long carriers exhibited significant differences whereas summer-born long carriers did not differ from short carriers. This is relevant because season of birth, and specifically variation in daylight exposure, has been hypothesized to play a role in dopamine–melatonin balance [112]. Ha et al. [113] found no main effects of *DRD4* VNTR but found an epistatic interaction such that the combination of two repeat genotype and 5-HTTLPR long allele genotype was associated with poorest performance. This is hard to interpret because the 4 (short) and 7 (long) repeat alleles are by far the most commonly studied and best understood in terms of functionality [85]. Finally, one study has examined APOE ε4 genotype, which has been implicated in Alzheimer's disease, in relation to IGT performance, but found no significant relationship [114].

As is probably apparent, the findings to date are at best somewhat suggestive that low functionality 5-HTTLPR polymorphism is associated with poorer performance on the IGT, but no truly coherent pattern emerges from these studies. Although significant associations are present, the relationships are inconsistent and often depend on moderating variables, reflecting interaction effects. These relationships may in fact be robust and could be replicated in future studies. Indeed, the current findings provide the basis for more precise a priori hypothesis testing. However, given this inconsistency and the modest sample sizes used in many of these studies, the significant associations reported may also be false positives and it would be imprudent to over-interpret their significance at this point.

Like the IGT, the BART [115] was not developed as a behavioral economic measure per se, but it certainly meets the fundamental characteristics of being one. The BART assesses risk-taking propensity using a series of electronic balloons that can be variably inflated. Each time the button is pressed to inflate the balloon, a monetary reward is earned, but each inflation also increases the risk the balloon will pop. If this happens, the earned money is lost. In this way, the task assays preferences for risk, as performance reflects a person's willingness to trade a larger reward for greater risk of loss. Importantly, BART performance has been associated with various forms of addictive behavior [116–118], suggesting it is a plausible intermediate phenotype for addiction. This is also supported by initial evidence of heritability. Anokhin et al. [119] examined BART performance at ages 12 and 14 in 752 identical and fraternal twins and found significant evidence of heritability at age 12 for males and females (28% and 17%, respectively). In addition, from age 12 to age 14, participants exhibited significant increases in risk-taking but relative performance on the BART closely corresponded across time. At age 14, heritability increased to 55% for males but was nonsignificant and was dropped from the model for females. This suggests that genetic influences may be moderated by sex over the course of adolescence. A role for genetic factors is also suggested by a second study that examined brain activity profiles during the BART using electroencephalography. Specifically, Fein and Chiang [120] examined feedback error-related negativity (F-ERN), a cognitive index of the valence of negative consequence, during BART performance in a sample of alcohol-dependent adults. A significant inverse relationship was observed between F-ERN and density of alcohol FH, suggesting that one mechanism of risk may be a cognitive insensitivity to negative consequences, in turn leading to more risky behavior. This is clearly an incipient area of research, but these two studies are both suggestive that BART performance may be influenced by genetic factors.

In addition to findings using validated behavioral tasks, a number of genetic associations have been reported with performance on behavioral economic tasks that have not undergone extensive validation or necessarily been examined in relation to addictive behavior. For example, Dreber et al. [121] used an investment task in which participants allocated some portion of $250 into an investment that could substantially increase or decrease based on the coin toss. Two loci were characterized, the DRD2/ANKK1 TaqIA SNP (rs1800497) and DRD4 VNTR, and, although no ANKK1 association was present, individuals with at least one long version of the DRD4 polymorphism exhibited significantly greater willingness to take a risk. Also examining DRD4 VNTR, Carpenter et al. [122] administered experimental lotteries involving various levels of risk and ambiguity, finding that long allele carriers selected more risk under escalating levels of ambiguity and possible loss conditions. Finally, Kuhnen and Chiang [123] examined both DRD4 VNTR and 5-HTTLPR in relation to how much money participants were willing to allocate to a risky or riskless outcome. In this case, both polymorphisms were significantly associated with performance. Possession of a long DRD4 allele was associated with being willing to take 25% more risk, and S/L or L/L 5-HTTLPR individuals took 28% more risk than S/S individuals. Although these are only a small number of studies conducted in this domain, the findings do consistently implicate the long version of DRD4 VNTR with greater risk-taking, a finding that is consistent with previous associations between this polymorphism and both addictive disorders and other relevant phenotypes [85].

Taken together, the studies using the IGT, BART, and other measures provide additional support for the hypothesis that behavioral economics may provide complementary phenotypes for understanding the genetic basis of addictive behavior. Given the heterogeneity of measures and findings, definitive conclusions cannot be drawn at this point, but these studies are certainly supportive of the need for further investigation of these phenotypes.

8.4 Novel Candidate Behavioral Economic Phenotypes

The focus so far has been on existing findings connecting genetic variation with behavioral economic phenotypes. In this section we will review two existing behavioral economic paradigms that may also generate useful phenotypes but have not been the focus of studies to date. The first is probability discounting, which indexes risk sensitivity, and the second is drug demand, which indexes the incentive value of a given drug. In both cases, these measures afford objective and precise assessment of decision-making preferences, and performance has been associated with addictive behavior, making them potentially high-quality phenotypes.

In the first case, probability discounting refers to how much a person devalues (discounts) a reward based on its relative uncertainty (probability) [124]. Its measurement is similar to delay discounting, in that it is typically assessed using decision-making tasks that pit smaller guaranteed rewards (e.g., $50 guaranteed) against larger

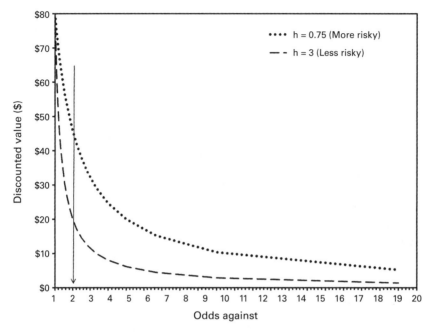

Figure 8.2
Prototypic probability discounting curves. The lines depict the subjective value (y-axis) of a certain monetary reward compared to $80 at escalating levels of risk (x-axis; odds against). Two levels of probability discounting functions (*h*) are depicted, with the steeper function reflecting a less risky profile. Two-to-one odds are included for illustrative purposes.

uncertain rewards (e.g., an 80% chance for $75). Then, over permuted trials that vary the amounts and probabilities of the uncertain reward, a person's probability discounting function can be generated, reflecting how much the person devalues the certain reward across varying risks. Prototypic probability discounting curves are presented in figure 8.2 and are also very similar to temporal discounting functions, but with important differences. Most obviously, the x-axis is odds-against, not duration of delay, reflecting the cost's being in terms of risk, not time. More importantly, though, the interpretation of decision-making preferences is reversed between delay discounting and probability discounting. In the first case, the higher the rate of delay discounting, the more the individual devalues a reward based on its delay; in the second case, the larger the individual's probability discounting function, denoted *h*, the more the individual is deterred by an increasingly uncertain reward, reflecting risk aversion. In other words, whereas high delay discounting reflects impulsivity, low probability discounting reflects risk proneness. For example, treating the two lines in figure 8.2 as individual people, for a choice between a smaller certain amount of money and one-third chance of \$80 (two-to-one odds, odds against = 2), the $h = 3$ person (less risky) would only take the uncertain choice when the certain reward was below \$20, but the $h = 0.75$ person (more risky) would take the uncertain reward when the certain reward was below \$46. For the choice of \$30 guaranteed versus a one-third chance for \$80, the former would take the smaller certain reward and the latter would take the larger riskier reward.

Although delay and probability discounting clearly tap similar processes and are significantly associated with one another in some studies, the aggregate evidence suggests they are distinct [19]. In addition, probability discounting and delay discounting are differentially related to addictive behavior. For example, in a recent study, significant differences were only present between pathological gamblers and controls on a measure of probability discounting, but not delay discounting [125], and probability discounting, but not delay discounting, has been found to predict treatment response for pathological gamblers [126]. Significant differences in probability discounting have also been detected for smokers versus controls [127], although these findings are less consistent [128, 129]. Importantly, probability discounting has also been determined to be temporally stable [130] and to systematically vary between isogenetic rodent strains [68], suggesting heritability. However, it is also worth noting that alcohol exposure has been shown to induce more risky probability discounting in an animal model [131]. Thus, probability discounting represents a complementary decision-making profile to delay discounting and may be sensitive to genetic variation that contributes to cognitive and affective processing of risk versus reward.

The second behavioral economic candidate phenotype is drug demand [8], which refers to consumption of a given drug under escalating levels of cost. There are diverse methods for assessing drug demand, including various forms of consumption (e.g., in vivo, hypothetical) and costs (e.g., operant output, money) (for a review, see [132]), but the approach that maximally permits assessment of individual variation in drug demand is the purchase task methodology. Typically, purchase tasks assess estimated consumption of a substance under various levels of price, either using hypothetical commodities [133, 134] or various prices and actual receipt of one outcome [135, 136]. These tasks permit a multifaceted assessment of an individual's incentive value of the drug, or how much the person values the substance as a reinforcer. Specifically, these tasks permit demand curve analysis of price effects on consumption, a powerful analytic approach that generates several putatively distinct facets of incentive value. These are depicted in figure 8.3 and include intensity (i.e., consumption under conditions of minimal cost), O_{max} (maximum expenditure), P_{max} (i.e., the price at which consumption becomes elastic), breakpoint (i.e., the price that first reduces consumption to zero), and elasticity (i.e., aggregate price sensitivity). Importantly, these variables have been consistently associated with drug consumption and associated severity [26, 134, 137–140] and response to psychosocial and pharmacological interventions [141–143]. In addition, demand indices have been found to be stable over time [137, 144] and

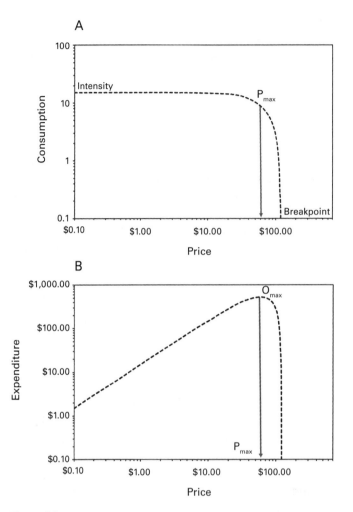

Figure 8.3

Prototypic behavioral economic demand and expenditure curves. (A) Consumption of a drug or any other commodity across escalating levels of price. Intensity refers to consumption at minimal levels of cost, P_{max} refers to the price at which demand becomes elastic (i.e., consumption becomes disproportionately sensitive to increasing price), and breakpoint refers to the first price to reduce consumption to zero. The overall price elasticity for the demand curve can also be derived. (B) The commensurate expenditure curve, reflecting consumption in panel A. O_{max} refers to the maximum expenditure on the commodity and reflects expenditure at P_{max}.

elasticity of cocaine demand has been shown to systematically vary by rodent strain [145], providing initial evidence of heritability.

Importantly, demand indices as phenotypes would probably better fit in the category of intermediate phenotypes rather than endophenotypes (see chapter 1). From a theoretical perspective, demand for a drug reflects its incentive value and is not an innate latent risk characteristic that is independent of the condition. Demand as a characteristic cannot be present in the absence of experience with the drug in the same way that delay discounting or the other domain-general tasks can. In this way, the phenotype of high (or low) incentive value of the drug is theorized to be sensitive to underlying genetic variation that is related to the reward value of the drug and, in turn, to confer risk, but not necessarily expected to meet the endophenotype criterion of being present in the absence of the condition. In this way, demand indices are akin to other informative phenotypes that are aspects of addictive behavior, such as drug use motives [146–148] or smoking topography [149], for example. These characteristics are not fully independent of the conditions but may nonetheless be informative about genetic influences. Although no published studies to date have reported significant differences in drug demand based on genetic variation, these indices of incentive value represent high-quality candidates for future work in this area.

8.5 Conclusions and Future Directions

The aim of this chapter was to provide an overview of the prospect of using behavioral economics to contribute novel phenotypes for improving the understanding of genetic influences on addictive behavior. By any measure, this is a novel domain; the earliest studies cited are less than ten years old, and the majority of studies were published within the last five years. The most supportive evidence comes from the study of delay discounting as an endophenotype, but studies of IGT and BART performance also suggest that genetic factors may underlie performance on those measures, and they are at least worthy of consideration as candidate intermediate phenotypes.

Going forward, there is clearly a need for further study in this area. The first priorities are relatively clear; replication and disambiguation of the existing findings will be critical to establish valid findings from false positives. To this end, an advantage to the behavioral economic measure discussed is that relatively short paper-and-pencil versions have been validated and may be viable in large-scale genome-wide association studies. Indeed, even the extended tasks are not excessively onerous and may be similarly viable. Second, it will be important for future studies to carefully characterize the role of additional factors that come into play in behavioral economic decision making. Although these measures may be relatively narrow assays of preferences, they are nonetheless contextualized within developmental and socioeconomic factors, both of which should be considered. In addition, the observed preferences on these tasks are partially subserved by potentially even more focal underlying cognitive processes, including abstraction, attention, and working memory. As such, it will be important to assess these variables concurrently to avoid inadvertently identifying genetic influences on a behavioral economic measure that are actually better understood as influencing cognition.

Related to this, a further future priority will be incorporating neuroeconomics, the intersection of behavioral economics and cognitive neuroscience, into studies of the genetic basis for behavioral economic variables. Choice behavior is ultimately a manifestation of underlying cognitive and affective brain activity, and neuroeconomic phenotypes provide a portal into those processes. Neuroeconomic phenotypes have the potential to reveal both genetic influences on decision-making preferences and also how those effects are manifested in terms of brain mechanisms. All of these prospects have the potential to validate the utility of behavioral economic phenotypes and also substantively extend our understanding of these mechanisms.

Finally, although one common theme throughout this chapter is the promise of behavioral economic phenotypes, a second was that caution is nonetheless warranted. For example, although these phenotypes may be comparatively narrower than clinical diagnosis, they may nonetheless have a complex genetic architecture. As discussed, significant environmental contributions, epistatic interactions, and gene-by-environment interactions have all been reported thus far and suggest the role of genetic factors in behavioral economic phenotypes will not be a simple one. Critical thinking and systematic inquiry will be essential in future work pursuing the prospect of behavioral economic intermediate phenotypes.

References

1. Camerer, C. (1999). Behavioral economics: Reunifying psychology and economics. *Proceedings of the National Academies of Sciences, 96*, 10575–10577.

2. Kahneman, D., & Tversky, A. (2000). *Choices, Values and Frames*. New York: Cambridge University Press and the Russell Sage Foundation.

3. Hursh, S. R. (1980). Economic concepts for the analysis of behavior. *Journal of the Experimental Analysis of Behavior, 34*, 219–238.

4. Bickel, W. K., Green, L., & Vuchinich, R. E. (1995). Behavioral economics [Editorial]. *Journal of the Experimental Analysis of Behavior, 64*, 257–262.

5. Camerer, C. (2003). *Behavioral Game Theory: Experiments in Strategic Interaction*. Princeton, NJ: Princeton University Press.

6. Bickel, W. K., Miller, M. L., Yi, R., et al. (2007). Behavioral and neuroeconomics of drug addiction: Competing neural systems and temporal discounting processes. *Drug and Alcohol Dependence, 90*(Suppl 1), S85–S91.

7. Higgins, S. T., Heil, S. H., & Lussier, J. P. (2004). Clinical implications of reinforcement as a determinant of substance use disorders. *Annual Review of Psychology, 55,* 431–461.

8. Hursh, S. R., Galuska, C. M., Winger, G., & Woods, J. H. (2005). The economics of drug abuse: A quantitative assessment of drug demand. *Molecular Interventions, 5,* 20–28.

9. Bickel, W. K., Mueller, E. T., MacKillop, J., & Yi, R. (in press). Behavioral economic and neuroeconomic perspectives on addiction. In K. J. Sher (Ed.), *Oxford Handbook of Substance Use Disorders.* Oxford, UK: Oxford University Press.

10. Madden, G. J., Petry, N. M., Badger, G. J., & Bickel, W. K. (1997). Impulsive and self-control choices in opioid-dependent patients and non-drug-using control participants: drug and monetary rewards. *Experimental and Clinical Psychopharmacology, 5,* 256–262.

11. Vuchinich, R. E., & Simpson, C. A. (1998). Hyperbolic temporal discounting in social drinkers and problem drinkers. *Experimental and Clinical Psychopharmacology, 6,* 292–305.

12. Mischel, W., Shoda, Y., & Peake, P. K. (1988). The nature of adolescent competencies predicted by preschool delay of gratification. *Journal of Personality and Social Psychology, 54,* 687–696.

13. Bradford, W. D. (2010). The association between individual time preferences and health maintenance habits. *Medical Decision Making, 30,* 99–112.

14. Kirby, K. N., Petry, N. M., & Bickel, W. K. (1999). Heroin addicts have higher discount rates for delayed rewards than non-drug-using controls. *Journal of Experimental Psychology. General, 128,* 78–87.

15. Reynolds, B., & Schiffbauer, R. (2004). Measuring state changes in human delay discounting: an experiential discounting task. *Behavioural Processes, 67,* 343–356.

16. Dougherty, D. M., Mathias, C. W., Marsh, D. M., & Jagar, A. A. (2005). Laboratory behavioral measures of impulsivity. *Behavior Research Methods, 37,* 82–90.

17. Frederick, S., Loewenstein, G., & O'Donoghue, T. (2003). Time discounting and time preference: A critical review. In Loewenstein, G., Read, D., & Baumeister, R. F. (Eds.), *Time and Decision: Economic and Psychological Perspectives on Intertemporal Choice* (pp. 13–86). New York: Russell Sage Foundation.

18. Madden, G. J., Bickel, W. K., & Jacobs, E. A. (1999). Discounting of delayed rewards in opioid-dependent outpatients: Exponential or hyperbolic discounting functions? *Experimental and Clinical Psychopharmacology, 7,* 284–293.

19. Green, L., & Myerson, J. (2004). A discounting framework for choice with delayed and probabilistic rewards. *Psychological Bulletin, 130,* 769–792.

20. Vuchinich, R. E., & Simpson, C. A. (1998). Hyperbolic temporal discounting in social drinkers and problem drinkers. *Experimental and Clinical Psychopharmacology, 6,* 292–305.

21. Petry, N. M. (2001). Delay discounting of money and alcohol in actively using alcoholics, currently abstinent alcoholics, and controls. *Psychopharmacology, 154,* 243–250.

22. Bickel, W. K., Odum, A. L., & Madden, G. J. (1999). Impulsivity and cigarette smoking: Delay discounting in current, never, and ex-smokers. *Psychopharmacology, 146*, 447–454.

23. Coffey, S. F., Gudleski, G. D., Saladin, M. E., & Brady, K. T. (2003). Impulsivity and rapid discounting of delayed hypothetical rewards in cocaine-dependent individuals. *Experimental and Clinical Psychopharmacology, 11*, 18–25.

24. MacKillop, J., Anderson, E. J., Castelda, B. A., Mattson, R. E., & Donovick, P. J. (2006). Divergent validity of measures of cognitive distortions, impulsivity, and time perspective in pathological gambling. *Journal of Gambling Studies, 22*, 339–354.

25. MacKillop, J., Amlung, M. T., Few, L. R., et al. (2011). Delayed reward discounting and addictive behavior: A meta-analysis. *Psychopharmacology, 216*, 305–321.

26. MacKillop, J., Miranda, J. R., Monti, P. M., et al. (2010). Alcohol demand, delayed reward discounting, and craving in relation to drinking and alcohol use disorders. *Journal of Abnormal Psychology, 119*, 115–125.

27. Bobova, L., Finn, P. R., Rickert, M. E., & Lucas, J. (2009). Disinhibitory psychopathology and delay discounting in alcohol dependence: personality and cognitive correlates. *Experimental and Clinical Psychopharmacology, 17*, 51–61.

28. Sweitzer, M. M., Donny, E. C., Dierker, L. C., Flory, J. D., & Manuck, S. B. (2008). Delay discounting and smoking: Association with the Fagerstrom Test for Nicotine Dependence but not cigarettes smoked per day. *Nicotine & Tobacco Research, 10*, 1571–1575.

29. Alessi, S. M., & Petry, N. M. (2003). Pathological gambling severity is associated with impulsivity in a delay discounting procedure. *Behavioural Processes, 64*, 345–354.

30. Audrain-McGovern, J., Rodriguez, D., Epstein, L. H., et al. (2009). Does delay discounting play an etiological role in smoking or is it a consequence of smoking? *Drug and Alcohol Dependence, 103*, 99–106.

31. Ayduk, O., Mendoza-Denton, R., Mischel, W., et al. (2000). Regulating the interpersonal self: strategic self-regulation for coping with rejection sensitivity. *Journal of Personality and Social Psychology, 79*, 776–792.

32. Simpson, C. A., & Vuchinich, R. E. (2000). Reliability of a measure of temporal discounting. *Psychological Record, 50*, 3–16.

33. Baker, F., Johnson, M. W., & Bickel, W. K. (2003). Delay discounting in current and never-before cigarette smokers: Similarities and differences across commodity, sign, and magnitude. *Journal of Abnormal Psychology, 112*, 382–392.

34. Clare, S., Helps, S., & Sonuga-Barke, E. J. (2010). The quick delay questionnaire: A measure of delay aversion and discounting in adults. *Attention Deficit and Hyperactivity Disorders, 2*, 43–48.

35. Weatherly, J. N., Derenne, A., & Terrell, H. K. (2011). Testing the reliability of delay discounting of ten commodities using the fill-in-the-blank method. *Psychological Record, 61*, 113–126.

36. Beck, R. C., & Triplett, M. F. (2009). Test–retest reliability of a group-administered paper–pencil measure of delay discounting. *Experimental and Clinical Psychopharmacology*, *17*, 345–355.

37. Takahashi, T., Furukawa, A., Miyakawa, T., Maesato, H., & Higuchi, S. (2007). Two-month stability of hyperbolic discount rates for delayed monetary gains in abstinent inpatient alcoholics. *Neuroendocrinology Letters*, *28*, 131–136.

38. Anokhin, A. P., Golosheykin, S., Grant, J. D., & Heath, A. C. (2011). Heritability of delay discounting in adolescence: A longitudinal twin study. *Behavior Genetics*, *41*, 175–183.

39. Kirby, K. N. (2009). One-year temporal stability of delay-discount rates. *Psychonomic Bulletin & Review*, *16*, 457–462.

40. Odum, A. L. (2011). Delay discounting: Trait variable? *Behavioural Processes*, *87*, 1–9.

41. Romer, D., Duckworth, A. L., Sznitman, S., & Park, S. (2010). Can adolescents learn self-control? Delay of gratification in the development of control over risk taking. *Prevention Science*, *11*, 319–330.

42. Dom, G., D'Haene, P., Hulstijn, W., & Sabbe, B. (2006). Impulsivity in abstinent early- and late-onset alcoholics: Differences in self-report measures and a discounting task. *Addiction*, *101*, 50–59.

43. Kollins, S. H. (2003). Delay discounting is associated with substance use in college students. *Addictive Behaviors*, *28*, 1167–1173.

44. Anker, J. J., Perry, J. L., Gliddon, L. A., & Carroll, M. E. (2009). Impulsivity predicts the escalation of cocaine self-administration in rats. *Pharmacology, Biochemistry, and Behavior*, *93*, 343–348.

45. Marusich, J. A., & Bardo, M. T. (2009). Differences in impulsivity on a delay-discounting task predict self-administration of a low unit dose of methylphenidate in rats. *Behavioural Pharmacology*, *20*, 447–454.

46. Perry, J. L., Larson, E. B., German, J. P., Madden, G. J., & Carroll, M. E. (2005). Impulsivity (delay discounting) as a predictor of acquisition of IV cocaine self-administration in female rats. *Psychopharmacology*, *178*, 193–201.

47. Perry, J. L., Nelson, S. E., Anderson, M. M., Morgan, A. D., & Carroll, M. E. (2007). Impulsivity (delay discounting) for food and cocaine in male and female rats selectively bred for high and low saccharin intake. *Pharmacology, Biochemistry, and Behavior*, *86*, 822–837.

48. Setlow, B., Mendez, I. A., Mitchell, M. R., & Simon, N. W. (2009). Effects of chronic administration of drugs of abuse on impulsive choice (delay discounting) in animal models. *Behavioural Pharmacology*, *20*, 380–389.

49. Krishnan-Sarin, S., Reynolds, B., Duhig, A. M., et al. (2007). Behavioral impulsivity predicts treatment outcome in a smoking cessation program for adolescent smokers. *Drug and Alcohol Dependence*, *88*, 79–82.

50. MacKillop, J., & Kahler, C. W. (2009). Delayed reward discounting predicts treatment response for heavy drinkers receiving smoking cessation treatment. *Drug and Alcohol Dependence, 104,* 197–203.

51. Yoon, J. H., Higgins, S. T., Heil, S. H., et al. (2007). Delay discounting predicts postpartum relapse to cigarette smoking among pregnant women. *Experimental and Clinical Psychopharmacology, 15,* 176–186.

52. Sheffer, S., MacKillop, J., McGeary, J., et al. (2012). Delay discounting, locus of control, and cognitive impulsivity independently predict tobacco dependence treatment outcomes in a highly dependent, lower socioeconomic group of smokers. *American Journal on Addictions, 21,* 221–232.

53. Tucker, J. A., Vuchinich, R. E., & Rippens, P. D. (2002). Predicting natural resolution of alcohol-related problems: A prospective behavioral economic analysis. *Experimental and Clinical Psychopharmacology, 10,* 248–257.

54. Tucker, J. A., Vuchinich, R. E., Black, B. C., & Rippens, P. D. (2006). Significance of a behavioral economic index of reward value in predicting drinking problem resolution. *Journal of Consulting and Clinical Psychology, 74,* 317–326.

55. Tucker, J. A., Foushee, H. R., & Black, B. C. (2008). Behavioral economic analysis of natural resolution of drinking problems using IVR self-monitoring. *Experimental and Clinical Psychopharmacology, 16,* 332–340.

56. Tucker, J. A., Roth, D. L., Vignolo, M. J., & Westfall, A. O. (2009). A behavioral economic reward index predicts drinking resolutions: Moderation revisited and compared with other outcomes. *Journal of Consulting and Clinical Psychology, 77,* 219–228.

57. Washio, Y., Higgins, S. T., Heil, S. H., et al. (2011). Delay discounting is associated with treatment response among cocaine-dependent outpatients. *Experimental and Clinical Psychopharmacology, 19,* 243–248.

58. Stanger, C., Ryan, S. R., Fu, H., et al. (2012). Delay discounting predicts adolescent substance abuse treatment outcome. *Experimental and Clinical Psychopharmacology, 20,* 205–212.

59. Passetti, F., Clark, L., Davis, P., et al. (2011). Risky decision-making predicts short-term outcome of community but not residential treatment for opiate addiction. Implications for case management. *Drug and Alcohol Dependence, 118,* 12–18.

60. Anderson, K. G., & Woolverton, W. L. (2005). Effects of clomipramine on self-control choice in Lewis and Fischer 344 rats. *Pharmacology, Biochemistry, and Behavior, 80,* 387–393.

61. Madden, G. J., Smith, N. G., Brewer, A. T., Pinkston, J. W., & Johnson, P. S. (2008). Steady-state assessment of impulsive choice in Lewis and Fischer 344 rats: Between-condition delay manipulations. *Journal of the Experimental Analysis of Behavior, 90,* 333–344.

62. Wilhelm, C. J., & Mitchell, S. H. (2009). Strain differences in delay discounting using inbred rats. *Genes, Brain, and Behavior, 8,* 426–434.

63. Isles, A. R., Humby, T., Walters, E., & Wilkinson, L. S. (2004). Common genetic effects on variation in impulsivity and activity in mice. *Journal of Neuroscience, 24*, 6733–6740.

64. Helms, C. M., Gubner, N. R., Wilhelm, C. J., Mitchell, S. H., & Grandy, D. K. (2008). D4 receptor deficiency in mice has limited effects on impulsivity and novelty seeking. *Pharmacology, Biochemistry, and Behavior, 90*, 387–393.

65. Carroll, M. E., Morgan, A. D., Lynch, W. J., Campbell, U. C., & Dess, N. K. (2002). Intravenous cocaine and heroin self-administration in rats selectively bred for differential saccharin intake: phenotype and sex differences. *Psychopharmacology, 161*, 304–313.

66. Fattore, L., Piras, G., Corda, M. G., & Giorgi, O. (2009). The Roman high- and low-avoidance rat lines differ in the acquisition, maintenance, extinction, and reinstatement of intravenous cocaine self-administration. *Neuropsychopharmacology, 34*, 1091–1101.

67. Moreno, M., Cardona, D., Gomez, M. J., et al. (2010). Impulsivity characterization in the Roman high- and low-avoidance rat strains: Behavioral and neurochemical differences. *Neuropsychopharmacology, 35*, 1198–1208.

68. Wilhelm, C. J., & Mitchell, S. H. (2008). Rats bred for high alcohol drinking are more sensitive to delayed and probabilistic outcomes. *Genes, Brain, and Behavior, 7*, 705–713.

69. Oberlin, B. G., & Grahame, N. J. (2009). High-alcohol preferring mice are more impulsive than low-alcohol preferring mice as measured in the delay discounting task. *Alcoholism, Clinical and Experimental Research, 33*, 1294–1303.

70. Wilhelm, C. J., Reeves, J. M., Phillips, T. J., & Mitchell, S. H. (2007). Mouse lines selected for alcohol consumption differ on certain measures of impulsivity. *Alcoholism, Clinical and Experimental Research, 31*, 1839–1845.

71. Helms, C. M., Reeves, J. M., & Mitchell, S. H. (2006). Impact of strain and D-amphetamine on impulsivity (delay discounting) in inbred mice. *Psychopharmacology, 188*, 144–151.

72. Goldman, D., Oroszi, G., & Ducci, F. (2005). The genetics of addictions: Uncovering the genes. *Nature Reviews. Genetics, 6*, 521–532.

73. Petry, N. M., Kirby, K. N., & Kranzler, H. R. (2002). Effects of gender and family history of alcohol dependence on a behavioral task of impulsivity in healthy subjects. *Journal of Studies on Alcohol, 63*, 83–90.

74. Crean, J., Richards, J. B., & de Wit, H. (2002). Effect of tryptophan depletion on impulsive behavior in men with or without a family history of alcoholism. *Behavioural Brain Research, 136*, 349–357.

75. Herting, M. M., Schwartz, D., Mitchell, S. H., & Nagel, B. J. (2010). Delay discounting behavior and white matter microstructure abnormalities in youth with a family history of alcoholism. *Alcoholism, Clinical and Experimental Research, 34*, 1590–1602.

76. Acheson, A., Vincent, A. S., Sorocco, K. H., & Lovallo, W. R. (2011). Greater discounting of delayed rewards in young adults with family histories of alcohol and drug use disorders: Studies

from the Oklahoma Family Health Patterns Project. *Alcoholism, Clinical and Experimental Research*, 35, 1607–1613.

77. Eisenberg, D. T., MacKillop, J., Modi, M., et al. (2007). Examining impulsivity as an endophenotype using a behavioral approach: A DRD2 TaqI A and DRD4 48-bp VNTR association study. *Behavioral and Brain Functions*, 3, 2.

78. Garcia, J. R., MacKillop, J., Aller, E. L., et al. (2010). Associations between dopamine D4 receptor gene variation with both infidelity and sexual promiscuity. *PLoS ONE*, 5, e14162.

79. Sweitzer, M. M., Halder, I., Flory, J. D., et al. (2012). Polymorphic variation in the dopamine D4 receptor predicts delay discounting as a function of childhood socioeconomic status: Evidence for differential susceptibility. *Social Cognitive and Affective Neuroscience*. [epub ahead of print]

80. Paloyelis, Y., Asherson, P., Mehta, M. A., Faraone, S. V., & Kuntsi, J. (2010). DAT1 and COMT effects on delay discounting and trait impulsivity in male adolescents with attention deficit/hyperactivity disorder and healthy controls. *Neuropsychopharmacology*, 35, 2414–2426.

81. Boettiger, C. A., Mitchell, J. M., Tavares, V. C., et al. (2007). Immediate reward bias in humans: fronto–parietal networks and a role for the catechol-O-methyltransferase 158(Val/Val) genotype. *Journal of Neuroscience*, 27, 14383–14391.

82. Smith, C. T., & Boettiger, C. A. (2012). Age modulates the effect of COMT genotype on delay discounting behavior. *Psychopharmacology*, 222, 609–617.

83. White, M. J., Lawford, B. R., Morris, C. P., & Young, R. M. (2009). Interaction between DRD2 C957T polymorphism and an acute psychosocial stressor on reward-related behavioral impulsivity. *Behavior Genetics*, 39, 285–295.

84. Munafo, M. R., Timpson, N. J., David, S. P., Ebrahim, S., & Lawlor, D. A. (2009). Association of the DRD2 gene Taq1A polymorphism and smoking behavior: A meta-analysis and new data. *Nicotine & Tobacco Research*, 11, 64–76.

85. McGeary, J. (2009). The DRD4 exon 3 VNTR polymorphism and addiction-related phenotypes: A review. *Pharmacology, Biochemistry, and Behavior*, 93, 222–229.

86. Tammimaki, A. E., & Mannisto, P. T. (2010). Are genetic variants of COMT associated with addiction? *Pharmacogenetics and Genomics*, 20, 717–741.

87. Bechara, A., Damasio, A. R., Damasio, H., & Anderson, S. W. (1994). Insensitivity to future consequences following damage to human prefrontal cortex. *Cognition*, 50, 7–15.

88. Bechara, A., Damasio, H., Tranel, D., & Anderson, S. W. (1998). Dissociation of working memory from decision making within the human prefrontal cortex. *Journal of Neuroscience*, 18, 428–437.

89. Bechara, A., Dolan, S., Denburg, N., et al. (2001). Decision-making deficits, linked to a dysfunctional ventromedial prefrontal cortex, revealed in alcohol and stimulant abusers. *Neuropsychologia*, 39, 376–389.

90. Bechara, A., Dolan, S., & Hindes, A. (2002). Decision-making and addiction: II. Myopia for the future or hypersensitivity to reward? *Neuropsychologia, 40*, 1690–1705.

91. Dolan, S. L., Bechara, A., & Nathan, P. E. (2008). Executive dysfunction as a risk marker for substance abuse: The role of impulsive personality traits. *Behavioral Sciences & the Law, 26*, 799–822.

92. Bechara, A., & Damasio, H. (2002). Decision-making and addiction: I. Impaired activation of somatic states in substance dependent individuals when pondering decisions with negative future consequences. *Neuropsychologia, 40*, 1675–1689.

93. Hammers, D. B., & Suhr, J. A. (2010). Neuropsychological, impulsive personality, and cerebral oxygenation correlates of undergraduate polysubstance use. *Journal of Clinical and Experimental Neuropsychology, 32*, 599–609.

94. Buelow, M. T., & Suhr, J. A. (2009). Construct validity of the Iowa Gambling Task. *Neuropsychology Review, 19*, 102–114.

95. Lovallo, W. R., Yechiam, E., Sorocco, K. H., Vincent, A. S., & Collins, F. L. (2006). Working memory and decision-making biases in young adults with a family history of alcoholism: studies from the Oklahoma Family Health Patterns Project. *Alcoholism, Clinical and Experimental Research, 30*, 763–773.

96. Viswanath, B., Janardhan Reddy, Y. C., Kumar, K. J., Kandavel, T., & Chandrashekar, C. R. (2009). Cognitive endophenotypes in OCD: A study of unaffected siblings of probands with familial OCD. *Progress in Neuro-Psychopharmacology & Biological Psychiatry, 33*, 610–615.

97. Acheson, A., Robinson, J. L., Glahn, D. C., Lovallo, W. R., & Fox, P. T. (2009). Differential activation of the anterior cingulate cortex and caudate nucleus during a gambling simulation in persons with a family history of alcoholism: Studies from the Oklahoma Family Health Patterns Project. *Drug and Alcohol Dependence, 100*, 17–23.

98. Lesch, K. P., Bengel, D., Heils, A., et al. (1996). Association of anxiety-related traits with a polymorphism in the serotonin transporter gene regulatory region. *Science, 274*, 1527–1531.

99. Canli, T., & Lesch, K. P. (2007). Long story short: The serotonin transporter in emotion regulation and social cognition. *Nature Neuroscience, 10*, 1103–1109.

100. van den Bos, R., Homberg, J., Gijsbers, E., den Heijer, E., & Cuppen, E. (2009). The effect of COMT Val158 Met genotype on decision-making and preliminary findings on its interaction with the 5-HTTLPR in healthy females. *Neuropharmacology, 56*, 493–498.

101. Lage, G. M., Malloy-Diniz, L. F., Matos, L. O., et al. (2011). Impulsivity and the 5-HTTLPR polymorphism in a non-clinical sample. *PLoS ONE, 6*, e16927.

102. Homberg, J. R., van den Bos, R., den Heijer, E., Suer, R., & Cuppen, E. (2008). Serotonin transporter dosage modulates long-term decision-making in rat and human. *Neuropharmacology, 55*, 80–84.

103. Stoltenberg, S. F., Lehmann, M. K., Anderson, C., Nag, P., & Anagnopoulos, C. (2011). Serotonin transporter (5-HTTLPR) genotype and childhood trauma are associated with individual differences in decision making. *Front Genet*, *2*, 33.

104. He, Q., Xue, G., Chen, C., et al. (2010). Serotonin transporter gene-linked polymorphic region (5-HTTLPR) influences decision making under ambiguity and risk in a large Chinese sample. *Neuropharmacology*, *59*, 518–526.

105. Stoltenberg, S. F., & Vandever, J. M. (2010). Gender moderates the association between 5-HTTLPR and decision-making under ambiguity but not under risk. *Neuropharmacology*, *58*, 423–428.

106. Must, A., Juhasz, A., Rimanoczy, A., et al. (2007). Major depressive disorder, serotonin transporter, and personality traits: Why patients use suboptimal decision-making strategies? *Journal of Affective Disorders*, *103*, 273–276.

107. da Rocha, F. F., Malloy-Diniz, L., Lage, N. V., et al. (2008). Decision-making impairment is related to serotonin transporter promoter polymorphism in a sample of patients with obsessive–compulsive disorder. *Behavioural Brain Research*, *195*, 159–163.

108. Maurex, L., Zaboli, G., Wiens, S., et al. (2009). Emotionally controlled decision-making and a gene variant related to serotonin synthesis in women with borderline personality disorder. *Scandinavian Journal of Psychology*, *50*, 5–10.

109. Kang, J. I., Namkoong, K., Ha, R. Y., et al. (2010). Influence of BDNF and COMT polymorphisms on emotional decision making. *Neuropharmacology*, *58*, 1109–1113.

110. Roussos, P., Giakoumaki, S. G., Pavlakis, S., & Bitsios, P. (2008). Planning, decision-making and the COMT rs4818 polymorphism in healthy males. *Neuropsychologia*, *46*, 757–763.

111. Roussos, P., Giakoumaki, S. G., & Bitsios, P. (2010). Cognitive and emotional processing associated with the season of birth and dopamine D4 receptor gene. *Neuropsychologia*, *48*, 3926–3933.

112. Eisenberg, D. T., Campbell, B., Mackillop, J., Lum, J. K., & Wilson, D. S. (2007). Season of birth and dopamine receptor gene associations with impulsivity, sensation seeking and reproductive behaviors. *PLoS ONE*, *2*, e1216.

113. Ha, R. Y., Namkoong, K., Kang, J. I., Kim, Y. T., & Kim, S. J. (2009). Interaction between serotonin transporter promoter and dopamine receptor D4 polymorphisms on decision making. *Progress in Neuro-Psychopharmacology & Biological Psychiatry*, *33*, 1217–1222.

114. Caselli, R. J., Dueck, A. C., Locke, D. E., et al. (2011). Longitudinal modeling of frontal cognition in APOE epsilon4 homozygotes, heterozygotes, and noncarriers. *Neurology*, *76*, 1383–1388.

115. Lejuez, C. W., Read, J. P., Kahler, C. W., et al. (2002). Evaluation of a behavioral measure of risk taking: the Balloon Analogue Risk Task (BART). *Journal of Experimental Psychology. Applied*, *8*, 75–84.

116. Lejuez, C. W., Aklin, W. M., Jones, H. A., et al. (2003). The Balloon Analogue Risk Task (BART) differentiates smokers and nonsmokers. *Experimental and Clinical Psychopharmacology, 11,* 26–33.

117. Lejuez, C. W., Aklin, W., Bornovalova, M., & Moolchan, E. T. (2005). Differences in risk-taking propensity across inner-city adolescent ever- and never-smokers. *Nicotine & Tobacco Research, 7,* 71–79.

118. Bornovalova, M. A., Daughters, S. B., Hernandez, G. D., Richards, J. B., & Lejuez, C. W. (2005). Differences in impulsivity and risk-taking propensity between primary users of crack cocaine and primary users of heroin in a residential substance-use program. *Experimental and Clinical Psychopharmacology, 13,* 311–318.

119. Anokhin, A. P., Golosheykin, S., Grant, J., & Heath, A. C. (2009). Heritability of risk-taking in adolescence: A longitudinal twin study. *Twin Research and Human Genetics, 12,* 366–371.

120. Fein, G., & Chang, M. (2008). Smaller feedback ERN amplitudes during the BART are associated with a greater family history density of alcohol problems in treatment-naive alcoholics. *Drug and Alcohol Dependence, 92,* 141–148.

121. Dreber, A., Apicella, C. L., Eisenberg, D. T. A., et al. (2009). The 7R polymorphism in the dopamine receptor D4 gene (DRD4) is associated with financial risk taking in men. *Evolution and Human Behavior, 30,* 85–92.

122. Carpenter, J. P., Garcia, J. R., & Lum, J. K. (2011). Dopamine receptor genes predict risk preferences, time preferences, and related economic choices. *Journal of Risk and Uncertainty, 42,* 233–261.

123. Kuhnen, C. M., & Chiao, J. Y. (2009). Genetic determinants of financial risk taking. *PLoS ONE, 4,* e4362.

124. Rachlin, H., Raineri, A., & Cross, D. (1991). Subjective probability and delay. *Journal of the Experimental Analysis of Behavior, 55,* 233–244.

125. Madden, G. J., Petry, N. M., & Johnson, P. S. (2009). Pathological gamblers discount probabilistic rewards less steeply than matched controls. *Experimental and Clinical Psychopharmacology, 17,* 283–290.

126. Petry, N. M. (2012). Discounting of probabilistic rewards is associated with gambling abstinence in treatment-seeking pathological gamblers. *Journal of Abnormal Psychology, 121,* 151–159.

127. Reynolds, B., Richards, J. B., Horn, K., & Karraker, K. (2004). Delay discounting and probability discounting as related to cigarette smoking status in adults. *Behavioural Processes, 65,* 35–42.

128. Reynolds, B., Karraker, K., Horn, K., & Richards, J. B. (2003). Delay and probability discounting as related to different stages of adolescent smoking and non-smoking. *Behavioural Processes, 64,* 333–344.

129. Reynolds, B., Patak, M., Shroff, P., et al. (2007). Laboratory and self-report assessments of impulsive behavior in adolescent daily smokers and nonsmokers. *Experimental and Clinical Psychopharmacology*, *15*, 264–271.

130. Ohmura, Y., Takahashi, T., Kitamura, N., & Wehr, P. (2006). Three-month stability of delay and probability discounting measures. *Experimental and Clinical Psychopharmacology*, *14*, 318–328.

131. Nasrallah, N. A., Yang, T. W., & Bernstein, I. L. (2009). Long-term risk preference and suboptimal decision making following adolescent alcohol use. *Proceedings of the National Academy of Sciences of the United States of America*, *106*, 17600–17604.

132. MacKillop, J., & Murphy, C. M. (2013). Drug self-administration paradigms: methods for quantifying motivation in experimental research. In J. MacKillop & H. de Wit (Eds.), *The Wiley–Blackwell Handbook of Addiction Psychopharmacology* (pp. 315–344). Oxford, UK: Wiley–Blackwell.

133. Jacobs, E. A., & Bickel, W. K. (1999). Modeling drug consumption in the clinic using simulation procedures: demand for heroin and cigarettes in opioid-dependent outpatients. *Experimental and Clinical Psychopharmacology*, *7*, 412–426.

134. Murphy, J. G., & MacKillop, J. (2006). Relative reinforcing efficacy of alcohol among college student drinkers. *Experimental and Clinical Psychopharmacology*, *14*, 219–227.

135. Amlung, M. T., Acker, J., Stojek, M. K., Murphy, J. G., & Mackillop, J. (2012). Is talk "cheap"? An initial investigation of the equivalence of alcohol purchase task performance for hypothetical and actual rewards. *Alcoholism, Clinical and Experimental Research*, *36*, 716–724.

136. MacKillop, J., Brown, C. L., Stojek, M. K., et al. (2012). Behavioral economic analysis of withdrawal- and cue-elicited craving for tobacco: an initial investigation. *Nicotine & Tobacco Research*, *14*, 1426–1434.

137. Murphy, J. G., MacKillop, J., Skidmore, J. R., & Pederson, A. A. (2009). Reliability and validity of a demand curve measure of alcohol reinforcement. *Experimental and Clinical Psychopharmacology*, *17*, 396–404.

138. Murphy, J. G., Mackillop, J., Tidey, J. W., Brazil, L. A., & Colby, S. M. (2011). Validity of a demand curve measure of nicotine reinforcement with adolescent smokers. *Drug and Alcohol Dependence*, *113*, 207–214.

139. MacKillop, J., & Tidey, J. W. (2011). Cigarette demand and delayed reward discounting in nicotine-dependent individuals with schizophrenia and controls: an initial study. *Psychopharmacology*, *216*, 91–99.

140. MacKillop, J., Murphy, J. G., Ray, L. A., et al. (2008). Further validation of a cigarette purchase task for assessing the relative reinforcing efficacy of nicotine in college smokers. *Experimental and Clinical Psychopharmacology*, *16*, 57–65.

141. MacKillop, J., & Murphy, J. G. (2007). A behavioral economic measure of demand for alcohol predicts brief intervention outcomes. *Drug and Alcohol Dependence, 89*, 227–233.

142. Madden, G. J., & Kalman, D. (2010). Effects of bupropion on simulated demand for cigarettes and the subjective effects of smoking. *Nicotine & Tobacco Research, 12*, 416–422.

143. Bujarski, S., Mackillop, J., & Ray, L. A. (2012). Understanding naltrexone mechanism of action and pharmacogenetics in Asian Americans via behavioral economics: aA preliminary study. *Experimental and Clinical Psychopharmacology, 20*, 181–190.

144. Few, L. R., Acker, J., Murphy, C., & Mackillop, J. (2012). Temporal stability of a cigarette purchase task. *Nicotine & Tobacco Research, 14*, 761–765.

145. Christensen, C. J., Kohut, S. J., Handler, S., Silberberg, A., & Riley, A. L. (2009). Demand for food and cocaine in Fischer and Lewis rats. *Behavioral Neuroscience, 123*, 165–171.

146. Kristjansson, S. D., Agrawal, A., Littlefield, A. K., et al. (2011). Drinking motives in female smokers: Factor structure, alcohol dependence, and genetic influences. *Alcoholism, Clinical and Experimental Research, 35*, 345–354.

147. Miranda, R., Ray, L., Justus, A., et al. (2010). Initial evidence of an association between OPRM1 and adolescent alcohol misuse. *Alcoholism, Clinical and Experimental Research, 34*, 112–122.

148. van der Zwaluw, C. S., Kuntsche, E., & Engels, R. C. (2011). Risky alcohol use in adolescence: the role of genetics (DRD2, SLC6A4) and coping motives. *Alcoholism, Clinical and Experimental Research, 35*, 756–764.

149. Strasser, A. A., Malaiyandi, V., Hoffmann, E., Tyndale, R. F., & Lerman, C. (2007). An association of CYP2A6 genotype and smoking topography. *Nicotine & Tobacco Research, 9*, 511–518.

9 Using Functional Magnetic Resonance Imaging to Develop Intermediate Phenotypes for Substance Use Disorders

Rachel E. Thayer and Kent E. Hutchison

Studies over the past decade have suggested that substance use disorders, and in particular alcohol use disorders, are highly heritable, with as much as half of observed variance related to genetic factors [1]. Attempts to link addiction phenotypes to genetic variations have largely failed to produce consistent findings, perhaps because of the complexity of many genetic variations that each account for a very small amount of variance [2–4]. Clearly, one reason for the difficulty in identifying associations may be related to the fact that phenotypes that have traditionally been studied are based on diagnostic criteria from the *Diagnostic and Statistical Manual of Mental Disorders* (4th ed., text revision) [5]. These phenotypes are less likely to reflect functional effects of genetic variation than biological phenotypes. Translational brain-based phenotypes measured with neuroimaging modalities such as functional magnetic resonance imaging (fMRI) offer greater potential to elucidate risk for substance use and addictive disorders. Further, neuroimaging phenotypes serve as a bridge between basic preclinical research using animal models and applied clinical research. Intermediate phenotypes of basic mechanisms closely tied to the neurobiology of addiction may also provide targets for improving treatment interventions, relapse prevention, and medication development.

The scope of neuroimaging studies has traditionally been limited to testing separate bivariate associations between genetic variables and fMRI variables or fMRI variables and clinical measures. In these models, the clinical phenotype is influenced by a constellation of core components with known biological substrates, such as changes in the incentive value of drug-related stimuli, control and inhibitory processes, and the balance between reward drive processes and executive control [6–9]. These constructs may be measured using specific cognitive tasks performed during scanning. Most recently, scientists have advocated for testing genetic factors within mediational models that include both neuroimaging phenotypes (i.e., the intermediate phenotype) and more complex clinical phenotypes [2].

This chapter focuses on the application of fMRI to probing neural networks implicated in the development and maintenance of addiction in adolescent and adult populations and on how measures of these functional networks, in turn, may serve as intermediate phenotypes that link genetic variation with clinical phenotypes related to substance use disorders. In particular, the substance use disorders literature supports the importance of the incentive reward and executive control networks, both of which show particular promise in terms of generating translational phenotypes related to addiction. The chapter discusses specific fMRI tasks designed to probe relevant processes and regions of interest, followed by the importance of resting state fMRI, an emerging approach to examine network connectivity. Genetic associations with fMRI and functional connectivity are also briefly reviewed before conclusions and likely directions of future research are discussed.

9.1   A Conceptual Framework: Reward Learning and Executive Control

A key point of interest in current addiction research is the mechanism that leads substance users to transition from light or occasional use to problematic use and finally to lose control over using substances. Neurobiological research has focused on pleasure and reward systems in the brain that putatively change in response to chronic substance abuse and mechanisms that underlie chronic relapse even after prolonged abstinence.[1] Consistent with this model, studies have shown that anticipation of receiving a drug significantly impacts the brain's reward regions [10]. Thus, prolonged use leads to potentially irreversible changes in how the brain processes reward [11]. Conversely, executive control processes are instrumental for adaptive behaviors including delay of immediate gratification and emotion regulation [12–14]. Populations at risk for developing substance use disorders may display initial tendencies to discount risks while overemphasizing rewards, especially during adolescence [15]. In addition, substance use itself may exacerbate problems with executive control, such that chronic abusers suffer from neural adaptations in the reward and incentive salience networks as well as networks involved in executive control, making it very difficult to control their substance use [2, 14].

Thus, brain measures that reflect reliable patterns of activity among particular regions associated with reward or control processes may serve as useful intermediate phenotypes related to addiction. Subcortical "older" brain regions involved in basic drives, which project to the frontal areas involved in executive control, include the insula, ventral tegmental area (VTA), putamen, and caudate. Areas related to executive control such as the orbitofrontal cortex (OFC), dorsolateral prefrontal cortex (DLPFC), inferior frontal gyrus (IFG), and anterior cingulate cortex (ACC) have also been a focus of addiction research. These areas are broadly implicated in inhibition and impulsive decisions. For example, the OFC has been associated with control of craving and decision making [16] and, when damaged, is related to increased preference for immediate over delayed, larger rewards [17]. The DLPFC has also been identified in regulation of reward regions and executive functioning [18]. Frontal areas do not fully mature until well into young adulthood, so drug use emerging in adolescence offers a window into the interaction of brain development and substance use, while substance-dependent adults inform more specific associations between long-term drug use and functional brain changes. The following sections describe several common fMRI tasks currently used in addiction research. In general, substance users show a number of abnormalities in activation patterns in these tasks compared to nonusing controls.

9.2 Using fMRI Tasks to Generate Intermediate Phenotypes

9.2.1 Cue Reactivity

The most common approach to date in fMRI studies of substance use disorders uses cue reactivity tasks, where activation during presentation of substance-related stimuli is compared to activation during presentation of neutral or control stimuli. Similar approaches have used a variety of stimulus modalities, including sight, taste, smell, and tactile cues. This basic platform is the foundation for substance use brain–behavior associations and further explorations of how substance use influences brain function.

Greater blood-oxygen-level-dependent (BOLD) response in key regions related to craving and addiction has been frequently observed in substance users compared to healthy controls for a variety of drugs of abuse. In particular, robust activation has been found in regions related to reward processing, such as the striatum and areas of the prefrontal cortex (PFC) [19–21]. In a large sample of heavy drinkers, alcohol taste cues elicited activations in the dorsal striatum, insula, OFC, ACC, and VTA, and alcohol use disorder symptom severity was significantly associated with activation in the insula, precuneus, and striatum [19]. Similar effects have been found in adolescents as well as adults. For example, visual alcoholic beverage stimuli produced greater activations in anterior and limbic regions related to reward, desire, and positive affect among adolescents who reported consuming more drinks per month and greater craving [22]. Cue reactivity has distinguished between light and preclinical heavy drinkers when alcohol cues were compared to images related to life goals or food stimuli [23]. In chronic cocaine abusers, increased activation has been found compared to that in matched controls in the left DLPFC and occipital cortex when viewing cocaine stimuli but not appetitive control stimuli such as food [24]. Further, exposure to tactile cannabis cues was associated with BOLD response in reward regions, including the VTA, thalamus, ACC, insula, and amygdala [25]. A meta-analysis of smoking cue reactivity found reliable activations in the precuneus, posterior and anterior cingulate gyri, dorsal and medial PFC, insula, dorsal striatum, and extended visual system across studies [26]. More specific findings related to marijuana use, gambling and heavy smoking, and abstinence from substance use are detailed below.

One example is a study conducted by Cousijn and colleagues [27], who found that cue reactivity among frequent cannabis users was associated with high severity of problematic use but not with quantity of use. Marijuana-related images produced higher activation than neutral images in the VTA of frequent users compared to sporadic users and controls, and activation in the OFC, ACC, and striatum was only greater in a subset of frequent users who endorsed severe problems related to their substance use. Frequent marijuana users also showed negative correlations between craving and activation in the right putamen and right DLPFC.

Similarly, a study by Goudriaan and colleagues [28] examined cue reactivity in a sample of problem gamblers and heavy smokers compared to healthy controls. Increased craving in nonsmoking, treatment-seeking problem gamblers was associated with activation in the left ventrolateral PFC and left insula during viewing of gambling pictures. This pattern was not observed in heavy smokers, where craving was correlated with left prefrontal and left amygdala activation during viewing of smoking stimuli. Reactivity to smoking cues was not significantly different between heavy smokers and gamblers or controls, but heavy smokers did show within-group differences when stratified by level of nicotine dependence. Smokers who endorsed increased dependence symptoms had greater activation than their smoking peers in the ventromedial PFC, ACC, insula, and temporal gyrus when viewing smoking compared to neutral cues. Interestingly, the PFC was associated with craving in both the gamblers and heavy smokers, but activation in the insula accompanied craving specifically in the group of gamblers, whose problems were severe enough that participants had sought treatment. The PFC, ACC, and insula were also associated with dependence symptoms.

Cue reactivity has also been examined in relation to abstinence from substance use, which is of clinical relevance with respect to the aim of reducing relapse rates. In particular, stressors may activate incentive reward regions and thus trigger relapse to drug seeking [29]. One pilot study in a sample of heroin abusers found widespread differences in brain activity, including greater activation to heroin cues in the posterior and anterior cingulate under short-term abstinence and decreased activation in the posterior cingulate, insula, and dorsal striatum under long-term abstinence [30]. Activity in prefrontal areas was significantly greater in smokers who had been abstinent for longer than 12 months compared to controls and current smokers when presented with smoking cues versus neutral stimuli, while current smokers exhibited less cortical but more subcortical activation compared to the other groups [31]. Another study suggested that smoking cues elicited the greatest activation in moderately dependent smokers, possibly because highly dependent smokers may be driven by internal rather than external cues [32]. In attempting to remain abstinent, findings suggest that, until heightened responses to drug cues have diminished, substance users must continue to exercise increased cognitive control.

In summary, presentation of drug cues reliably elicits activation in widespread regions of interest and is associated with behavioral measures of problematic substance use such as craving. The simplicity of these paradigms is also noteworthy, given that analysis often involves straightforward contrasts between drug-related and neutral stimuli. External validity is likely strong, especially in the case of users making quit attempts facing exposure to substance use cues. However, other fMRI tasks have been developed to more explicitly model cognitive constructs related to addiction such as decision making, impulsivity, and inhibition. Several of these tasks are described below.

9.2.2 Balloon Analogue Risk Task

The Balloon Analogue Risk Task (BART) [33] assesses behavioral risk taking and has been modified for use in neuroimaging [19]. Participants pump up successive balloons to collect money or points, with the goal of saving points before each balloon pops. The balloons explode at varied intervals, so participants must weigh the increasing risk of losing points against a potential reward from continuing to pump the balloon. High mean number of pumps before saving points, adjusted for number of explosions, and high total number of popped balloons reflect greater risk taking. Behavioral studies have shown that BART performance is associated with general risk taking in young adults including substance use, unprotected sex, and theft [34], and also distinguishes substance users from controls [35].

Responses during the BART may broadly tap into both the incentive reward and executive control networks. Claus and Hutchison [36] assessed risk taking on a modified BART in a sample of adults with an alcohol use disorder by contrasting risky versus nonrisky decisions and cash-out responses versus explosions. The task used three different colors of balloons, two of which exploded after five or eight pumps on average and one control, which did not explode. Neural response in the ACC, insula, brain stem, and striatum increased as risk increased within a given trial, but ACC, striatum, and brain stem activation was negatively associated with hazardous drinking as measured by the Alcohol Use Disorders Identification Test (AUDIT) [37]. AUDIT scores were also associated with activation in the ACC and insula following negative feedback from explosions. These findings suggest that risk taking related to substance use occurs as a function of increased reward sensitivity and decreased ability to predict negative outcomes [36]. In a modified BART task, adolescent males with combined conduct and substance diagnoses showed widespread differences compared to controls, with hypoactivation in the ACC, OFC, DLPFC, insula, amygdala, and basal ganglia when making risky decisions and activation differences during wins and losses resembling patterns of decreased reward sensitivity in older, chronic substance abusers [38]. Increased weekly alcohol consumption has also been negatively associated with activation in the medial PFC during risky decisions but positively associated with PFC activation during successful outcomes on the BART [39], further suggesting that substance users experience greater impulsivity (i.e., less control) and increased response to rewards during decision making.

9.2.3 Iowa Gambling Task

The Iowa Gambling Task (IGT) models real-world decision making by requiring participants to learn a reward–punishment rule over many trials. Participants choose between decks in a simulated card game based on monetary rewards and losses. Healthy adults have shown increased activation in the lateral prefrontal cortex and ACC during early decisions and activation in the OFC during later decisions, suggesting that the PFC and ACC are involved in directing behavior during evaluation of long-term goals while the OFC responds to values that should be pursued [40]. Interestingly, participants with a family history of alcoholism but no current abuse or history of dependence showed greater activation in the left dorsal ACC and left caudate nucleus during the IGT than family history negative participants, despite similar behavioral performance overall [41]. Observations of neuronal response differences on a decision-making task even in nonusing participants at elevated risk for substance use disorders support use of the ACC response during the IGT as an intermediate phenotype.

Studies of substance users' performance on the IGT have shown widespread differences in activation compared to healthy controls but may specifically provide a demonstration of the interaction between reward and control processes. For example, substance-dependent participants showed decreased activation in ventral medial and superior frontal regions during decision making compared to controls and to substance users with gambling problems [42]. Furthermore, a comparison by Wesley and colleagues [43] of chronic marijuana users and controls during the strategy development phase of the IGT showed no differences in performance but significant differences in activation. Following losses, controls showed greater activation in widespread areas including the ACC, medial frontal cortex, and precuneus. In addition, activation in the ACC and PFC was positively correlated with later performance for controls but not chronic users; however, no group differences were observed during win evaluation. This implies that marijuana use is associated with less sensitivity to negative feedback when approaching a new task [43], and further that nonusers may first activate reward then control processes to recover from loss and redirect behavior toward long-term goals. In comparison, substance users may not initially respond to negative feedback, thus delaying or possibly preventing later initiation of control processes.

9.2.4 Go/No-Go

In go/no-go paradigms, participants are instructed to respond with a button press to every presented stimulus unless a stimulus is followed by a cue to inhibit their response. For example, go trials with arrows pointing left or right should elicit a response, while in no-go trials participants should stop themselves from responding to left or right arrows followed immediately by an upward arrow or a beep. Interestingly, an alcohol administration study in healthy controls found decreased activation during false alarms on no-go trials in the ACC, PFC, insula, and parietal regions [44]. In other words, participants under the influence of alcohol showed decreased activity in those regions compared to placebo when incorrectly responding to no-go stimuli. In addition, former smokers showed greater prefrontal activation than controls during a go/no-go paradigm while current heavy smokers demonstrated less PFC activation compared to controls [31]. Repeated exposure to alcohol, and other substances in general, may blunt responses to negative consequences and negatively impact the ability to inhibit responses over long-term use.

Studies of the go/no-go paradigm in adolescents typically describe differing activation patterns but not task performance between youth at risk for developing substance use disorders and nonusing controls. Adolescents with a family history of alcoholism demonstrated less prefrontal activation when inhibiting responses on no-go trials than family history negative controls despite no performance differences overall [45]. A sample of adolescents with very limited substance use histories who later transitioned into heavy alcohol use showed decreased activation at baseline on no-go trials in widespread regions including left dorsal and medial frontal areas, cingulate gyrus, and left putamen [46]. After nearly a month of monitored abstinence, adolescent marijuana users showed increased BOLD response during no-go trials in right DLPFC, medial frontal, and parietal regions, as well as increased activation during go trials in right prefrontal, insula, and parietal regions, despite similar overall task performance as nonusing controls [47]. These findings indicate that substance users with limited histories maintain task performance levels by recruiting additional neural resources, but given long-term use, these activation patterns, particularly decreases in control regions, may become more prominent until performance is no longer maintained and eventually reduced.

9.2.5   Variations of the Stroop Task

Classic Stroop tasks involve measuring reaction time when reading ink colors aloud that are either congruent or incongruent with printed words, such as being required to say blue when the word red is printed in blue ink, but the task also has several variations which all examine reaction to incongruence. A common variant is emotional Stroop tasks, which measure differing reaction times between neutral words and relevant negative- or positive-valence words. The Stroop task is often modified for use during scanning by instructing participants to press one of several buttons that correspond to different colors or valences in response to word presentations. The Stroop task is particularly relevant to addiction research as a measure of cognitive control.

In general, substance users demonstrate less activation and perform poorly compared to controls on incongruent trials. Methamphetamine-dependent participants showed decreased activation in the right IFG, ACC, and anterior insula during incongruent trials of a color-word Stroop, and no regions of increased activation throughout the task compared to controls [48]. Similarly, a cocaine-dependent group did not show any areas of greater activation than controls but did exhibit decreased activation in the right IFG, right inferior parietal gyrus, and right superior temporal gyrus during a counting Stroop [49]. Another study in participants who met criteria for cocaine abuse or dependence showed hypoactivation in the ACC and medial OFC when presented with drug-related versus matched neutral words [50]. Other fMRI tasks have not typically evoked group differences in the IFG, so Stroop tasks may offer unique advantages to studies of executive control. These findings support deficits in cognitive control associated with severe substance abuse and dependence.

9.2.6   Brain Regions as Functional Intermediate Phenotypes

The patterns of activation across the fMRI tasks reviewed above support the possibility of specific brain regions' serving as intermediate phenotypes. The insula, striatum, ACC, and regions of the frontal cortex including the DLPFC, OFC, and medial frontal cortex have shown the most consistent activation differences between substance users and nonusing controls across substances and tasks. Many of these regions have also been identified through recent resting state functional connectivity studies, further emphasizing their role in addiction.

9.3   Emerging Evidence from Resting State Network Connectivity Studies

A growing body of research has emerged that describes the formation of functional connectivity in the brain using fMRI measures of BOLD response during a resting state, which allows for an assessment of correlated activity across select regions of interest when the brain is not otherwise engaged in task-specific activity [51]. Networks identified in this manner are thought to underlie aspects of functional domains such as sensation and cognition. Several recent studies have examined the typical developmental trajectory of network connectivity. An analysis of connections suggested that networks were strengthened with increasing maturity and the boundaries between networks were sharpened [52]. During development, long-range connections increase while short-range connections decrease, increasing the overall efficiency of major networks [53]. This research suggests that as the brain develops, nearby functional regions are segregated and the long-term connections between regions and networks are strengthened [52]. One large study examined functional connectivity in a sample of 210 subjects ages 7 to 31 [54]. Consistent with previous reports, the study noted that functional networks transition from a local to distributed architecture during adolescence [54]. This pattern reflects the development of more efficient neural pathways, which in turn facilitate more effective cognitive processing. Disruption in the evolution of major networks may result in inefficient cognitive processing and eventually lead to problematic behaviors downstream [53, 55].

Applications of functional connectivity to addiction are still relatively limited, despite the vastly increasing number of published resting state functional connectivity studies in recent years. Limitations of this emerging field of study must be noted, including small sample sizes and potential interactions with acute drug effects or withdrawal symptoms [56]. On a basic level, functional connectivity as an emerging modality must overcome similar obstacles as did fMRI until basic brain–behavior alterations in substance use are understood and accounted for in sufficiently powered studies with large sample sizes. Nevertheless, functional connectivity represents enormous potential for exploring neural mechanisms of brain function, especially when considered in concert with structural or functional imaging techniques. Current research has described the intrinsic properties of networks related to attention [57], cognitive performance [58], and control [53, 55, 59]. The following sections focus on two networks likely central to addiction, namely the incentive salience/reward and executive control networks. The salience network includes the dorsal ACC and insula, with strong connections to limbic structures, while the executive control network involves links between the DLPFC and the parietal cortex and appears to act on salience [55]. Thus, these two networks may be strongly related to the identification of salience cues linked with emotion and reward and acting on those cues.

9.3.1 The Incentive Salience/Reward Network

The incentive salience/reward network was identified in an independent component analysis which used the right orbital frontoinsula as a seed region, previously identified in an fMRI working memory task and chosen for its role in autonomic processing [55]. This extensive network primarily includes the orbital frontoinsula and anterior insula, dorsal ACC, and superior temporal pole, with additional connections to the DLPFC, supplementary motor area, portions of the amygdala, thalamus, and hypothalamus, and the VTA. A measure of anticipatory anxiety correlated with increased dorsal ACC connectivity within this network, and overall the identified regions have been shown to coactivate in response to emotional drives, stress, hunger, and receiving rewards [55]. Thus, it is an ideal candidate for a primary network related to reward learning and addiction.

Preliminary studies in substance users have observed alterations in connectivity between regions associated with the salience/reward network, including both increased and decreased connectivity compared to nonusers. Chronic heroin users showed increased connectivity compared to controls between the accumbens and portions of the ACC, between the accumbens and OFC, and between the amygdala and OFC [60] while chronic cocaine abusers have shown increased correlations between ventral striatum and OFC [24]. Increased strength of dorsal-ACC-to-striatum connections has been associated with severity of nicotine addiction [61]. Current prescription opioid dependence was associated with decreased functional connectivity for seed regions including the anterior insula, nucleus accumbens, and amygdala, with greater changes occurring with longer duration of use [62]. Compared to nonusing controls, chronic heroin users showed decreased connectivity of the ACC, which was related to increased cue reactivity and craving [63]. Individuals with cocaine dependence have shown decreased connectivity within subcircuits of a broader mesocorticolimbic system, including between the VTA and thalamus/nucleus accumbens, between the amygdala and medial PFC, and between the hippocampus and dorsal medial PFC, with the VTA to thalamus/nucleus accumbens connection in particular negatively correlated with years of cocaine use [64]. Another study in a cocaine-dependent group found alterations in ACC, middle frontal gyrus, and middle temporal gyrus connectivity in comparison to controls, which were associated with poor performance on a delay discounting task [65]. Finally, the ACC and insula have both been implicated in impaired awareness, which is relevant to addiction as it may relate to recognition of negative consequences of use, need for treatment, and strong desire to use [17]. These studies have all been conducted with chronic users, possibly indicating that decreased connectivity between reward learning nodes and control nodes plays a pivotal role in addiction.

9.3.2   The Executive Control Network

The executive control network was also identified by Seeley and colleagues [55] but used the right DLPFC as its seed region due to its involvement in control processes including working memory. This network is comprised of the DLPFC, ventrolateral and dorsomedial PFC, lateral parietal cortices, and a portion of the left frontoinsula, with additional functional connections to the dorsal caudate and anterior thalamus. Interestingly, no connectivity was observed between these areas and limbic, hypothalamic, or midbrain regions. The executive control network likely operates on cues identified by the salience/reward network and incorporates regions important for sustained attention and response inhibition [55]. Frontal areas, particularly the DLPFC, are responsible for adjusting control in response to feedback from the environment [59]. Functional deficits in this network could therefore be particularly indicative of substance use that has transitioned to loss of behavioral control inherent in severe abuse or dependence.

Two studies have examined prefrontal functional connectivity in adolescents with family histories of alcoholism. At-risk youth exhibited reduced connectivity between anterior PFC and cerebellar regions compared to family history negative controls, which correlated with reduced white matter integrity in the internal capsule and superior longitudinal fasciculus [66]. Decreased DLPFC and posterior parietal connectivity was also observed in family history positive adolescents, but no group differences in white matter architecture in frontoparietal tracts were observed [67]. Alterations in functional connectivity that occur prior to substance use in at-risk youth offer potential intermediate phenotypes and suggest that these differences may contribute to later development of substance use disorders because of altered control processing [66].

Similar to salience/reward network findings, both increased and decreased functional connectivity has been shown in substance users. Schulte and colleagues [68] found increased connectivity among middle cingulate, posterior cingulate, and medial PFC in control compared to alcoholic participants, and greater midbrain to OFC connectivity in alcoholics relative to controls. Structural fiber integrity was associated with increased functional connectivity overall [68]. In contrast, chronic heroin users have shown decreased connectivity between PFC and OFC and between PFC and ACC [69]. Reduced prefrontal connectivity was observed in cocaine-dependent participants relative to controls, even in the absence of any group differences in underlying microstructural white matter integrity [70]. Connectivity strength between DLPFC and striatum was negatively associated with performance on a reinforcement learning task and increased craving in a sample of male alcohol-dependent participants [71]. At the least, these findings describe functional differences in control processing in substance users, with deficits in executive control associated with increasing severity of use. These differences may be particularly useful as markers of transitioning toward problematic use.

9.4   Summary and Future Directions

Cue reactivity paradigms have identified regions of the brain involved in addiction and are the most commonly used tasks in fMRI substance use studies to date. Substance users evidence heightened activation toward drug cues in areas related to reward processing, which is often enhanced with severity of use. Likewise, cognitive fMRI tasks have shown widespread differences in activation patterns between substance users and controls, even when behavioral task performance is similar overall. The current literature broadly suggests that at-risk populations (e.g., family history positive, light to moderate users) display differences in reward-related activations, as well as differences in cognitive control-related brain activity. Finally, severe substance use at the level of abuse or dependence is associated with reduced activation in executive control areas accompanied by poor performance compared to nonusing controls. Responses to specific fMRI tasks in particular regions thus offer potential for use as intermediate phenotypes predictive of severity of substance use, especially given activation differences observed independently of current use in populations at risk for developing substance use disorders.

Thus, a neuroimaging approach may be useful for identifying neurobiological phenotypes that connect genetic variation with variation in clinical phenotypes. Studies in populations with positive family histories of substance use disorders offer a glimpse into genetic contributions to brain functioning, but recent research has linked fMRI abnormalities in substance users to specific genetic variations. For example, published studies have linked BOLD response to cannabis cues with variation in the cannabinoid receptor 1 gene (CNR1) and the fatty acid amide hydrolase (FAAH) gene [72]. In a group of dependent smokers, cue reactivity in the right superior frontal gyrus and right insula differed by polymorphism of a dopamine receptor D4 (DRD4) variable number tandem repeat [73]. Studies have also linked variation in the opioid receptor mu 1 (OPRM1) and DRD4 genes [74] and CNR1 gene [75] with increased BOLD response to alcohol cues in mesocorticolimbic structures.

Resting state functional connectivity is currently emerging as the imaging modality favored in studying neural associations underlying widespread brain functioning. Applications to addiction research remain limited, but substance use has been associated with broad alterations in functional connectivity. Importantly, differences have been observed in substance-naive adolescents with positive family histories, indicating particular promise for functional connectivity results as intermediate phenotypes of substance use disorders. Abnormalities in the salience/reward network likely indicate frequent but not necessarily problematic use, as even moderate substance users may demonstrate increased reward responsiveness. Studies of connectivity between regions related to executive control indicate that substance abuse and dependence are associated with general deficits. However, examinations to date of the executive control network have been related to structural integrity more often than functional differences during fMRI tasks, and further combined approaches are needed to describe cognitive mechanisms of severe substance use. Finally, few studies have suggested that functional connectivity measures may serve as useful phenotypes in substance abusing samples. One study recently found an association between variation in nicotinic acetylcholine receptor alpha 5 (CHRNA5) and connectivity in a sample of smokers [76]. The CHRNA5 has been linked more broadly with smoking in both genome-wide [77] and candidate gene studies [78], and it is possible that the association with functional connectivity may partially explain the influence of this gene on risk for tobacco dependence.

The strongest predictor of transitioning to disordered substance use may be the difference in activation between the salience reward and executive control networks [13, 79, 80]. To the extent that greater cue reactivity in the incentive salience network occurs with less activation of the control network, it is likely that an individual experiences lessened control over drug use behavior. It is also possible that increasing imbalance between reward and control systems is related to lack of insight into problem severity and decreasing self-awareness [7]. Future studies should incorporate both cue exposure tasks as well as tasks that probe inhibitory control. Ideally, future studies should find a way to probe inhibitory control during the presentation of drug-related stimuli, as this approach may be most closely related to how these processes operate in the real world and thus may have greater external validity. For example, features of cue reactivity and go/no-go tasks could be combined to examine response inhibition toward various stimuli (i.e., contrasting no-go during drug cues vs. other appetitive stimuli), or a decision-making task could incorporate weighing possible positive and negative outcomes of substance use to mirror real-life experiences. Given that the interplay between these two networks may be partially driven by changes in functional connectivity, future studies should also incorporate connectivity analyses that are focused on identifying the loss of connectivity between reward-based network nodes and executive control network nodes. This type of study might be particularly useful in the context of a longitudinal design, which is intended to tease apart underlying interactions between possible premorbid neural risks and neural effects of substance use.

Note

1. For a review, see G. F. Koob and N. D. Volkow (2010), Neurocircuitry of addiction, *Neuropsychopharmacology*, *35*, 217–238.

References

1. Stacey, D., Clarke, T. K., & Schumann, G. (2009). The genetics of alcoholism. *Current Psychiatry Reports*, *11*, 364–369.

2. Hutchison, K. E. (2010). Substance use disorders: realizing the promise of pharmacogenomics and personalized medicine. *Annual Review of Clinical Psychology*, *6*, 577–589.

3. Liu, J. Z., Tozzi, F., Waterworth, D. M., et al. (2010). Meta-analysis and imputation refines the association of 15q25 with smoking quantity. *Nature Genetics*, *42*, 436–440.

4. Manolio, T. A., Collins, F. S., Cox, N. J., et al. (2009). Finding the missing heritability of complex diseases. *Nature*, *461*, 747–753.

5. American Psychiatric Association. (2000). *Diagnostic and Statistical Manual of Mental Disorders* (4th ed., text rev.). Washington, DC: American Psychiatric Association.

6. Filbey, F. M., Claus, E., Audette, A. R., et al. (2008). Exposure to the taste of alcohol elicits activation of the mesocorticolimbic neurocircuitry. *Neuropsychopharmacology*, *33*, 1391–1401.

7. Goldstein, R. Z., Craig, A. D., Bechara, A., et al. (2009). The neurocircuitry of impaired insight in drug addiction. *Trends in Cognitive Sciences*, *13*, 372–380.

8. Leland, D. S., Arce, E., Miller, D. A., & Paulus, M. P. (2008). Anterior cingulate cortex and benefit of predictive cueing on response inhibition in stimulant dependent individuals. *Biological Psychiatry*, *63*, 184–190.

9. Potenza, M. N. (2007). To do or not to do? The complexities of addiction, motivation, self-control, and impulsivity. *American Journal of Psychiatry, 164,* 4–6.

10. van Hell, H. H., Vink, M., Ossewaarde, L., et al. (2010). Chronic effects of cannabis use on the human reward system: an fMRI study. *European Neuropsychopharmacology, 20,* 153–163.

11. Koob, G. F. (2006). The neurobiology of addiction: a neuroadaptational view relevant for diagnosis. *Addiction, 101*(Suppl 1), 23–30.

12. Bechara, A. (2005). Decision making, impulse control and loss of willpower to resist drugs: a neurocognitive perspective. *Nature Neuroscience, 8,* 1458–1463.

13. Goldstein, R. Z., & Volkow, N. D. (2002). Drug addiction and its underlying neurobiological basis: neuroimaging evidence for the involvement of the frontal cortex. *American Journal of Psychiatry, 159,* 1642–1652.

14. Kalivas, P. W., & Volkow, N. D. (2005). The neural basis of addiction: a pathology of motivation and choice. *American Journal of Psychiatry, 162,* 1403–1413.

15. Bava, S., & Tapert, S. F. (2010). Adolescent brain development and the risk for alcohol and other drug problems. *Neuropsychology Review, 20,* 398–413.

16. London, E. D., Ernst, M., Grant, S., Bonson, K., & Weinstein, A. (2000). Orbitofrontal cortex and human drug abuse: functional imaging. *Cerebral Cortex, 10,* 334–342.

17. Volkow, N. D., Wang, G. J., Fowler, J. S., & Tomasi, D. (2012). Addiction circuitry in the human brain. *Annual Review of Pharmacology and Toxicology, 52,* 321–336.

18. Goldstein, R. Z., & Volkow, N. D. (2011). Dysfunction of the prefrontal cortex in addiction: neuroimaging findings and clinical implications. *Nature Reviews. Neuroscience, 12,* 652–669.

19. Claus, E. D., Ewing, S. W., Filbey, F. M., Sabbineni, A., & Hutchison, K. E. (2011). Identifying neurobiological phenotypes associated with alcohol use disorder severity. *Neuropsychopharmacology, 36,* 2086–2096.

20. Kareken, D. A., Claus, E. D., Sabri, M., et al. (2004). Alcohol-related olfactory cues activate the nucleus accumbens and ventral tegmental area in high-risk drinkers: preliminary findings. *Alcoholism, Clinical and Experimental Research, 28,* 550–557.

21. Myrick, H., Anton, R. F., Li, X., et al. (2004). Differential brain activity in alcoholics and social drinkers to alcohol cues: relationship to craving. *Neuropsychopharmacology, 29,* 393–402.

22. Tapert, S. F., Cheung, E. H., Brown, G. G., et al. (2003). Neural response to alcohol stimuli in adolescents with alcohol use disorder. *Archives of General Psychiatry, 60,* 727–735.

23. Ihssen, N., Cox, W. M., Wiggett, A., Fadardi, J. S., & Linden, D. E. (2011). Differentiating heavy from light drinkers by neural responses to visual alcohol cues and other motivational stimuli. *Cerebral Cortex, 21,* 1408–1415.

24. Wilcox, C. E., Teshiba, T. M., Merideth, F., Ling, J., & Mayer, A. R. (2011). Enhanced cue reactivity and fronto-striatal functional connectivity in cocaine use disorders. *Drug and Alcohol Dependence, 115,* 137–144.

25. Filbey, F. M., Schacht, J. P., Myers, U. S., Chavez, R. S., & Hutchison, K. E. (2009). Marijuana craving in the brain. *Proceedings of the National Academy of Sciences of the United States of America, 106*, 13016–13021.

26. Engelmann, J. M., Versace, F., Robinson, J. D., et al. (2012). Neural substrates of smoking cue reactivity: a meta-analysis of fMRI studies. *NeuroImage, 60*, 252–262.

27. Cousijn, J., Goudriaan, A. E., Ridderinkhof, K. R., et al. (2012). Neural responses associated with cue-reactivity in frequent cannabis users. *Addiction Biology, 18*, 570–580.

28. Goudriaan, A. E., de Ruiter, M. B., van den Brink, W., Oosterlaan, J., & Veltman, D. J. (2010). Brain activation patterns associated with cue reactivity and craving in abstinent problem gamblers, heavy smokers and healthy controls: an fMRI study. *Addiction Biology, 15*, 491–503.

29. Koob, G. F., & Volkow, N. D. (2010). Neurocircuitry of addiction. *Neuropsychopharmacology, 35*, 217–238.

30. Lou, M., Wang, E., Shen, Y., & Wang, J. (2012). Cue-elicited craving in heroin addicts at different abstinent time: an fMRI pilot study. *Substance Use & Misuse, 47*, 631–639.

31. Nestor, L., McCabe, E., Jones, J., Clancy, L., & Garavan, H. (2011). Differences in "bottom-up" and "top-down" neural activity in current and former cigarette smokers: evidence for neural substrates which may promote nicotine abstinence through increased cognitive control. *NeuroImage, 56*, 2258–2275.

32. Vollstadt-Klein, S., Kobiella, A., Buhler, M., et al. (2011). Severity of dependence modulates smokers' neuronal cue reactivity and cigarette craving elicited by tobacco advertisement. *Addiction Biology, 16*, 166–175.

33. Lejuez, C. W., Read, J. P., Kahler, C. W., et al. (2002). Evaluation of a behavioral measure of risk taking: the Balloon Analogue Risk Task (BART). *Journal of Experimental Psychology. Applied, 8*, 75–84.

34. Lejuez, C. W., Aklin, W. M., Zvolensky, M. J., & Pedulla, C. M. (2003). Evaluation of the Balloon Analogue Risk Task (BART) as a predictor of adolescent real-world risk-taking behaviours. *Journal of Adolescence, 26*, 475–479.

35. Lejuez, C. W., Aklin, W. M., Jones, H. A., et al. (2003). The Balloon Analogue Risk Task (BART) differentiates smokers and nonsmokers. *Experimental and Clinical Psychopharmacology, 11*, 26–33.

36. Claus, E. D., & Hutchison, K. E. (2012). Neural mechanisms of risk taking and relationships with hazardous drinking. *Alcoholism, Clinical and Experimental Research, 36*, 932–940.

37. Babor, T. F., Higgins-Biddle, J. C., Saunders, J. B. & Monteiro, M. G. (2001). *The Alcohol Use Disorders Identification Test: Guidelines for Use in Primary Care.* World Health Organization.

38. Crowley, T. J., Dalwani, M. S., Mikulich-Gilbertson, S. K., et al. (2010). Risky decisions and their consequences: neural processing by boys with antisocial substance disorder. *PLoS ONE, 5*, e12835.

39. Bogg, T., Fukunaga, R., Finn, P. R., & Brown, J. W. (2012). Cognitive control links alcohol use, trait disinhibition, and reduced cognitive capacity: evidence for medial prefrontal cortex dysregulation during reward-seeking behavior. *Drug and Alcohol Dependence, 122,* 112–118.

40. Hartstra, E., Oldenburg, J. F., Van Leijenhorst, L., Rombouts, S. A., & Crone, E. A. (2010). Brain regions involved in the learning and application of reward rules in a two-deck gambling task. *Neuropsychologia, 48,* 1438–1446.

41. Acheson, A., Robinson, J. L., Glahn, D. C., Lovallo, W. R., & Fox, P. T. (2009). Differential activation of the anterior cingulate cortex and caudate nucleus during a gambling simulation in persons with a family history of alcoholism: studies from the Oklahoma Family Health Patterns Project. *Drug and Alcohol Dependence, 100,* 17–23.

42. Tanabe, J., Thompson, L., Claus, E., et al. (2007). Prefrontal cortex activity is reduced in gambling and nongambling substance users during decision-making. *Human Brain Mapping, 28,* 1276–1286.

43. Wesley, M. J., Hanlon, C. A., & Porrino, L. J. (2011). Poor decision-making by chronic marijuana users is associated with decreased functional responsiveness to negative consequences. *Psychiatry Research, 191,* 51–59.

44. Anderson, B. M., Stevens, M. C., Meda, S. A., et al. (2011). Functional imaging of cognitive control during acute alcohol intoxication. *Alcoholism, Clinical and Experimental Research, 35,* 156–165.

45. Schweinsburg, A. D., Paulus, M. P., Barlett, V. C., et al. (2004). An fMRI study of response inhibition in youths with a family history of alcoholism. *Annals of the New York Academy of Sciences, 1021,* 391–394.

46. Norman, A. L., Pulido, C., Squeglia, L. M., et al. (2011). Neural activation during inhibition predicts initiation of substance use in adolescence. *Drug and Alcohol Dependence, 119,* 216–223.

47. Tapert, S. F., Schweinsburg, A. D., Drummond, S. P., et al. (2007). Functional MRI of inhibitory processing in abstinent adolescent marijuana users. *Psychopharmacology, 194,* 173–183.

48. Nestor, L. J., Ghahremani, D. G., Monterosso, J., & London, E. D. (2011). Prefrontal hypoactivation during cognitive control in early abstinent methamphetamine-dependent subjects. *Psychiatry Research, 194,* 287–295.

49. Barros-Loscertales, A., Bustamante, J. C., Ventura-Campos, N., et al. (2011). Lower activation in the right frontoparietal network during a counting Stroop task in a cocaine-dependent group. *Psychiatry Research, 194,* 111–118.

50. Goldstein, R. Z., Tomasi, D., Rajaram, S., et al. (2007). Role of the anterior cingulate and medial orbitofrontal cortex in processing drug cues in cocaine addiction. *Neuroscience, 144,* 1153–1159.

51. Raichle, M. E., MacLeod, A. M., Snyder, A. Z., et al. (2001). A default mode of brain function. *Proceedings of the National Academy of Sciences of the United States of America, 98,* 676–682.

52. Dosenbach, N. U., Nardos, B., Cohen, A. L., et al. (2010). Prediction of individual brain maturity using fMRI. *Science, 329,* 1358–1361.

53. Fair, D. A., Dosenbach, N. U., Church, J. A., et al. (2007). Development of distinct control networks through segregation and integration. *Proceedings of the National Academy of Sciences of the United States of America, 104,* 13507–13512.

54. Fair, D. A., Cohen, A. L., Power, J. D., et al. (2009). Functional brain networks develop from a "local to distributed" organization. *PLoS Computational Biology, 5,* e1000381.

55. Seeley, W. W., Menon, V., Schatzberg, A. F., et al. (2007). Dissociable intrinsic connectivity networks for salience processing and executive control. *Journal of Neuroscience, 27,* 2349–2356.

56. Sutherland, M. T., McHugh, M. J., Pariyadath, V., & Stein, E. A. (2012). Resting state functional connectivity in addiction: lessons learned and a road ahead. *NeuroImage, 62,* 2281–2295.

57. Fox, M. D., Corbetta, M., Snyder, A. Z., Vincent, J. L., & Raichle, M. E. (2006). Spontaneous neuronal activity distinguishes human dorsal and ventral attention systems. *Proceedings of the National Academy of Sciences of the United States of America, 103,* 10046–10051.

58. Andrews-Hanna, J. R., Snyder, A. Z., Vincent, J. L., et al. (2007). Disruption of large-scale brain systems in advanced aging. *Neuron, 56,* 924–935.

59. Dosenbach, N. U., Fair, D. A., Miezin, F. M., et al. (2007). Distinct brain networks for adaptive and stable task control in humans. *Proceedings of the National Academy of Sciences of the United States of America, 104,* 11073–11078.

60. Ma, N., Liu, Y., Li, N., et al. (2010). Addiction related alteration in resting-state brain connectivity. *NeuroImage, 49,* 738–744.

61. Hong, L. E., Gu, H., Yang, Y., et al. (2009). Association of nicotine addiction and nicotine's actions with separate cingulate cortex functional circuits. *Archives of General Psychiatry, 66,* 431–441.

62. Upadhyay, J., Maleki, N., Potter, J., et al. (2010). Alterations in brain structure and functional connectivity in prescription opioid-dependent patients. *Brain, 133,* 2098–2114.

63. Liu, J., Liang, J., Qin, W., et al. (2009). Dysfunctional connectivity patterns in chronic heroin users: an fMRI study. *Neuroscience Letters, 460,* 72–77.

64. Gu, H., Salmeron, B. J., Ross, T. J., et al. (2010). Mesocorticolimbic circuits are impaired in chronic cocaine users as demonstrated by resting-state functional connectivity. *NeuroImage, 53,* 593–601.

65. Camchong, J., MacDonald, A. W., III, Nelson, B., et al. (2011). Frontal hyperconnectivity related to discounting and reversal learning in cocaine subjects. *Biological Psychiatry, 69,* 1117–1123.

66. Herting, M. M., Fair, D., & Nagel, B. J. (2011). Altered fronto-cerebellar connectivity in alcohol-naive youth with a family history of alcoholism. *NeuroImage, 54,* 2582–2589.

67. Wetherill, R. R., Bava, S., Thompson, W. K., et al. (2012). Frontoparietal connectivity in substance-naive youth with and without a family history of alcoholism. *Brain Research, 1432*, 66–73.

68. Schulte, T., Muller-Oehring, E. M., Sullivan, E. V., & Pfefferbaum, A. (2012). Synchrony of corticostriatal–midbrain activation enables normal inhibitory control and conflict processing in recovering alcoholic men. *Biological Psychiatry, 71*, 269–278.

69. Ma, N., Liu, Y., Fu, X. M., et al. (2011). Abnormal brain default-mode network functional connectivity in drug addicts. *PLoS ONE, 6*, e16560.

70. Kelly, C., Zuo, X. N., Gotimer, K., et al. (2011). Reduced interhemispheric resting state functional connectivity in cocaine addiction. *Biological Psychiatry, 69*, 684–692.

71. Park, S. Q., Kahnt, T., Beck, A., et al. (2010). Prefrontal cortex fails to learn from reward prediction errors in alcohol dependence. *Journal of Neuroscience, 30*, 7749–7753.

72. Filbey, F. M., Schacht, J. P., Myers, U. S., Chavez, R. S., & Hutchison, K. E. (2010). Individual and additive effects of the CNR1 and FAAH genes on brain response to marijuana cues. *Neuropsychopharmacology, 35*, 967–975.

73. McClernon, F. J., Hutchison, K. E., Rose, J. E., & Kozink, R. V. (2007). DRD4 VNTR polymorphism is associated with transient fMRI–BOLD responses to smoking cues. *Psychopharmacology, 194*, 433–441.

74. Filbey, F. M., Ray, L., Smolen, A., et al. (2008). Differential neural response to alcohol priming and alcohol taste cues is associated with DRD4 VNTR and OPRM1 genotypes. *Alcoholism, Clinical and Experimental Research, 32*, 1113–1123.

75. Hutchison, K. E., Haughey, H., Niculescu, M., et al. (2008). The incentive salience of alcohol: translating the effects of genetic variant in CNR1. *Archives of General Psychiatry, 65*, 841–850.

76. Hong, L. E., Hodgkinson, C. A., Yang, Y., et al. (2010). A genetically modulated, intrinsic cingulate circuit supports human nicotine addiction. *Proceedings of the National Academy of Sciences of the United States of America, 107*, 13509–13514.

77. Saccone, S. F., Hinrichs, A. L., Saccone, N. L., et al. (2007). Cholinergic nicotinic receptor genes implicated in a nicotine dependence association study targeting 348 candidate genes with 3713 SNPs. *Human Molecular Genetics, 16*, 36–49.

78. Bierut, L. J., Stitzel, J. A., Wang, J. C., et al. (2008). Variants in nicotinic receptors and risk for nicotine dependence. *American Journal of Psychiatry, 165*, 1163–1171.

79. Baler, R. D., & Volkow, N. D. (2006). Drug addiction: the neurobiology of disrupted self-control. *Trends in Molecular Medicine, 12*, 559–566.

80. Bickel, W. K., Miller, M. L., Yi, R., et al. (2007). Behavioral and neuroeconomics of drug addiction: competing neural systems and temporal discounting processes. *Drug and Alcohol Dependence, 90*(Suppl 1), S85–S91.

10   Implicit Cognition: An Intermediate Phenotype for Addiction?

Reinout W. Wiers, Eske M. Derks, and Thomas E. Gladwin

In this chapter we will introduce the topic of implicit cognition in relation to addiction and discuss our current knowledge regarding the question of whether implicit cognitive processes may be regarded as an intermediate phenotype ("endophenotype"). We discuss the scarce genetic research in relation to implicit cognition and addiction, as well as recent research, aimed at modifying implicit cognitive processes, which has shown promising first findings regarding congruent changes in addictive behaviors. This line of research may therefore eventually meet some of the promises of endophenotype research: while it is difficult to change genetic risk factors, it may be possible to change the intermediate processes and thus reduce the risk for relapse or onset of addiction.

Before we set out to define implicit cognition as a potential endophenotype, it is important to briefly discuss the *phenotype* of interest: addiction or addictive behaviors. First, there is much discussion about the question of whether a psychoactive substance is necessary in definitions of addiction (currently defined as "substance dependence," as in the *Diagnostic and Statistical Manual of Mental Disorders*, 4th ed.; *DSM–IV*) and about the directly related question of whether behavioral addictions should be included such as excessive gambling, gaming, sex, or even excessive cases of more mundane behaviors such as work, eating, or buying. Current discussions regarding (re)definitions for the fifth edition (*DSM–V*) appear to resolve with including gambling, which would end the definition of addiction as substance dependence while restricting the behavioral addictions to only this category of excessive behaviors [1]. We note here that all of the relevant research we discuss in this chapter pertains to addictive behaviors including a psychoactive substance, but similar processes could well play a role in a variety of behavioral addictions [2, 3]. Second, it is important to realize that the outcome phenotype (addiction) is very heterogeneous. As already observed some 25 years ago, based on decades of research on personality and addiction, it is essential to distinguish between at least two broad subtypes of addicted people in order to find any meaningful relationship between a risk factor (personality) and the outcome of interest (addiction). The reason is that both individuals scoring high on internalizing characteristics (anxiety, depression) and individuals scoring high on externalizing characteristics (impulsivity, sensation seeking) are at increased risk for addiction, but through different mechanisms [4–9]. To illustrate the difference, in the "typical" externalizing pathway, addiction is preceded by externalizing problem behaviors, substance use, and related problems that begin at an early age and develop into addiction with antisocial personality characteristics, a type which primarily characterizes males. In the internalizing pathway, by contrast, substance use typically escalates later in life, as a means to cope with other (internalizing) problems, both in males and in females (although it should be noted that there is some preliminary evidence for an internalizing pathway at early age in adolescents who suffered from depression during childhood [10]). It is important to keep this rough distinction in mind (note that more subtle typologies have been developed [11, 12]) when considering any potential endophenotype for addiction. In line with this distinction, there are indications that the heritability of addiction is affected by both age of onset of problems and gender, with strongest heritability for men with early age of onset of dependence [13, 14]), although findings have not always been consistent [15].

10.1 Implicit Cognition

The term "implicit cognition" has been used both with reference to a measurement procedure and with reference to the outcome of the measurement procedure: cognitive processes [16]. To avoid confusion, we refer to the first use with the term "indirect measurement" [17]. Implicit cognitive processes are typically assessed with indirect measures: rather than asking participants to reflect upon reasons for their (addictive) behavior, behavioral measures are used that should indirectly reveal underlying processes. In relation to cognitive processes, the term "implicit" refers to a set of criteria that indicate relatively automatic, spontaneous cognitive processes [18]. These criteria should not be regarded as absolute or perfectly correlated [19–21]. Implicit cognitive processes are relatively automatic, require minimal cognitive effort, and are not dependent upon verbal abilities and IQ. These processes can be contrasted with reflective cognitive processes, which are deliberate, require cognitive effort, and are dependent on verbal abilities and IQ [18, 21–23]. One way to interpret these different cognitive processes is that they constitute different systems [24], with unique underlying neural architecture [25, 26]. Other researchers view implicit and explicit processes more as different processing streams [22, 27], which might be subserved by the same neural architecture (e.g., through reprocessing). Finally, there are researchers who only recognize the difference in assessment strategy but conceptualize the measured outcomes as different ways to assess a single underlying cognitive process (in the field of addiction represented by Goldman and colleagues, who have argued that "expectancies"—either measured with indirect or direct measures—constitute a single unifying construct [28, 29]).

We distinguish three relevant implicit cognitive processes: an attentional bias for a substance, automatically activated memory associations, and automatically activated action tendencies [18]. We now describe the related measurement procedures and main findings, discuss interactions with an important moderator (executive control), and then turn to what is known on the role of genetic factors in these measures.

10.1.1 Attentional Bias

Regarding attentional bias, most studies have used either a variety of the addiction Stroop or a variety of the visual probe task (for a review, see [30]). In the addiction Stroop task, attentional bias is inferred from a *slowing* in reaction times (RTs) when participants name the ink color of words referring to their substance of abuse as compared with neutral words. Note that the addiction Stroop task should not be confused with the classical Stroop task (presenting conflicting color names while the participant has to name the ink color of the words), which is regarded as a measure of executive control, response inhibition, or control over interfering information [31, 32]. Importantly, the tasks are structurally different: while the content of the words directly interferes with the task at hand (naming the ink color) in a classical Stroop, the content in an emotional Stroop is merely distracting [33]. Note that in some studies assessing a substance-related attentional bias, a classical Stroop task is used as a comparison [34]. In some other recent studies, a classical Stroop task was used as a global index of executive control, a moderator of the impact of implicit cognitive processes on behavior [35, 36].

In the visual probe task, two pictures or words are presented simultaneously for a brief period of time, one representing the substance of abuse and the other a matched neutral stimulus. This is followed by the presentation of a probe (e.g., an arrow pointing up or down), to which the participant has to react. Attentional bias is inferred if participants react *faster* to the probe when it replaces a representation of the substance, compared with when it replaces the neutral picture or word. Using these two measures, researchers have fairly consistently found that heavier substance use is related to a stronger attentional bias, both in student and in general population samples [30, 37]. One notorious problem with the visual probe task is its poor reliability [38]. An alternative dependent variable used with the same task is to study eye movements, which can be used to distinguish between different underlying processes, that is, engagement and disengagement [39]. Attentional bias as assessed with eye movements or brain potentials correlates higher with subjective craving than the RT-based measure [40]. Other measures have been used as well–for example, a variety of the change detection paradigm, where it has been found that substance abusers detect quick changes in a complex visual scene faster when they occur in substance-related stimuli than when they occur in non-substance-related stimuli [41]. More recently, researchers have started to use the "attentional blink" paradigm, where stimuli are presented at a very rapid pace which permits the study of encoding versus later postattentive processing stages [42].

Across different methods and substances of abuse, there is converging evidence that substance abusers show an attentional bias and that this is most pronounced in the relatively slow disengagement component of attention [30, 37]. Whether a fast engagement attentional bias is also present in some stages of addiction is more controversial [39, 42, 43]. It is also unclear which role an attentional bias plays in the etiology of addictive behaviors: does it prelude or follow heavy substance use, and does it play a role in the acceleration from use to problematic use [30]? And does it stabilize or decline after long periods of addiction? Some recent studies have indicated that an attentional bias may decrease after prolonged addiction [44, 45]. These findings could be interpreted as support of incentive habit models of addiction [46]: an attentional bias, as a marker of incentive salience of drug cues [47, 48], is of primary importance in the first escalation phase of addiction, but habit processes become more pronounced in later stages of addiction. Of importance, from a clinical perspective, there are indications that an attentional bias for alcohol and drugs may increase during treatment [49, 50], and this increase has been found to predict treatment outcome [49, 51, 52]. Furthermore, an attentional bias can be retrained in addiction, which may help to reduce alcohol use, as indicated by one study in problem drinkers, which lacked a control group [53], and by one small randomized clinical trial (RCT; including a sham-training group), which indicated that the trained group of alcoholic patients had a significantly longer time to relapse [50].

10.1.2 Memory Associations

Memory associations can be assessed both with open-ended word association tests and with RT tests. Common tests of word association have used free word association, in which the participant lists the first word that comes to mind in response to a cue word, phrase, or picture. In these controlled association tests, a first response of some type (e.g., verb) is requested using "top of mind" instructions. For example, an instruction is given to write down the first thing that comes to mind for "feeling good, Friday night…" If such tests do not directly inquire about the target concept (e.g., drug associations), then the tests are indirect and may have the capability of assessing implicit processes as basic memory research has demonstrated [18, 54]. Stacy and colleagues developed different versions of word association methods and found that they prospectively predict substance use [18, 54–56].

In addition to open-ended memory association measures, researchers have used RT tests to assess memory associations, with varieties of the Implicit Association Test (IAT; [57]) being by far the most widely used test. The IAT is a RT measure used to probe individual differences in associations between a drug and two attribute categories (e.g., "positive" vs. "negative" if one assesses implicit attitudes). The target category (alcohol or another substance in addiction research) also requires a contrast category (often soft drinks or water for alcohol) although some research has used a version with only one target category (the Single Target or Single Category IAT [58]). On each trial of the task, participants rapidly categorize visually presented stimuli (pictures or words) by pressing one of two response keys. For example, they may be instructed to press the left response key when an alcohol-related word or a positive word is presented and to press the right response key in response to alcohol-unrelated or negative words. The rationale for the task is that if participants automatically evaluate alcohol as positive rather than negative, they should be quicker to respond when alcohol-related words and positive words share the same response key (as in the example), compared to another block of the task where alcohol-related words and negative words share the same response key. The IAT has a number of strengths, which explain its popularity: it is a flexible tool (different associations can be assessed), easy to use and much more reliable than many other implicit measures, with test–retest correlations around 0.70 [59, 60]. However, the validity of the measure has been criticized, with much ongoing debate [16, 19, 33, 61–70].

Many studies have now reported correlations between memory associations assessed with an IAT and alcohol or drug use, in most cases after controlling for explicit attitudes or expectancies (for reviews and meta-analyses, see [18, 71–75]). Perhaps surprisingly, most studies overall found strong negative associations with the concept "alcohol," which tended to be weaker in heavy drinkers compared with light drinkers [71, 72, 76]. Note that researchers did not assess only the positive–negative dimension; heavy drinkers were also found to more strongly associate alcohol with (positive) arousal than light drinkers [71, 72] and to more strongly associate alcohol with approach than with avoidance [77, 78]. The latter variety could be considered a measure of automatically activated action tendencies (discussed below). Note further that the IAT can be used with bipolar attributes (e.g., one response button associated with positive, the other with negative), but also unipolar versions have been used, in which positive and negative associations are assessed separately, against neutral categories [71]. Although using unipolar versions is theoretically appealing, a direct comparison found better predictive validity for a bipolar IAT [79]. Finally, partly in response to criticisms concerning the validity of the IAT, different varieties have been proposed, such as a personalized IAT [80] and the single-block or recode-free IAT [81]. Both have been successfully applied to alcohol research by Houben and colleagues [82–84]. One recent variant is the Brief IAT, in which one category is present but never explicitly categorized [85].

Other RT measures have also been used in the domain of addiction, such as varieties of semantic priming [86, 87] and the Extrinsic Affective Simon Task (EAST [88]). Although these measures have theoretical advantages over the IAT (e.g., when one wants to assess unidirectional associations, which is of particular relevance in negative reinforcement [89, 90]), the problem of these alternatives is their far more modest reliability [19, 59, 91].

A small number of studies have directly compared the predictive validity of open-ended word association measures and RT-based measures [92, 93] and found stronger predictive power for the open-ended word association measures, which was confirmed by the meta-analyses which included these measures [75]. Finally, regarding the prediction of treatment outcome and possibilities for direct manipulation, there is very little work yet. Wiers and colleagues found that a cognitive behavioral intervention in which alcohol expectancies are challenged [94] affected explicit expectancies but hardly affected implicit associations [60]. Similarly, motivational interviewing was not found to affect implicit alcohol associations or their correlation with drinking [95]. More positively, two studies found that affective memory associations can be influenced in social drinkers, with short-term reductions of drinking behavior [96, 97], and one as-yet unpublished study found that mindfulness training can reduce the impact of implicit associations on drinking behavior [98].

10.1.3 Action Tendencies

Action tendencies have been conceptualized as an important aspect of emotion and motivation [99, 100]. Different tasks have been used to assess an automatically activated tendency to approach or avoid a class of substances (or other stimuli). First, as noted above, the IAT has been used to assess associations between alcohol with approach versus avoid words [77, 78]. Second, a measure with a symbolic approach or avoid movement has been used, in which a manikin is presented above or under a picture of a substance or a control picture, and participants are to move the manikin toward the substance in one block (and away from control pictures) and away from the substance (and toward control pictures) in another block. The difference in RT between the blocks gives a measure of a tendency to approach or avoid the substance, and the resulting measure (approach bias) has been found to correlate positively with use of a variety of substances, including cigarettes [45], alcohol [101, 102], and cannabis [103]. Another measure, the alcohol Approach Avoidance Task (alcohol-AAT) uses an actual approach or avoid response with a joystick [104]. With this measure it was found that heavy but not light drinkers have a faster reaction to pull (approach) than to push (avoid) the joystick to alcohol cues, even when they reacted to the format of the picture (landscape vs. portrait). Note that the alcohol-AAT is an irrelevant-feature task (participants react to another feature of the stimuli than the contents), similar to a (pictorial) EAST [105], while the manikin is structurally similar to the IAT (relevant-feature task: one block approach substance, one block avoid substance, hence more explicit categorization). As noted above, relevant-feature tasks like the IAT usually have a better reliability than irrelevant-feature tasks [19, 33, 61], and this was also found for the manikin task [106]. However, a recent study with an irrelevant-feature cannabis-AAT found a reasonably good reliability (0.68) and found that only heavy cannabis users show an approach bias for cannabis stimuli and that this approach bias uniquely predicted escalation of cannabis use in the six months to follow [107].

Regarding intervention possibilities, the AAT is interesting, because it can easily be manipulated, in a similar fashion as in attentional retraining, by changing the percentage of alcohol pictures that come in the format which is to be approached or avoided [108]. In a first preclinical study using this methodology, Wiers and colleagues found that relatively heavily drinking students who had pushed in response to most of the alcohol pictures drank less alcohol in a subsequent taste test than did heavily drinking students who had pulled in response to most of the alcohol pictures [108]. In addition, the manipulation of the approach bias (with the pictorial AAT) generalized to an approach–avoid IAT. In a subsequent clinical study (RCT), Wiers and colleagues trained alcoholic patients to avoid alcohol, using the alcohol-AAT [109]. Again, generalized effects were found on the IAT: patients who were trained to push the joystick in response to alcohol cues changed their relatively strong alcohol-approach associations to relatively strong alcohol-avoidance associations while no such change was found in the control groups. Moreover, trained patients were less likely to have relapsed one year later [109].

10.1.4 Executive Control as Moderator of the Impact of Implicit Cognition

Theoretically, the impact of implicit cognitive processes on behavior should be moderated by a combination of ability to control (executive control capacity) and motivation to control [110–112]. A number of recent studies have investigated the interaction between implicit cognitive processes (mostly memory associations) and executive control functions (ECFs; mostly working memory or interference control) and found that in individuals with relatively poor executive control, implicit memory associations are a better predictor of alcohol use and smoking [113], alcohol use and problems [35, 114, 115], an attentional bias for alcohol [116], aggression after alcohol [36], with similar findings for eating and other social behaviors with an impulsive component [117, 118] (for reviews, see [111, 119, 120]). Note that these effects hold up after controlling for measures of impulsivity and sensation seeking [121]. Note further that some of these studies also found that for individuals with relatively strong executive functions, explicit attitudes and expectancies better predicted the behavior under study [114, 118], but a recent large study found no evidence for an interaction between executive control and explicit expectancies [122]. In addition to stable individual differences in ECFs, the relative weight of implicit cognitive processes in the prediction of behavior appears to be also stronger when executive control is temporarily weakened, for example through ego depletion [123, 124] or alcohol consumption [125–127].

Regarding the motivation to control behavior, there is one aforementioned study [96] in which the goal was to reduce the impact of implicit memory associations on addictive behavior through motivational interviewing, which was not found. One other recent study found that external motivation (parental rule setting) moderated the impact of an approach bias for alcohol on drinking behavior in adolescents: in boys with permissive parents, an approach bias for alcohol predicted heavy alcohol involvement, which was not the case for boys with more strict parents or for girls [128].

Finally, regarding possibilities for intervention, it is important to note that working memory can also be trained, with positive results in children with attention-deficit/ hyperactivity disorder and related problems [129, 130]. One recent study trained problem drinkers' working memory (real training or sham training) and found that those participants in the experimental condition who had strong positive alcohol associations successfully reduced their alcohol intake as a result of the training [131]. This finding suggests that participants with strong automatically activated memory associations profited most from increased control over their associations. In another study, Houben and colleagues trained inhibitory control using an adapted go/no-go task [132] in which participants in the experimental group never responded to alcohol pictures (always paired with a no-go sign) while no such contingency was introduced in the control condition [133]. This manipulation increased negative alcohol associations and decreased drinking behavior in the week after the experiment [133].

Hence, all three implicit cognitive processes distinguished here (attentional bias, memory associations, and approach bias) as well as an important moderator (executive control) appear to be trainable, with first positive results on the corresponding (addictive) behaviors. We now turn to the question of to what extent these phenomena are genetically moderated.

10.2 Genetics and Implicit Cognition (and Executive Control)

10.2.1 Findings on Genetic Variation and Implicit Cognition

Any relationship between genetic factors and implicit cognition would be expected to be indirect—for instance, it seems unlikely that a genetic variant would be associated with automatically attending to bottles of beer. Various types of implicit cognition may well involve common genetic influences, and different genetic pathways may result in common cognitive biases. Genetic variants in several genes have been found to be associated with addiction. Although the functional pathways via which these genetic variants increase the risk for addiction remain to be further elucidated, we will propose a theoretical framework for the role of the genes involved in addiction-related implicit cognition (see figure 10.1). We identify and briefly discuss three broadly defined, interconnected pathways: (1) variations related to the acute hedonic value[1]

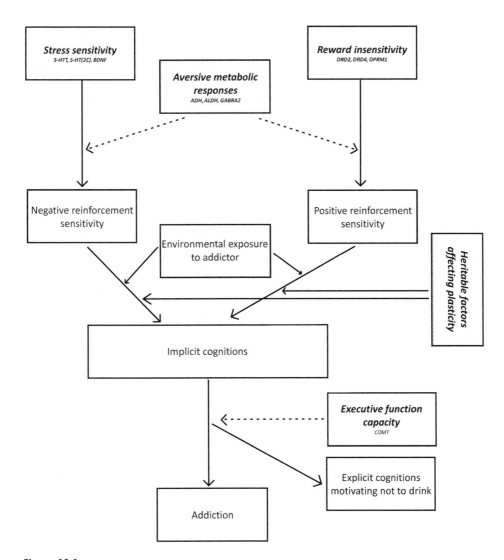

Figure 10.1

A heuristic model illustrating possible pathways to addiction, involving implicit cognitive processes.

of potentially addictive substances or behavior (or "addictors" for short), (2) variations related to emotion regulation that indirectly affect negative reinforcement value, and (3) variations related to reward (in)sensitivity that affect positive reinforcement value. These pathways may provide a foundation for understanding findings concerning genetic variation and addiction-related implicit cognition. Subsequently, we tentatively link these pathways to the few existing studies directly relating genetic variation to implicit cognition.

10.2.2   Genetic Variations Related to the Acute Hedonic Value of Addictors

First, genetic variation can impact the metabolism of addictive substances and, thereby, the extent to which their use will be reinforcing, which will impact the associations between cues signaling the drug and reward. A well-known example is the set of *ADH* and *ALDH* genes, which code for enzymes involved with alcohol metabolism. Certain alleles of these genes are associated with facial flushing and other aversive effects as a consequence of drinking alcohol and, accordingly, have been shown to be protective against alcoholism in various populations [134, 135]. Single nucleotide polymorphisms in the *GABRA2* gene associated with alcoholism [136] are also associated with weaker subjective sensations (such as "getting high") due to alcohol consumption [137], possibly due to relatively weak signals to stop drinking.

10.2.3   Genetic Variants Related to Negative Reinforcement

The second pathway concerns negative reinforcement. Genetic variants related to the serotonin transporter gene *SLC6A4* (also referred to as 5-HTT) [138] and the brain-derived neurotrophic factor (*BDNF*) [139] gene appear to be related to general stress sensitivity—for instance, such that they influence the chance that childhood adversity will lead to depression [140, 141]. The "short" repeat of a stress-related polymorphism of the promotor region of the SCL6A4 gene (5-HTTLPR) results in slower reuptake of serotonin, and hence desensitization of serotonin receptors, and is also related to higher sensitivity to nicotine withdrawal [142], which would seem likely to be due to an enhancement of the acute effect of smoking in terms of negative reinforcement. Indeed, the 5-HTT short allele has been shown to interact with stress to predict alcohol and drug use in college students [143]. Further, two different polymorphisms related to serotonin, which are located in the *HTR2C* (also referred to as *5-HT(2C)*) gene, have been found to affect the odds of smoking initiation [144]. The *BDNF* gene is involved in neuronal growth and differentiation and has a Val/Met polymorphism (Val66Met) in which the Met allele is associated with more robust fear and anxiety behavior in mice [145, 146] and, anatomically, with smaller amygdala and hippocampus volume in humans [147], possibly due to reduced BDNF production associated with the Met allele. Concerning its relationship to addiction, low *BDNF* serum levels are associated with an increased chance of addiction to cocaine [148] and methamphetamine [149] and with the severity of alcohol withdrawal [150]. In line with these findings, the MET variant of the *BDNF* gene is associated with violence and delirium tremens in addiction [151]. We note that changes in BDNF protein levels over time, as opposed to variation over subjects, are also related to addiction: craving is associated with increases in *BDNF* protein levels, and *BDNF* protein levels rise after abstinence [152], possibly reflecting the activation of mechanisms related to drug seeking [153].

10.2.4 Genetic Variants Related to Positive Reinforcement Value

Genetic variation may affect addictor sensitivity via acute effects involving enhanced positive reinforcement. Such enhancement has been argued to result from dopaminergic deficiencies, for which the stimulatory effects of addictive drugs on the endogenous opioid system can compensate [154]. Low dopamine (DA) receptor density is associated with various clinical disorders as well as with personality characteristics such as novelty seeking [155] and approach tendencies [156]. Genetic polymorphisms in the *ANKK1* gene (TaqI A and B) have been associated with low receptor density as well as with alcohol addiction, specifically a severe form of addiction, obesity, and addiction to other substances including cocaine [154, 157]. In addition, the Taq1A polymorphism has been suggested to interact with parental rule setting regarding adolescent alcohol use [158, 159]. The Taq1A and B polymorphisms are also in strong linkage disequilibrium (LD) with the D2 dopamine receptor (*DRD2*) gene and were previously considered to be in the promotor region of *DRD2*. Due to the strong LD in this region, it is unclear whether associations between clinical disorders and Taq1A and B polymorphisms indicate the involvement of *ANKK1* or *DRD2*.

A polymorphism (-521 C/T) in the promotor region of the dopamine D4 receptor (*DRD4*) gene has been found to affect novelty seeking in a meta-analysis [155]. Reduced receptor density may result in increased novelty seeking or related personality changes, which has been proposed as a consequence of DA genetic variation that may mediate increased risk for drug use [160, 161]. Some studies have found evidence for a role of the 7-repeat variable number of tandem repeats (VNTR) polymorphism in *DRD4* in novelty seeking and approach behavior, but these relationships were found to have relatively weak evidence in meta-analyses [155, 156]. However, there are indications that the *DRD4* 7-repeat VNTR may involve more general processes, such as tendency to imitate, which would only lead to heavy drinking in the presence of heavy drinking peers [162], in line with the idea of plasticity genes [163].

A similar pathway involving positive reinforcement involves genetic variation in a gene coding for the mu-opioid receptor (*OPRM1*). The A118G polymorphism of the *OPRM1* gene has been shown to confer functional differences to mu-opioid receptors, such that the G allele binds beta-endorphin three times more strongly than the A allele. The G allele of this polymorphism has been shown to be associated with relatively strong craving for alcohol [164, 165] and with relatively strong automatically activated approach tendencies to alcohol and other appetitive stimuli [104], assessed with the alcohol-AAT (as discussed above). A functional magnetic resonance imaging study showed relatively strong reactions to alcohol in mesolimbic areas, both for carriers of the *OPRM1* G allele and for carriers of the *DRD4* 7+ repeat VNTR [166]. In a recent positron-emission tomography study, smokers with the G allele of the *OPRM1* gene were found to have reduced receptor availability of opioid receptors in brain regions involved with reinforcement [167]. The relationship between *OPRM1* and addiction may thus, similarly to *DRD* genes, reflect an enhanced positive reinforcement of drug use due to a baseline deficiency in the reward system. Of possible interest is the disinhibitory effect of mu-opioid receptors on dopamine release, suggesting a possible interaction with dopamine-related aspects of addiction.

10.2.5 Modulating Influences via Executive Functions

Modulating effects involving the above biological network concern the ECFs. As discussed above, relatively strong ECFs concern a protective factor in addiction: less impulsive individuals are less likely to develop addiction, and stronger ECFs appear to "uncouple" automatic addictor-related associations from behavior [35, 111, 113, 114, 117, 118]. ECFs are highly heritable [168, 169]. There are indications that the Val158Met polymorphism of the catechol-O-methyltransferase (*COMT*) gene may explain part of the genetic variation. *COMT* is involved with the clearance of prefrontal dopamine; the Met allele is associated with relatively low enzyme activity, and hence an increase in tonic dopamine levels and decrease in dopamine sensitivity compared to the Val allele. Individuals carrying a Met versus Val allele have better ECFs [170] and more efficient prefrontal processing (i.e., more deactivation) during tasks that tax ECFs [171]. The effect of the *COMT* variations appears to be specific to ECFs as opposed to component processes such as attention or short-term memory capacity without executive components [170, 172]. Interestingly, the negative effects of the Val allele of the *COMT* gene appear to be modifiable via methylation [173], which could be an interesting epigenetic effect from the present perspective (for a review on epigenetic effects in addiction, see [174]).

10.2.6 Implicit Cognition as an Endophenotype?

While current evidence is scarce and indirect, we suggest that the pathways described above may connect various genetic variations' influences to implicit cognition in addiction. Initial findings have connected some of the above genes to some forms of addiction-related implicit cognitive processes. Genetic influences involving attentional bias for alcohol have been found in a study with young adolescents and young adults [175]. A dot probe task with alcoholic drinks and soft drinks was used to assess attentional bias to alcohol pictures. The risk variant in the *OPRM1* gene (the G allele) was associated with a relatively strong attentional bias for alcohol in young adolescents while the *DRD4 DRD4* 7+ repeat VNTR was found to be associated with an attentional bias for alcohol in young adults. While we know of no other study relating attentional bias for alcohol to genetic factors, it should be noted that in research on anxiety, one of the above sources of genetic variation—serotonin-related polymorphisms related to stress—has been associated with attentional bias for emotional stimuli [176–179]; for a meta-analysis, see [180]. A stronger attentional bias could be explained from increased emotional salience of stimuli and thereby a stronger influence on learning processes. It is entirely possible that in a subgroup of individuals with an attentional bias for threatening stimuli, this bias is coupled with a bias to approach alcohol (as a way to cope with the stress). Similar processes could influence the development of automatic associations, in two ways matching the basic dichotomy of internalizing and externalizing pathways to addiction. Interactions involving genes and implicit cognition have also been found for implicit memory associations [181]. Only for subgroups with high-risk variants in *ALDH2* or *COMT* were alcohol-coping associations (assessed with an IAT; see above) predictive of drinking behavior. Finally, as mentioned above, the risk variant of a polymorphism in the *OPRM1* gene has been found to be related to relatively strong automatic approach tendencies for alcohol and other appetitive stimuli in male heavy drinkers [104]. Hence, individuals for whom drug use compensates for a genetic lack of sensitivity to normal reinforcement would form a group at risk for developing addiction-related approach tendencies. Individuals with a genetic tendency to experience strong negative emotions could experience greater reinforcement for drug use because it helps them avoid those emotions. In both cases, individuals with strong executive function would be better able to overcome the effects of these tendencies (see figure 10.1).

To account for interactions between the various influences that are likely involved in addictive behaviors, an endophenotype for addiction would have to be defined in terms of constellations of pathways, with the common result of a tendency to develop difficult-to-control implicit cognitions directed toward addictive behaviors. There are two such constellations which we argue should be distinguished: the positive and negative reinforcement endophenotype. The positive endophenotype revolves around heritable biological deficits that drug use ameliorates (e.g., decreased ability for dopaminergic stimulation) while the core of the negative endophenotype is a general stress vulnerability that confers added negative reinforcement value on the acute effects of drugs. For example, coping drinking would be considered an expression of the latter pathway and would be expected to be associated with different genes than individuals with different drinking motivation. On the other hand, some genetic factors would be expected to have common effects in either endophenotype: genetic variation that renders drug metabolism aversive should always be protective, as would factors that increase executive control.

10.3   Conclusion

From an implicit cognition or dual-process perspective, the endophenotype for addiction vulnerability would consist of a set of interrelated genetically influenced factors: (1) nonaversive metabolism of the addictor, (2) enhanced (positive or negative) reinforcement of acute effects, and (3) relatively weak ECFs. The "perfect storm" would involve all factors working toward a tendency to addiction. This vulnerability would be expected to be expressed at a behavioral level as a common endophenotype, defined in terms of automaticity development: the ease with which hard-to-control cognitive biases arise from addictor-related reinforcement. The measurement of such an endophenotype would require novel methods aimed at detecting this kind of vulnerability, possibly involving experimentally controlled tests of automaticity development (for an example of such an approach, see [182]). From the perspective sketched above, research focusing on genetic variation and implicit cognition, and in particular the development of implicit cognitive processes, may be highly relevant to understanding the biological basis of the vulnerability to addiction.

We briefly note that genetic influences on neural plasticity may well modulate many of the pathways discussed above. The hippocampus has been linked to associative processes supporting addiction in animal studies: stimulation of the hippocampus reinstates cocaine seeking after extinction [183], and suppression of hippocampal activity reduces cue-evoked cocaine seeking [184]. As neural plasticity in the hippocampus is heritable, related genes may influence vulnerability for addiction via associative processes. Finally, variations in dopamine transporter (*DAT1*) and receptor (*DRD4*) genes [185] and *COMT* [186] may play a role in plasticity via effects on error-related feedback processing; variants leading to either very high or very low dopamine availability appear to cause increased reactivity to errors [186]. If these effects on errors could be generalized to include unexpected consequences in general, such variation may enhance learning due to the acute effects of addictors.

A particularly important application of the development of an endophenotype in terms of implicit cognition would be targeted interventions. If it is known what type of vulnerability has likely led to an individual's addiction, this may indicate which type of implicit cognition to target in interventions. For example, the patients with relatively strong automatically activated approach tendencies may benefit most from approach-bias retraining [109], and it is an interesting question whether patients could better be selected with respect to the strength of their approach tendencies or with respect to the associated genetic factor (presence of the *OPRM1* G allele). One issue in this discussion is measurement: while the reliability of the AAT is reasonably good [107], the reliability of a commonly used measure of attentional bias, the visual probe task, is poor [187]. Therefore, depending on the qualities of the measures, it may be more beneficial to select directly for the endophenotype (e.g., strong implicit cognitive process) or an associated gene. Similarly, patients with weak working memory capacity could benefit more from a specific training, either aimed at increasing their working memory [131] or aimed at changing their dysfunctional automatic cognitive processes, as has been found in the domain of anxiety [188]. In this domain, a first study also found genetic moderation of trainability: individuals with a low-expression form (S/S, S/Lg, or Lg/Lg) of the *SCL6A4* (5-HTTLPR) gene developed stronger biases for both negative and positive affective pictures through attentional retraining compared to those with the high-expression (La/La) form of the gene [189]. These first findings can be interpreted as initial support that implicit cognitive processes may constitute a malleable endophenotype in addiction and related disorders. However, more research is clearly needed to critically test this claim and its associated therapeutic consequences. We believe research into implicit cognitive processes as a potentially malleable endophenotype of addiction is an intriguing avenue for further research.

Note

1. While the term "hedonic" suggests that the strength of positive feelings following the addictor is crucial, research on expectancies [29] and implicit associations [71, 72] suggests that addictors are associated with a combination of positive and arousing feelings.

References

1. O'Brien, C. (2011). Addiction and dependence in DSM–V. *Addiction (Abingdon, England), 106,* 866–867.

2. Grant, J. E., Potenza, M. N., Weinstein, A., & Gorelick, D. A. (2010). Introduction to behavioral addictions. *American Journal of Drug and Alcohol Abuse, 36,* 233–241.

3. Orford, J. (2001). Addiction as excessive appetite. *Addiction, 96,* 15–31.

4. Babor, T. F., Hofmann, M., DelBoca, F. K., Hesselbrock, V., Meyer, R. E., Dolinsky, Z. S., et al. (1992). Types of alcoholics: I. evidence for an empirically derived typology based on indicators of vulnerability and severity. *Archives of General Psychiatry, 49,* 599–608.

5. Cloninger, C. R. (1987). Neurogenetic adaptive mechanisms in alcoholism. *Science, 236,* 410–416.

6. Conrod, P. J., Pihl, R. O., Stewart, S. H., & Dongier, M. (2000). Validation of a system of classifying female substance abusers on the basis of personality and motivational risk factors for substance abuse. *Psychology of Addictive Behaviors, 14,* 243–256.

7. Cox, W. M. (1987). Personality theory and research. In H. T. Blane & K. E. Leonard (Eds.), *Psychological Theories of Drinking and Alcoholism* (pp. 55–89). New York: Guilford.

8. Sher, K. J. (1991). *Children of Alcoholics: A Critical Appraisal of Theory and Research.* Chicago: University of Chicago Press.

9. Wiers, R. W., Sergeant, J. A., & Gunning, W. B. (1994). Psychological mechanisms of enhanced risk of addiction in children of alcoholics: a dual pathway? *Acta Paediatrica. Supplement, 404,* 9–13.

10. Saraceno, L., Heron, J., Munafó, M., Craddock, N., & van den Bree, M. B. (2012). The relationship between childhood depressive symptoms and problem alcohol use in early adolescence: findings from a large longitudinal population-based study. *Addiction, 107,* 567–577.

11. Zucker, R. A. (1987). The four alcoholisms: a developmental account of the etiological process. In P. C. Rivers (Ed.), Nebraska Symposium on Motivation, 1986: Alcohol and addictive behavior (pp. 27–83). Lincoln: University of Nebraska Press.

12. Lesch, O. M., & Walter, H. (1996). Subtypes of alcoholism and their role in therapy. *Alcohol and Alcoholism. Supplement, 1,* 63–67.

13. Liu, I. C., Blacker, D. L., Xu, R., Fitzmaurice, G., Tsuang, M. T., & Lyons, M. J. (2004). Genetic and environmental contributions to age of onset of alcohol dependence symptoms in male twins. *Addiction, 99,* 1403–1409.

14. McGue, M., Pickens, R. W., & Svikis, D. S. (1992). Sex and age effects on the inheritance of alcohol problems: a twin study. *Journal of Abnormal Psychology, 101,* 3–17.

15. Ehlers, C. L., Gizer, I. R., Vieten, C., Gilder, A., Gilder, D.A., Stouffer, G. M., et al. (2010). Age at regular drinking, clinical course, and heritability of alcohol dependence in the San Francisco family study: a gender analysis. *The American Journal on Addictions, 19,* 101–10.

16. De Houwer, J. (2006). What are implicit measures and why are we using them? In R. W. Wiers & A. W. Stacy (Eds.), *Handbook of Implicit Cognition and Addiction* (pp. 11–28). Thousand Oaks, CA: Sage.

17. Fazio, R. H., & Olson, M. A. (2003). Implicit measures in social cognition research: their meaning and use. *Annual Review of Psychology, 54,* 297–327.

18. Stacy, A. W., & Wiers, R. W. (2010). Implicit cognition and addiction: a tool for explaining paradoxical behavior. *Annual Review of Clinical Psychology, 6,* 551–575.

19. De Houwer, J., Teige-Mocigemba, S., Spruyt, A., & Moors, A. (2009). Implicit measures: a normative analysis and review. *Psychological Bulletin, 135,* 347–368.

20. Keren, G., & Schul, Y. (2009). Two is not always better than one: A critical evaluation of two-system theories. *Perspectives on Psychological Science, 4,* 533–550.

21. Evans, J. S. B. T. (2008). Dual-processing accounts of reasoning, judgment, and social cognition. *Annual Review of Psychology, 59,* 255–278.

22. Strack, F., & Deutsch, R. (2004). Reflective and impulsive determinants of social behavior. *Personality and Social Psychology Review, 8,* 220–247.

23. Kahneman, D. (2003). A perspective on judgment and choice: mapping bounded rationality. *American Psychologist, 58,* 697–720.

24. Evans, J. S. B. T. (2003). In two minds: dual-process accounts of reasoning. *Trends in Cognitive Sciences*, *7*, 454–459.

25. Satpute, A. B., & Lieberman, M. D. (2006). Integrating automatic and controlled processes into neurocognitive models of social cognition. *Brain Research*, *1079*(1), 86–97.

26. Heatherton, T. F., & Wagner, D. D. (2011). Cognitive neuroscience of self-regulation failure. *Trends in Cognitive Sciences*, *15*, 132–139.

27. Gladwin, T. E., Figner, B., Crone, E. A., & Wiers, R. W. (2011). Addiction, adolescence, and the integration of control and motivation. *Developmental Cognitive Neuroscience*, *1*, 364–376.

28. Goldman, M. S., Reich, R. R., & Darkes, J. (2006). Expectancy as a unifying construct in alcohol-related cognition. In R. W. Wiers & A. W. Stacy (Eds.), *Handbook on Implicit Cognition and Addiction* (pp. 105–119). Thousand Oaks, CA: Sage.

29. Goldman, M. S., Del Boca, F. K., & Darkes, J. (1999). Alcohol expectancy theory: the application of cognitive neuroscience. In K. E. Leonard & H. T. Blane (Eds.), *Psychological Theories of Drinking and Alcoholism* (2nd ed., pp. 203–246). New York: Guilford.

30. Field, M., & Cox, W. M. (2008). Attentional bias in addictive behaviors: a review of its development, causes, and consequences. *Drug and Alcohol Dependence*, *97*, 1–20.

31. MacLeod, C. M. (1991). Half a century of research on the Stroop effect: an integrative review. *Psychological Bulletin*, *109*, 163–203.

32. Miyake, A., Friedman, N. P., Emerson, M. J., Witzki, A. H., Howerter, A., & Wager, T. D. (2000). The unity and diversity of executive functions and their contributions to complex "frontal lobe" tasks: a latent variable analysis. *Cognitive Psychology*, *41*, 49–100.

33. De Houwer, J. (2003). A structural analysis of indirect measures of attitudes. In J. Musch & K. C. Klauer (Eds.), *The Psychology of Evaluation: Affective Processes in Cognition and Emotion* (pp. 219–244). Mahwah, NJ: Erlbaum.

34. Fadardi, J. S., & Cox, W. M. (2006). Alcohol attentional bias: drinking salience or cognitive impairment? *Psychopharmacology*, *185*, 169–178.

35. Houben, K., & Wiers, R. W. (2009). Response inhibition moderates the relationship between implicit associations and drinking behavior. *Alcoholism, Clinical and Experimental Research*, *33*, 626–633.

36. Wiers, R. W., Beckers, L., Houben, K., & Hofmann, W. (2009). A short fuse after alcohol: implicit power associations predict aggressiveness after alcohol consumption in young heavy drinkers with limited executive control. *Pharmacology, Biochemistry, and Behavior*, *93*, 300–305.

37. Cox, W. M., Fadardi, J. S., & Pothos, E. M. (2006). The addiction-Stroop test: theoretical considerations and procedural recommendations. *Psychological Bulletin*, *132*, 443–476.

38. Tull, M. T., McDermott, M. J., Gratz, K. L., Coffey, S. F., & Lejuez, C. W. (2011). Cocaine-related attentional bias following trauma cue exposure among cocaine dependent in-patients with and without post-traumatic stress disorder. *Addiction*, *106*, 1810–1818.

39. Field, M., Mogg, K., & Bradley, B. P. (2006). Attention to drug-related cues in drug abuse and addiction: component processes. In R. W. Wiers & A. W. Stacy (Eds.), *Handbook on Implicit Cognition and Addiction* (pp. 151–163). Thousand Oaks, CA: Sage.

40. Field, M., Munafo, M. R., & Franken, I. H. (2009). A meta-analytic investigation of the relationship between attentional bias and subjective craving in substance abuse. *Psychological Bulletin*, *135*, 589–607.

41. Jones, B. T., Jones, B. C., Smith, H., & Copley, N. (2003). A flicker paradigm for inducing change blindness reveals alcohol and cannabis information processing biases in social users. *Addiction*, *98*, 235–244.

42. Tibboel, H., De Houwer, J., & Field, M. (2010). Reduced attentional blink for alcohol-related stimuli in heavy social drinkers. *Journal of Psychopharmacology*, *24*, 1349–1356.

43. Leventhal, A. M., Waters, A. J., Breitmeyer, B. G., Miller, E. K., Tapia, E., & Li, Y. (2008). Subliminal processing of smoking-related and affective stimuli in tobacco addiction. *Experimental and Clinical Psychopharmacology*, *16*, 301–312.

44. Loeber, S., Duka, T., Welzel, H., Nakovics, H., Heinz, A., Flor, H., et al. (2009). Impairment of cognitive abilities and decision making after chronic use of alcohol: the impact of multiple detoxifications. *Alcohol and Alcoholism*, *44*, 372–381.

45. Mogg, K., Bradley, B. P., Field, M., & De Houwer, J. (2003). Eye movements to smoking-related pictures in smokers: relationship between attentional biases and implicit and explicit measures of stimulus valence. *Addiction*, *98*, 825–836.

46. Everitt, B. J., & Robbins, T. W. (2005). Neural systems of reinforcement for drug addiction: from actions to habits to compulsion. *Nature Neuroscience*, *8*, 1481–1489.

47. Robinson, T. E., & Berridge, K. C. (2003). Addiction. *Annual Review of Psychology*, *54*, 25–53.

48. Franken, I. H. A. (2003). Drug craving and addiction: integrating psychological and neuro-psychopharmacological approaches. *Progress in Neuro-Psychopharmacology & Biological Psychiatry*, *27*, 563–579.

49. Cox, W. M., Hogan, L. M., Kristian, M. R., & Race, J. H. (2002). Alcohol attentional bias as a predictor of alcohol abusers' treatment outcome. *Drug and Alcohol Dependence*, *68*, 237–243.

50. Schoenmakers, T., de Bruin, M., Lux, I. F., Goertz, A. G., Van Kerkhof, D. H., & Wiers, R. W. (2010). Clinical effectiveness of attentional bias modification training in abstinent alcoholic patients. *Drug and Alcohol Dependence*, *109*, 30–36.

51. Carpenter, K. M., Schreiber, E., Church, S., & McDowell, D. (2006). Drug Stroop performance: relationships with primary substance of use and treatment outcome in a drug-dependent outpatient sample. *Addictive Behaviors*, *31*, 174–181.

52. Marissen, M. A. E., Franken, I. H. A., Waters, A. J., Blanken, P., Van Den Brink, W., & Hendriks, V. M. (2006). Attentional bias predicts heroin relapse following treatment. *Addiction*, *101*, 1306–1312.

53. Fadardi, J. S., & Cox, W. M. (2009). Reversing the sequence: reducing alcohol consumption by overcoming alcohol attentional bias. *Drug and Alcohol Dependence, 101*, 137–145.

54. Stacy, A. W., Ames, S. L., & Grenard, J. (2006). Word association tests of associative memory and implicit processes: theoretical and assessment issues. In R. W. Wiers & A. W. Stacy (Eds.), *Handbook of Implicit Cognition and Addiction* (pp. 75–90). Thousand Oaks, CA: Sage.

55. Stacy, A. W. (1997). Memory activation and expectancy as prospective predictors of alcohol and marijuana use. *Journal of Abnormal Psychology, 106*, 61–73.

56. Stacy, A. W., Leigh, B. C., & Weingardt, K. R. (1997). An individual-difference perspective applied to word association. *Personality and Social Psychology Bulletin, 23*, 229–237.

57. Greenwald, A. G., McGhee, D. E., & Schwartz, J. L. K. (1998). Measuring individual differences in implicit cognition: the Implicit Association Test. *Journal of Personality and Social Psychology, 74*, 1464–1480.

58. Karpinski, A., & Steinman, R. B. (2006). The Single Category Implicit Association Test as a measure of implicit social cognition. *Journal of Personality and Social Psychology, 91*, 16–32.

59. Hofmann, W., Gawronski, B., Gschwendner, T., Le, H., & Schmitt, M. (2005). A meta-analysis on the correlation between the implicit association test and explicit self-report measures. *Personality and Social Psychology Bulletin, 31*, 1369–1385.

60. Wiers, R. W., van de Luitgaarden, J., van den Wildenberg, E., & Smulders, F. T. Y. (2005). Challenging implicit and explicit alcohol-related cognitions in young heavy drinkers. *Addiction (Abingdon, England), 100*, 806–819.

61. De Houwer, J., & De Bruycker, E. (2007). The Implicit Association Test outperforms the extrinsic affective Simon task as an implicit measure of inter-individual differences in attitudes. *British Journal of Social Psychology, 46*(Pt 2), 401–421.

62. De Houwer, J., Teige-Mocigemba, S., Spruyt, A., & Moors, A. (2009). Theoretical claims necessitate basic research: reply to Gawronski, Lebel, Peters, and Banse (2009) and Nosek and Greenwald (2009). *Psychological Bulletin, 135*, 377–379.

63. Blanton, H., & Jaccard, J. (2006). Arbitrary metrics in psychology. *American Psychologist, 61*, 27–41.

64. Blanton, H., Jaccard, J., Klick, J., Mellers, B., Mitchell, G., & Tetlock, P. E. (2009). Strong claims and weak evidence: reassessing the predictive validity of the IAT. *Journal of Applied Psychology, 94*, 567–582, discussion 83–603.

65. van Ravenzwaaij, D., van der Maas, H. L., & Wagenmakers, E. J. (2011). Does the name–race Implicit Association Test measure racial prejudice? *Experimental Psychology, 58*, 271–277.

66. Rothermund, K., & Wentura, D. (2004). Underlying processes in the Implicit Association Test (IAT): dissociating salience from associations. *Journal of Experimental Psychology. General, 133*, 139–165.

67. Rothermund, K., Wentura, D., & De Houwer, J. (2005). Validity of the salience asymmetry account of the Implicit Association Test: reply to Greenwald, Nosek, Banaji, and Klauer (2005). *Journal of Experimental Psychology. General, 134*, 426.

68. Greenwald, A. G., Poehlman, T. A., Uhlmann, E. L., & Banaji, M. R. (2009). Understanding and using the Implicit Association Test: III. meta-analysis of predictive validity. *Journal of Personality and Social Psychology, 97*, 17–41.

69. Greenwald, A. G., Nosek, B. A., & Sriram, N. (2006). Consequential validity of the Implicit Association Test: comment on Blanton and Jaccard (2006). *American Psychologist, 61*, 56–61, discussion 2–71.

70. Nosek, B. A., & Sriram, N. (2007). Faulty assumptions: a comment on Blanton, Jaccard, Gonzales, and Christie (2006). *Journal of Experimental Social Psychology, 43*, 393–398.

71. Houben, K., & Wiers, R. W. (2006). Assessing implicit alcohol associations with the Implicit Association Test: fact or artifact? *Addictive Behaviors, 31*, 1346–1362.

72. Wiers, R. W., Van Woerden, N., Smulders, F. T. Y., & De Jong, P. J. (2002). Implicit and explicit alcohol-related cognitions in heavy and light drinkers. *Journal of Abnormal Psychology, 111*, 648–658.

73. McCarthy, D. M., & Thompsen, D. M. (2006). Implicit and explicit measures of alcohol and smoking cognitions. *Psychology of Addictive Behaviors, 20*, 436–444.

74. Reich, R. R., Below, M. C., & Goldman, M. S. (2010). Explicit and implicit measures of expectancy and related alcohol cognitions: a meta-analytic comparison. *Psychology of Addictive Behaviors, 24*, 13–25.

75. Rooke, S. E., Hine, D. W., & Thorsteinsson, E. B. (2008). Implicit cognition and substance use: a meta-analysis. *Addictive Behaviors, 33*, 1314–1328.

76. Houben, K., & Wiers, R. W. (2006). A test of the salience asymmetry interpretation of the Alcohol-IAT. *Experimental Psychology, 53*, 292–300.

77. Ostafin, B. D., & Palfai, T. P. (2006). Compelled to consume: the Implicit Association Test and automatic alcohol motivation. *Psychology of Addictive Behaviors, 20*, 322–327.

78. Palfai, T. P., & Ostafin, B. D. (2003). Alcohol-related motivational tendencies in hazardous drinkers: assessing implicit response tendencies using the modified-IAT. *Behaviour Research and Therapy, 41*, 1149–1162.

79. Houben, K., Nosek, B., & Wiers, R. W. (2010). Seeing the forest through the trees: a comparison of different IAT variants measuring implicit alcohol associations. *Drug and Alcohol Dependence, 106*, 204–211.

80. Olson, M. A., & Fazio, R. H. (2004). Reducing the influence of extrapersonal associations on the Implicit Association Test: personalizing the IAT. *Journal of Personality and Social Psychology, 86*, 653–667.

81. Rothermund, K., Teige-Mocigemba, S., Gast, A., & Wentura, D. (2009). Minimizing the influence of recoding in the Implicit Association Test: the Recoding-Free Implicit Association Test (IAT-RF). *Quarterly Journal of Experimental Psychology, 62,* 84–98.

82. Houben, K., & Wiers, R. W. (2007). Personalizing the alcohol-IAT with individualized stimuli: relationship with drinking behavior and drinking-related problems. *Addictive Behaviors, 32,* 2852–2864.

83. Houben, K., & Wiers, R. W. (2007). Are drinkers implicitly positive about drinking alcohol? Personalizing the alcohol-IAT to reduce negative extrapersonal contamination. *Alcohol and Alcoholism (Oxford, Oxfordshire), 42,* 301–307.

84. Houben, K., Rothermund, K., & Wiers, R. W. (2009). Predicting alcohol use with a recoding-free variant of the Implicit Association Test. *Addictive Behaviors, 34,* 487–489.

85. Sriram, N., & Greenwald, A. G. (2009). The Brief Implicit Association Test. *Experimental Psychology, 56,* 283–294.

86. Ostafin, B. D., Palfai, T. P., & Wechsler, C. E. (2003). The accessibility of motivational tendencies toward alcohol: approach, avoidance, and disinhibited drinking. *Experimental and Clinical Psychopharmacology, 11,* 294–301.

87. Zack, M., Toneatto, T., & MacLeod, C. M. (1999). Implicit activation of alcohol concepts by negative affective cues distinguishes between problem drinkers with high and low psychiatric distress. *Journal of Abnormal Psychology, 108,* 518–531.

88. De Houwer, J. (2003). The Extrinsic Affective Simon Task. *Experimental Psychology, 50,* 77–85.

89. Wiers, R. W., Houben, K., Smulders, F. T. Y., Conrod, P. J., & Jones, B. T. (2006). To drink or not to drink: the role of automatic and controlled cognitive processes in the etiology of alcohol-related problems. In R. W. Wiers & A. W. Stacy (Eds.), *Handbook of Implicit Cognition and Addiction* (pp. 339–361). Thousand Oaks, CA: Sage.

90. Wiers, R. W. (2008). Alcohol and drug expectancies as anticipated changes in affect: negative reinforcement is not sedation. *Substance Use & Misuse, 43,* 429–444.

91. Roefs, A., Huijding, J., Smulders, F. T., Macleod, C. M., de Jong, P. J., Wiers, R. W., et al. (2011). Implicit measures of association in psychopathology research. *Psychological Bulletin, 137,* 149–193.

92. Ames, S. L., Grenard, J. L., Thush, C., Sussman, S., & Wiers, R. W. (2007). Comparison of indirect assessments of association as predictors of marijuana use among at-risk adolescents. *Experimental and Clinical Psychopharmacology, 15,* 204–218.

93. Thush, C., Wiers, R. W., Ames, S. L., Grenard, J., Sussman, S., & Stacy, A. W. (2007). Apples and oranges? Comparing indirect measures of alcohol-related cognition predicting alcohol use in at-risk adolescents. *Psychology of Addictive Behaviors, 21,* 587–591.

94. Darkes, J., & Goldman, M. S. (1993). Expectancy challenge and drinking reduction: experimental evidence for a mediational process. *Journal of Consulting and Clinical Psychology*, *61*, 344–353.

95. Thush, C., Wiers, R. W., Moerbeek, M., Ames, S. L., Grenard, J. L., Sussman, S., et al. (2009). Influence of motivational interviewing on explicit and implicit alcohol-related cognition and alcohol use in at-risk adolescents. *Psychology of Addictive Behaviors*, *23*, 146–151.

96. Houben, K., Schoenmakers, T. M., & Wiers, R. W. (2010). I didn't feel like drinking but I don't know why: the effects of evaluative conditioning on alcohol-related attitudes, craving and behavior. *Addictive Behaviors*, *35*, 1161–1163.

97. Houben, K., Havermans, R. C., & Wiers, R. W. (2010). Learning to dislike alcohol: conditioning negative implicit attitudes toward alcohol and its effect on drinking behavior. *Psychopharmacology*, *211*, 79–86.

98. Ostafin, B. D., Bauer, C., & Myxter, P. (2012). Mindfulness decouples the relation between automatic alcohol motivation and heavy drinking. *Journal of Social and Clinical Psychology*, *31*, 729–745.

99. Frijda, N. H. (2010). Impulsive action and motivation. *Biological Psychology*, *84*, 570–579.

100. Frijda, N. H. (Ed.). (1986). *The Emotions*. Cambridge, UK: Cambridge University Press.

101. Field, M., Mogg, K., & Bradley, B. P. (2005). Craving and cognitive biases for alcohol cues in social drinkers. *Alcohol and Alcoholism*, *40*, 504–510.

102. Field, M., Kiernan, A., Eastwood, B., & Child, R. (2008). Rapid approach responses to alcohol cues in heavy drinkers. *Journal of Behavior Therapy and Experimental Psychiatry*, *39*, 209–218.

103. Field, M., Eastwood, B., Bradley, B. P., & Mogg, K. (2006). Selective processing of cannabis cues in regular cannabis users. *Drug and Alcohol Dependence*, *85*, 75–82.

104. Wiers, R. W., Rinck, M., Dictus, M., & van den Wildenberg, E. (2009). Relatively strong automatic appetitive action-tendencies in male carriers of the OPRM1 G-allele. *Genes, Brain and Behavior*, *8*, 101–106.

105. Huijding, J., & de Jong, P. J. (2005). A pictorial version of the Extrinsic Affective Simon Task: sensitivity to generally affective and phobia-relevant stimuli in high and low spider fearful individuals. *Experimental Psychology*, *52*, 289–295.

106. Field, M., Caren, R., Fernie, G., & De Houwer, J. (2011). Alcohol approach tendencies in heavy drinkers: comparison of effects in a relevant stimulus–response compatibility task and an approach/avoidance Simon task. *Psychology of Addictive Behaviors*, *25*, 697–701.

107. Cousijn, J., Goudriaan, A. E., & Wiers, R. W. (2011). Reaching out towards cannabis: approach-bias in heavy cannabis users predicts changes in cannabis use. *Addiction*, *106*, 1667–1674.

108. Wiers, R. W., Rinck, M., Kordts, R., Houben, K., & Strack, F. (2010). Retraining automatic action-tendencies to approach alcohol in hazardous drinkers. *Addiction*, *105*, 279–287.

109. Wiers, R. W., Eberl, C., Rinck, M., Becker, E., & Lindenmeyer, J. (2011). Re-training automatic action tendencies changes alcoholic patients' approach bias for alcohol and improves treatment outcome. *Psychological Science, 22*, 490–497.

110. Fazio, R. H. (1990). Multiple processes by which attitudes guide behavior: the MODE model as an integrative framework. In M. P. Zanna (Ed.), *Advances in Experimental Social Psychology* (Vol. 23, pp. 75–109). New York: Academic Press.

111. Hofmann, W., Friese, M., & Wiers, R. W. (2008). Impulsive versus reflective influences on health behavior: a theoretical framework and empirical review. *Health Psychology Review, 2*, 111–137.

112. Wiers, R. W., Bartholow, B. D., van den Wildenberg, E., Thush, C., Engels, R. C., Sher, K. J., et al. (2007). Automatic and controlled processes and the development of addictive behaviors in adolescents: a review and a model. *Pharmacology, Biochemistry, and Behavior, 86*, 263–283.

113. Grenard, J. L., Ames, S. L., Wiers, R. W., Thush, C., Sussman, S., & Stacy, A. W. (2008). Working memory capacity moderates the predictive effects of drug-related associations on substance use. *Psychology of Addictive Behaviors, 22*, 426–432.

114. Thush, C., Wiers, R. W., Ames, S. L., Grenard, J. L., Sussman, S., & Stacy, A. W. (2008). Interactions between implicit and explicit cognition and working memory capacity in the prediction of alcohol use in at-risk adolescents. *Drug and Alcohol Dependence, 94*, 116–124.

115. Friese, M., Hofmann, W., & Wanke, M. (2008). When impulses take over: moderated predictive validity of explicit and implicit attitude measures in predicting food choice and consumption behaviour. *British Journal of Social Psychology, 47*(Pt 3), 397–419.

116. Friese, M., Bargas-Avila, J., Hofmann, W., & Wiers, R. W. (2010). Here's looking at you, bud: alcohol-related memory structures predict eye movements for social drinkers with low executive control. *Social Psychological and Personality Science, 1*, 143–151.

117. Hofmann, W., Friese, M., & Roefs, A. (2009). Three ways to resist temptation: the independent contributions of executive attention, inhibitory control, and affect regulation to the impulse control of eating behavior. *Journal of Experimental Social Psychology, 45*, 431–435.

118. Hofmann, W., Gschwendner, T., Friese, M., Wiers, R. W., & Schmitt, M. (2008). Working memory capacity and self-regulatory behavior: toward an individual differences perspective on behavior determination by automatic versus controlled processes. *Journal of Personality and Social Psychology, 95*, 962–977.

119. Wiers, R. W., Houben, K., Roefs, A., Hofmann, W., & Stacy, A. W. (2010). Implicit cognition in health psychology: why common sense goes out of the window. In B. Gawronski & B. K. Payne (Eds.), *Handbook of Implicit Social Cognition* (pp. 463–488). New York: Guilford.

120. Wiers, R. W., & Stacy, A. W. (2010). Are alcohol expectancies associations? Comment on Moss and Albery (2009). *Psychological Bulletin, 136*, 12–16.

121. Wiers, R. W., Ames, S. L., Hofmann, W., Krank, M., & Stacy, A. W. (2010). Impulsivity, impulsive and reflective processes and the development of alcohol use and misuse in adolescents and young adults. *Frontiers in Psychology, 1*, 1–12.

122. Littlefield, A. K., Verges, A., McCarthy, D. M., & Sher, K. J. (2011). Interactions between self-reported alcohol outcome expectancies and cognitive functioning in the prediction of alcohol use and associated problems: a further examination. *Psychology of Addictive Behaviors*, *25*, 542–546.

123. Christiansen, P., Cole, J. C., & Field, M. (2012). Ego depletion increases ad-lib alcohol consumption: investigating cognitive mediators and moderators. *Experimental and Clinical Psychopharmacology*, *20*, 118–128.

124. Ostafin, B. D., Marlatt, G. A., & Greenwald, A. G. (2008). Drinking without thinking: an implicit measure of alcohol motivation predicts failure to control alcohol use. *Behaviour Research and Therapy*, *46*, 1210–1219.

125. Hofmann, W., & Friese, M. (2008). Impulses got the better of me: alcohol moderates the influence of implicit attitudes toward food cues on eating behavior. *Journal of Abnormal Psychology*, *117*, 420–427.

126. Field, M., Wiers, R. W., Christiansen, P., Fillmore, M. T., & Verster, J. C. (2010). Acute alcohol effects on inhibitory control and implicit cognition: implications for loss of control over drinking. *Alcoholism, Clinical and Experimental Research*, *34*, 1346–1352.

127. Schoenmakers, T., Wiers, R. W., & Field, M. (2008). Effects of a low dose of alcohol on cognitive biases and craving in heavy drinkers. *Psychopharmacology*, *197*, 169–178.

128. Pieters, S., Burk, W., Van Der Vorst, H., Wiers, R. W., & Engels, R. C. M. E. (2012). The moderating role of working memory capacity and alcohol-specific rule-setting on the relation between approach tendencies and alcohol use in young adolescents. *Alcoholism, Clinical and Experimental Research*, *36*, 915–922. doi:10.1016/j.dcn.2011.07.008].

129. Klingberg, T., Fernell, E., Olesen, P. J., Johnson, M., Gustafsson, P., Dahlstrom, K., et al. (2005). Computerized training of working memory in children with ADHD—a randomized, controlled trial. *Journal of the American Academy of Child and Adolescent Psychiatry*, *44*, 177–186.

130. Klingberg, T. (2010). Training and plasticity of working memory. *Trends in Cognitive Sciences*, *14*, 317–324.

131. Houben, K., Wiers, R. W., & Jansen, A. (2011). Getting a grip on drinking behavior: training working memory to reduce alcohol abuse. *Psychological Science*, *44*, 968–975.

132. Veling, H., Holland, R. W., & van Knippenberg, A. (2008). When approach motivation and behavioral inhibition collide: behavior regulation through stimulus devaluation. *Journal of Experimental Social Psychology*, *44*, 1013–1019.

133. Houben, K., Nederkoorn, C., Wiers, R. W., & Jansen, A. (2011). Resisting temptation: decreasing alcohol-related affect and drinking behavior by training response inhibition. *Drug and Alcohol Dependence*, *116*, 132–136.

134. Kuo, P.-H., Kalsi, G., Prescott, C. A., Hodgkinson, C. A., Goldman, D., van den Oord, E. J., et al. (2008). Association of ADH and ALDH genes with alcohol dependence in the Irish Affected

Sib Pair Study of Alcohol Dependence (IASPSAD) sample. *Alcoholism, Clinical and Experimental Research, 32*, 785–795.

135. Eng, M. Y., Luczak, S. E., & Wall, T. L. (2007). ALDH2, ADH1B, and ADH1C genotypes in Asians: a literature review. *Alcohol Research & Health, 30*, 22–27.

136. Covault, J., Gelernter, J., Hesselbrock, V., Nellissery, M., & Kranzler, H. R. (2004). Allelic and haplotypic association of GABRA2 with alcohol dependence. *American Journal of Medical Genetics. Part B, Neuropsychiatric Genetics, 129B*, 104–109.

137. Roh, S., Matsushita, S., Hara, S., Maesato, H., Matsui, T., Suzuki, G., et al. (2010). Role of GABRA2 in moderating subjective responses to alcohol. *Alcoholism, Clinical and Experimental Research, 35*, 400–407.

138. Mueller, A., Strahler, J., Armbruster, D., Lesch, K.-P., Brocke, B., & Kirschbaum, C. (2012). Genetic contributions to acute autonomic stress responsiveness in children. *International Journal of Psychophysiology, 83*, 302–308.

139. Alexander, N., Osinsky, R., Schmitz, A., Mueller, E., Kuepper, Y., & Hennig, J. (2010). The BDNF Val66Met polymorphism affects HPA-axis reactivity to acute stress. *Psychoneuroendocrinology, 35*, 949–953.

140. Aguilera, M., Arias, B., Wichers, M., Barrantes-Vidal, N., Moya, J., Villa, H., et al. (2009). Early adversity and 5-HTT/BDNF genes: new evidence of gene–environment interactions on depressive symptoms in a general population. *Psychological Medicine, 39*, 1425–1432.

141. Wichers, M., Kenis, G., Jacobs, N., Mengelers, R., Derom, C., Vlietinck, R, et al. (2008). The BDNF Val(66)Met × 5-HTTLPR × child adversity interaction and depressive symptoms: an attempt at replication. *American Journal of Medical Genetics. Part B, Neuropsychiatric Genetics, 147B*, 120–123.

142. Watanabe, M. A. E., Nunes, S. O. V., Nunes, S. O. V., Amarante, M. K., Guembarovski, R. L., Oda, J. M. M., et al. (2011). Genetic polymorphism of serotonin transporter 5-HTTLPR: involvement in smoking behaviour. *Journal of Genetics, 90*, 179–185.

143. Covault, J., Tennen, H., Armeli, S., Conner, T. S., Herman, A. I., Cillessen, A. H. N., et al. (2007). Interactive effects of the serotonin transporter 5-HTTLPR polymorphism and stressful life events on college student drinking and drug use. *Biological Psychiatry, 61*, 609–616.

144. Iordanidou, M., Tavridou, A., Petridis, I., Kyroglou, S., Kaklamanis, L., Christakidis, D., et al. (2010). Association of polymorphisms of the serotonergic system with smoking initiation in Caucasians. *Drug and Alcohol Dependence, 108*, 70–76.

145. Chen, Z.-Y., Jing, D., Bath, K. G., Ieraci, A., Khan, T., Siao, C.-J., et al. (2006). Genetic variant BDNF (Val66Met) polymorphism alters anxiety-related behavior. *Science, 314*, 140–143.

146. Frielingsdorf, H., Bath, K. G., Soliman, F., Difede, J., Casey, B. J., & Lee, F. S. (2010). Variant brain-derived neurotrophic factor Val66Met endophenotypes: implications for posttraumatic stress disorder. *Annals of the New York Academy of Sciences, 1208*, 150–157.

147. Montag, C., Weber, B., Fliessbach, K., Elger, C., & Reuter, M. (2009). The BDNF Val66Met polymorphism impacts parahippocampal and amygdala volume in healthy humans: incremental support for a genetic risk factor for depression. *Psychological Medicine, 39,* 1831–1839.

148. Corominas, M., Roncero, C., Ribases, M., Castells, X., & Casas, M. (2007). Brain-derived neurotrophic factor and its intracellular signaling pathways in cocaine addiction. *Neuropsychobiology, 55,* 2–13.

149. Mendelson, J., Baggott, M. J., Flower, K., & Galloway, G. (2011). Developing biomarkers for methamphetamine addiction. *Current Neuropharmacology, 9,* 100–103.

150. Heberlein, A., Muschler, M., Wilhelm, J., Frieling, H., Lenz, B., Gröschl, M., et al. (2010). BDNF and GDNF serum levels in alcohol-dependent patients during withdrawal. *Progress in Neuro-Psychopharmacology & Biological Psychiatry, 34,* 1060–1064.

151. Matsushita, S., Kimura, M., Miyakawa, T., Yoshino, A., Murayama, M., Masaki, T., et al. (2004). Association study of brain-derived neurotrophic factor gene polymorphism and alcoholism. *Alcoholism, Clinical and Experimental Research, 28,* 1609–1612.

152. Grimm, J. W., Lu, L., Hayashi, T., Hope, B. T., Su, T.-P., & Shaham, Y. (2003). Time-dependent increases in brain-derived neurotrophic factor protein levels within the mesolimbic dopamine system after withdrawal from cocaine: implications for incubation of cocaine craving. *Journal of Neuroscience, 23,* 742–747.

153. Lu, L., Wang, X., Wu, P., Xu, C., Zhao, M., Morales, M., et al. (2009). Role of ventral tegmental area glial cell line-derived neurotrophic factor in incubation of cocaine craving. *Biological Psychiatry, 66,* 137–145.

154. Noble, E. P. (2000). Addiction and its reward process through polymorphisms of the D2 dopamine receptor gene: a review. *European Psychiatry, 15,* 79–89.

155. Schinka, J. A., Letsch, E. A., & Crawford, F. C. (2002). DRD4 and novelty seeking: results of meta-analyses. *American Journal of Medical Genetics, 114,* 643–648.

156. Munafò, M. R., Yalcin, B., Willis-Owen, S. A., & Flint, J. (2008). Association of the dopamine D4 receptor (DRD4) gene and approach-related personality traits: meta-analysis and new data. *Biological Psychiatry, 63,* 197–206.

157. Noble, E. P. (1998). The D2 dopamine receptor gene: a review of association studies in alcoholism and phenotypes. *Alcohol (Fayetteville, N.Y.), 16,* 33–45.

158. van der Zwaluw, C. S., Engels, R. C., Vermulst, A. A., Franke, B., Buitelaar, J., Verkes, R. J., et al. (2010). Interaction between dopamine D2 receptor genotype and parental rule-setting in adolescent alcohol use: evidence for a gene–parenting interaction. *Molecular Psychiatry, 15,* 727–735.

159. van der Zwaluw, C. S., Kuntsche, E., & Engels, R. C. (2011). Risky alcohol use in adolescence: the role of genetics (DRD2, SLC6A4) and coping motives. *Alcoholism, Clinical and Experimental Research, 35,* 756–764.

160. Laucht, M., Becker, K., Blomeyer, D., & Schmidt, M. H. (2007). Novelty seeking involved in mediating the association between the dopamine D4 receptor gene exon III polymorphism and heavy drinking in male adolescents: results from a high-risk community sample. *Biological Psychiatry, 61,* 87–92.

161. Laucht, M., Becker, K., El-Faddagh, M., Hohm, E., & Schmidt, M. H. (2005). Association of the DRD4 exon III polymorphism with smoking in fifteen-year-olds: a mediating role for novelty seeking? *Journal of the American Academy of Child and Adolescent Psychiatry, 44,* 477–484.

162. Larsen, H., van der Zwaluw, C. S., Overbeek, G., Granic, I., Franke, B., & Engels, R. C. (2010). A variable-number-of-tandem-repeats polymorphism in the dopamine D4 receptor gene affects social adaptation of alcohol use: investigation of a gene–environment interaction. *Psychological Science, 21,* 1064–1068.

163. Belsky, J., Jonassaint, C., Pluess, M., Stanton, M., Brummett, B., & Williams, R. (2009). Vulnerability genes or plasticity genes? *Molecular Psychiatry, 14,* 746–754.

164. Ray, L. A., & Hutchison, K. E. (2004). A polymorphism of the mu-opioid receptor gene (OPRM1) and sensitivity to the effects of alcohol in humans. *Alcoholism, Clinical and Experimental Research, 28,* 1789–1795.

165. van den Wildenberg, E., Wiers, R. W., Dessers, J., Janssen, R. G., Lambrichs, E. H., Smeets, H. J., et al. (2007). A functional polymorphism of the mu-opioid receptor gene (OPRM1) influences cue-induced craving for alcohol in male heavy drinkers. *Alcoholism, Clinical and Experimental Research, 31,* 1–10.

166. Filbey, F. M., Ray, L., Smolen, A., Claus, E. D., Audette, A., & Hutchison, K. E. (2008). Differential neural response to alcohol priming and alcohol taste cues is associated with DRD4 VNTR and OPRM1 genotypes. *Alcoholism, Clinical and Experimental Research, 32,* 1113–1123.

167. Ray, L. A. (2011). Stress-induced and cue-induced craving for alcohol in heavy drinkers: preliminary evidence of genetic moderation by the OPRM1 and CRH-BP genes. *Alcoholism, Clinical and Experimental Research, 35,* 166–174.

168. Friedman, N. P., Miyake, A., Young, S. E., Defries, J. C., Corley, R. P., & Hewitt, J. K. (2008). Individual differences in executive functions are almost entirely genetic in origin. *Journal of Experimental Psychology. General, 137,* 201–225.

169. Karlsgodt, K. H., Bachman, P., Winkler, A. M., Bearden, C. E., & Glahn, D. C. (2011). Genetic influence on the working memory circuitry: behavior, structure, function and extensions to illness. *Behavioural Brain Research, 225,* 610–622.

170. Goldberg, T. E., Egan, M. F., Gscheidle, T., Coppola, R., Weickert, T., Kolachana, B. S., et al. (2003). Executive subprocesses in working memory: relationship to catechol-O-methyltransferase Val158Met genotype and schizophrenia. *Archives of General Psychiatry, 60,* 889–896.

171. El-Hage, W., Phillips, M. L., Radua, J., Gohier, B., Zelaya, F. O., Collier, D. A., & Surguladze, S. A. (2013). Genetic modulation of neural response during working memory in healthy individuals:

interaction of glucocorticoid receptor and dopaminergic genes. *Molecular Psychiatry*, *18*, 174–182.

172. Blanchard, M. M., Chamberlain, S. R., Roiser, J., Robbins, T. W., & Müller, U. (2011). Effects of two dopamine-modulating genes (DAT1 9/10 and COMT Val/Met) on n-back working memory performance in healthy volunteers. *Psychological Medicine*, *41*, 611–618.

173. Ursini, G., Bollati, V., Fazio L., Porcelli A., Iacovelli, L., Catalani, A., et al. (2011). Stress-related methylation of the catechol-O-methyltransferase Val 158 allele predicts human prefrontal cognition and activity. *Journal of Neuroscience*, *31*, 6692–6698.

174. Wong, C. C., Mill, J., & Fernandes, C. (2011). Drugs and addiction: an introduction to epigenetics. *Addiction*, *106*, 480–489.

175. Pieters, S., Van Der Vorst, H., Burk, W. J., Schoenmakers, T., Van Den Wildenberg, E., Smeets, H. J., et al. (2011). The effect of the OPRM1 and DRD4 polymorphisms on the relation between attentional bias and alcohol use in adolescence and young adulthood. *Developmental Cognitive Neuroscience.*, *1*, 591–599. doi:10.1016/j.dcn.2011.07.008].

176. Antypa, N., Cerit, H., Kruijt, A. W., Verhoeven, F. E. A., & Van der Does, A. J. W. (2011). Relationships among 5-HTT genotype, life events and gender in the recognition of facial emotions. *Neuroscience*, *172*, 303–313.

177. Beevers, C. G., Wells, T. T., Ellis, A. J., & McGeary, J. E. (2009). Association of the serotonin transporter gene promoter region (5-HTTLPR) polymorphism with biased attention for emotional stimuli. *Journal of Abnormal Psychology*, *118*, 670–681.

178. Kwang, T., Wells, T. T., McGeary, J. E., Swann, W. B., & Beevers, C. G. (2010). Association of the serotonin transporter promoter region polymorphism with biased attention for negative word stimuli. *Depression and Anxiety*, *27*, 746–751.

179. Naudts, K. H., Azevedo, R. T., David, A. S., van Heeringen, K., & Gibbs, A. A. (2012). Epistasis between 5-HTTLPR and ADRA2B polymorphisms influences attentional bias for emotional information in healthy volunteers. *International Journal of Neuropsychopharmacology*, *15*, 1027–1036.

180. Pergamin-Hight, L., Bakermans-Kranenburg, M. J., van Ijzendoorn, M. H., & Bar-Haim, Y. (2012). Variations in the promoter region of the serotonin transporter gene and biased attention for emotional information: a meta-analysis. *Biological Psychiatry*, *71*, 373–379.

181. Hendershot, C. S., Lindgren, K. P., Liang, T., & Hutchison, K. E. (2012). COMT and ALDH2 polymorphisms moderate associations of implicit drinking motives with alcohol use. *Addiction Biology*, *17*, 192–201.

182. Gladwin, T. E., & Wiers, R. W. (2012). Alcohol-related effects on automaticity due to experimentally manipulated conditioning. *Alcoholism, Clinical and Experimental Research*, *36*, 895–899.

183. Vorel, S. R., Liu, X., Hayes, R. J., Spector, J. A., & Gardner, E. L. (2001). Relapse to cocaine-seeking after hippocampal theta burst stimulation. *Science*, *292*, 1175–1178.

184. Sun, W., & Rebec, G. V. (2003). Lidocaine inactivation of ventral subiculum attenuates cocaine-seeking behavior in rats. *Journal of Neuroscience, 23*, 10258–10264.

185. Biehl, S. C., Dresler, T., Reif, A., Scheuerpflug, P., Deckert, J., & Herrmann, M. J. (2011). Dopamine transporter (DAT1) and dopamine receptor D4 (DRD4) genotypes differentially impact on electrophysiological correlates of error processing. *PLoS ONE, 6*, e28396.

186. Mueller, E. M., Makeig, S., Stemmler, G., Hennig, J., & Wacker, J. (2011). Dopamine effects on human error processing depend on catechol-o-methyltransferase VAL158MET genotype. *Journal of Neuroscience, 31*, 15818–15825.

187. Schmukle, S. C. (2005). Unreliability of the dot probe task. *European Journal of Personality, 19*, 595–605.

188. Salemink, E., & Wiers, R. W. (2012). Adolescent threat-related interpretive bias and its modification: the moderating role of regulatory control. *Behaviour Research and Therapy, 50*, 40–46.

189. Fox, E., Zougkou, K., Ridgewell, A., & Garner, K. (2011). The serotonin transporter gene alters sensitivity to attention bias modification: evidence for a plasticity gene. *Biological Psychiatry, 70*, 1049–1054.

11   The Role of Genetics in Addiction and the Expectancy Principle

Mark S. Goldman and Richard R. Reich

Studies that connect specific gene variants with alcohol consumption have identified many candidate genes that may contribute to alcoholism. In rodent models in particular, 86 candidate genes have been identified [1, 2]. Some alleles of these genes are risk related, others are protective, and some regulate the expression of other genes in risky or protective ways. Although genetically informative studies in animals clearly have made headway in identifying biomarkers of alcohol-related behaviors, these findings are inherently limited in two key ways when extended to humans. First, when behavior genetic techniques are applied to humans, the association between gene candidates and excessive drinking is not evident until early adulthood. Its presence, therefore, only becomes apparent simultaneously with, or even after, the peak in developmental risk for alcohol problems [3, 4]. As a consequence, the nature of the influence of these genes on the behavioral expression of risky drinking remains masked during their time of presumably greatest activity. Second, it has increasingly become apparent that variance in alcohol consumption/risky drinking is best attributed to the combined influence of many genes, with the portion specifically attributable to any single gene vanishingly small, perhaps as little as 0.1% [5]. Such small effect sizes are notoriously difficult to replicate in systematic follow-up studies, leaving the specific pathway of genetic influence uncertain [6, 7]. Because no single gene has emerged that reliably accounts for large amounts of drinking variance, it has become evident that complex combinations of multiple gene variants must contribute to phenotypic alcohol consumption.

To further appreciate the scientific task researchers face, we must then consider that this genetic complexity is in operation even before the influence of environmental exposures is taken into account. The interplay within environments can be just as intricate as among genes (see [7] for discussion of the "evironome"), and genetic and environmental influences rarely act in isolation, but in complex interplay with each other. Finally, we must consider that genetic exposures, environmental exposures, and their interplay occur in the context of development; that is, the dynamics of gene × environment interactions may result in one outcome at one point in development, with a different outcome earlier or later [8, 9]. Sequencing of gene expression over development also matters; some genes do, or do not, express unless others have (or have not) previously expressed in the appropriate developmental sequence, at the appropriate point in maturation. As a result, patterns among dozens of genes, each with multiple possible alleles, interacting with the environment, and conferring different influence at different developmental time points, may lead to an exceedingly large number of possible etiological combinations. To identify key patterns among these possibilities in a way that would provide definitive insights into problematic drinking would require novel analytic methods and substantially greater statistical power [5, 9].

Because understanding of how phenotypes come to be depends on these as yet difficult-to-surmount issues of complexity, researchers have tried to simplify their focus by examining observable characteristics less far downstream than the eventual phenotypes, characteristics that have been called endophenotypes and intermediate phenotypes. As measureable characteristics of an individual that have been shown to be related to genetic profiles, and that are as predictive of drinking patterns as any variables that have been researched, alcohol expectancies certainly could be cataloged in this manner. Measured explicitly or implicitly, alcohol expectancies reflect the anticipated effects of drinking along with their associated reward/aversion value. Because what happens to a person when he or she drinks, and its associated payoff (positive or negative), is influenced by the person's genetic makeup as it has interplayed with environment up to the measurement point, the expectancy measurement profile of an individual could be understood as a developmentally appropriate, partial product of the gene–environment interaction that then affects alcohol consumption/ risky drinking. We shall review material in this chapter that relates to this understanding of alcohol expectancies.

However, because the general literature on expectancy has become so extensive in recent years, and has encompassed so many areas of biopsychological functioning, alcohol expectancy is best appreciated when placed in a larger context [8]. Framing this larger context recently in *The New York Times*, Andy Clark [10] said,

> ... the brain is complex, and builds new solutions upon old, perhaps resulting in a rather messy and hard-to-unravel edifice. Nonetheless, there can be fundamental processing ploys at work even in superficially much "messier" systems and the claim is indeed that the use of multilayer predictive routines is one such fundamental ploy.

Clark is not alone in identifying "multilayer predictive routines" as central to neurobehavioral functioning. Many decades after postulation of expectancies by Tolman [11], the noted philosopher Daniel Dennett [12] reiterated this idea when he characterized the brain as an "anticipatory machine" [12]. Even more recently [13], the "generation of predictions" was identified as a viable universal principle for organizing the multidisciplinary study of the brain and behavior. In the domain of this book, Redish, Jensen, and Johnson [14] identified a single common thread unifying the processes underlying addiction: "... decisions are based on *prediction* of value or expected utility ..." (p. 417). The central purpose of this chapter is to provide support for anticipatory influence at each step in the multilayer process from genes to the decision to drink alcohol. Although we emphasize work from our laboratory that has emphasized one kind of "window" on these operations, it needs to be recognized that what brings all the lines of research on expectancy/anticipatory processing together is the recognition that expectancy is a *principle* that characterizes the operation of many specific neurobehavioral mechanisms that have functional commonalities.

None of the words used to characterize this principle are preeminent, so the terms expectancy, anticipation, and prediction all may be used. Because, as we shall see, the expectancy principle has been at the heart of major theoretical efforts to explain the mechanism of motivation, the term motivation could also be included here (the term motivation is not, by itself, an explanation but has always called for further theory; e.g., see [15, 16]). In these characterizations, expectancy represents the linkage between the opportunity for reinforcement (context), the learned behavioral template for obtaining that reinforcement, and an anticipated outcome.

In addition, different expectancy mechanisms operate at many levels of biopsycho-social functioning. Each of these mechanisms, operating at different levels, therefore may be considered an endophenotype in its own right. As one example, we shall see later that the dopaminergic system has been characterized as the substrate for expectancy-based decisions [17]. It would not be much of a stretch to suggest that individual differences in dopaminergic functioning (an endophenotype) influence explicitly or implicitly measured verbal expectancies; anticipation of outcomes from substance use expressed as words likely reflects internal experiences of reward. Because each of these manifestations of anticipatory processes operates at a different level, with determinants that are both common and unique, each may be considered a separate endophenotype, albeit separate endophenotypes with some degree of coordinated function. Verbally expressed expectancies, whether measured explicitly or implicitly, may be labeled an intermediate phenotype because they represent the aggregated influence of multiple expectancy processes. And the influence of these aggregated processes is not trivial; verbal reports have been shown to account for large portions of the variance in drinking behavior, as much as 50% in error attenuated models [18].

11.1 Anticipation: The Central Principle

If organisms were only to react to external circumstances after the fact, the inevitable response delay would place them at an evolutionary disadvantage. Because time always moves forward, no context is static, and organisms must be prepared in advance to handle not-yet-encountered situations. As a result, the brain and nervous system have evolved to use information about past experiences to prepare for upcoming circumstances. In psychology, we refer to this capacity as memory but then make the mistake of viewing memory as a means of looking backward in time. Actually, memory is better understood as a means of looking forward, that is, applying information garnered from prior circumstances to continually adjust to the "stream" of environments/circumstances that we move into as our lives progress. And, as we know from memory research, this information is not a precise recording of experience. As Calvin and Bickerton [19] point out, "We ... make generalizations on inadequate evidence at very short notice, because that works better, in terms of evolutionary fitness, than making 100% correct generalization after a long period of cogitation" (p. 34). This is the essence of the expectancy principle.

In general, anticipatory decision making favors rewards and avoids adversity. It is critical to recognize, however, that rewards are not limited to the satisfaction of basic biological needs but may include any outcomes that enhance evolutionary fitness [20]. For example, jockeying for status in the social hierarchy is inherent to the motivational systems of social animals, because social status makes reproductive success more likely. The critical significance of social positioning has been underscored by the discovery of biological mechanisms of vicarious social learning such as mirror cells and brain regions that support mentalizing (taking the perspective of others). These discoveries provide mounting evidence that social information achieves privileged status in the brain [21]. That social information may be a substantial component in the computation of alcohol's reward value will be discussed in more detail below.

11.2 Domains of Expectancy Operation

11.2.1 Genetics

There is little question that genes express in anticipation of circumstances; consider the expression of genes that stimulate various phases of maturation. Puberty, for example, prepares the organism for procreation, an evolutionarily critical event. The information encoded in genes, however, is transmitted from generation to generation. Although beyond the scope of the current chapter, it is worth noting here that more recently described processes of epigenesis overlap considerably those of the expectancy principle noted above. Originally attached to the processes that convert pluripotent cells into specific cell lines (for specific organs, for example; [22]), epigenesis has come to have much wider meaning. Gottesman and Hanson [23] relate epigenetics to "...the complexities of how multiple genetic factors and multiple environmental factors become integrated over time through dynamic, often nonlinear, sometimes nonreversible, processes to produce behaviorally relevant endophenotypes and phenotypes" (p. 267). If we conceive of these endophenotypes and intermediate phenotypes as preparing the organism for likely-to-be-encountered situations, the concept overlaps considerably with expectancy.

In the domain of this chapter, genetic expression is the first step in biasing brain systems toward stimuli representing reward. Some mechanisms for this biasing include genetically based individual differences in sensitivity, reactivity, and metabolism (when substances are involved) (for review, see [24]). As an example in rodents, Katner et al. [25] compared genetically bred alcohol preferring (P) rats with Wistar rats, concluding that, "... the mere expectation of ethanol availability enhances the efflux of DA (dopamine) in the Nac (nucleus accumbens) of the P, but not the Wistar rat, which may play a role in the initiation or maintenance of ethanol seeking behavior in the P line" (p. 669). As an illustration of similar processes in humans, McCarthy et al. [26] assessed self-reported alcohol expectancies in individuals who differed genetically in the alcohol dehydrogenase allele (ALDH2; those with this allele metabolize alcohol less effectively and may find use more aversive). They reported that one mechanism by which the allelic variation may influence use is by lowering positive expectancies and reducing the expectancy–drinking relationship. A follow-up study [27] provided preliminary evidence that it is indeed the level of response to alcohol that mediates the relationship between ALDH2 status and expectancies. Among men, those with the ALDH2 allele had more adverse reaction to alcohol and increased expectancies for cognitive impairment. Statistical modeling showed that the ALDH2/ expectancy relationship was fully explained (mediated) by the level of response to alcohol. In females, having the ALDH2 allele decreased the likelihood that women would view alcohol as tension reducing, but a mediational relationship was not found. Undoubtedly, more gene expression patterns will be linked to verbal expectancies as research progresses.

11.2.2 Brain Systems

Because brain systems reflect gene expression patterns, it is perhaps evident that the expectancy principle would operate in this domain as well. In fact, expectancies have been widely implicated in the neurobiology of animal and human reward and reinforcement [17, 28–31]. Specifically, previous research has revealed brain circuitry that is differentially active in anticipation of rather than in response to rewards [29, 32–34]. In a very recent advance, Alexander and Brown [35] computationally modeled the multiple decision-making functions of the dorsal anterior cortex and surrounding medial prefrontal cortex as based on "forming expectations about actions and detecting surprising outcomes" (p. 1338).

Organisms have neural systems that appear designed to signal the availability of reward; that is, systems that use contextual signals to anticipate that reward is imminent [36]. As an example of this process, it has been noted that one of these pathways, embodied by dopaminergic neurons running from the striatum to the nucleus accumbens and on to the frontal lobes, is thought to convert "… an event or stimulus from a neutral 'cold' representation (mere information) into an attractive and 'wanted' incentive that can 'grab' attention" (p. 313, i.e., incentive salience [36]). Recasting the common understanding of dopaminergic pathways as the substrate for reward itself, Montague and Berns [37] described this system's function as reward prediction, contending specifically that the level of dopamine pathway activity signals, in advance of consummatory behaviors, that rewards available in a particular context differ from rewards anticipated. Such a mechanism would serve to adjust ongoing behavior toward the most beneficial expected outcomes. Further, Kupfermann et al. [17] note that "… dopaminergic neurons encode *expectations* (italics added) about external rewards" (p. 1010).

Apart from signaling reward availability, another means of identifying reward stimuli centers on the amygdaloid complex, an important area in the expression of emotion [38, 39]. Obviously, a neural system supporting the experience of emotion (pleasure or aversion) would be instrumental in encouraging certain behaviors and discouraging others. This emotion makes certain sensed patterns of information more salient and, therefore, more likely to be attended to, stored, and acted upon [30, 38, 40]. The linkage between neural processing of information and motivation/emotion is suggested by connections observed between areas such as the nucleus accumbens and amygdala, and information-processing areas such as the hippocampus and frontal cortex [40, 41]. Corticosteroids released by the hypothalamic–pituitary–adrenal (HPA) axis in response to threatening circumstances also influence hippocampal memory storage [42]. Researchers in this domain routinely apply expectancy in describing the function of the information–emotion link [38, 40].

To fully validate the expectancy model, brain reward systems must eventually be tied to behavior. Work by Matsumoto and Tanaka [43] indicates that the prefrontal cortex (anterior cingulate cortex) links signals of biologically important inputs *with actions* "… based on goal expectation and memory of action–outcome contingency" (p. 178). Given the conceptualization presented in these pages, memory of action–outcome contingency would be, of course, the essence of the expectancy. It is this "memory" that is reenacted in anticipation that the action will again lead to reward.

11.2.3 Complex Psychological Systems

Described at a different level (scale), these brain systems manifest themselves in terms we commonly refer to as psychological processes—for example, perception, learning, memory, and language (among others).

Perception

The human sensory system is constantly receiving a vast amount of sensory input. Not only is detailed information received by multiple senses in parallel, but incoming information is changing from moment to moment. It has been estimated that potentially billions of bits of information impinge on the human senses every second whereas, after processing, no more than 50 bits of information per second are available to consciousness [44]. Even allowing for substantial additional capacity to process information outside of consciousness, organisms must use selection processes to eliminate insignificant information while remaining responsive to information that matters for survival and adaptation.

One of the central purposes of a memory system may be to filter incoming stimuli. The choice of information to which the organism is biased (selectively responsive) is governed by the nature of our sensory systems as honed over evolutionary time and by the encoding of past sensory–perceptual configurations collected throughout the life experience of that organism. As one example, the human visual system has receptive elements configured to be sensitive to particular stimulus arrays (e.g., lines in various orientations, patterns, movement; [45]). However, because complex adaptations require plasticity, a good deal of perceptual selectivity in humans is based on past experience. Many receptive elements can be altered by experience [46]. Even basic elements of the visual system, once thought to be hardwired, have been found to be adaptable (plastic; [47, 48]). In addition, filtering mechanisms do not operate passively; they actively identify, process, and compare information to create whole perceptions from available information. Grossberg [46] offered an interesting example: suppose you hear a noise followed immediately by the words "eel is on the..." If that string of words is followed by the word "orange," you hear "peel is on the orange." If the word "wagon" completes the sentence, you hear "wheel is on the wagon." If the final word is "shoe," you hear "heel is on the shoe." In this example, Grossberg suggests that expectations activated by the final word in the sequence provide a means to fill in incomplete data and influence perception of the first word. It has been suggested [49] that because "the configuration of environmental stimuli and their behavioral context in daily life are unique and rarely repeated exactly" (p. 436), the brain must be able to recover a complete memory from stimulus conditions that supply only partial cues. Anticipatory processes occur not just with perception of visual or auditory stimuli, but with perception of motion [50], time perception [51], orienting behavior in early infancy [52], and the perception of music [53].

Learning/Memory

Associationist principles (e.g., [54]) provide an effective starting point for understanding how initially disconnected information may be "bonded" into a larger whole in the appropriate context. Such bonding ensures that relevant information can be reassembled efficiently and effectively repeated as needed. Because associations, long understood as central to memory encoding and retrieval, may serve as the "building blocks" of prediction, association-based memory models should not be seen as distinct from expectancy models, but instead as their foundation (both for explicit and implicit processing) [55]. Anticipatory processes have been use to describe operant [56] and classical conditioning [57, 58], comparative judgment [59, 60], models of memory [61], and the learning of complex motor skills [62].

Emotion/Motivation

A full explanation of the control of behavior in the real world, however, requires additional theoretical elements that encourage the translation of pure associational information into overt behavior. That is, associational principles provide the foundation for memory but do not determine what will be selected for remembering. As evolution progressed, selection pressures (e.g., adaptation, survival) influenced development of mechanisms that serve as signals for monitoring evolutionary success (fitness). We refer to these mechanisms when we discuss emotion, motivation, reward, incentive, reinforcement (positive and negative), and punishment. To govern real-world behavior, information that is accessed by associational search at some point must be associated with information about reward and punishment [30]. It is this final, critical information that specifies what searches will be undertaken and what behaviors will be carried out to consummate reward via approach or avoidance. Such information can also be used to create novel mental scenarios about not-yet-experienced situations from combinations of previous experiences to guide decisions [55]. The essence of the expectancy model is that context activates memories of objects/events that lead to anticipation/prediction of payoffs. These associations also are integrated with memories of neural and somatic reactions that give them emotional and motivational value [63, 64], along with the behaviors that are likely to lead to those rewards. In this way, expectancy serves as the theoretical amalgam of cognition (association) and motivation/emotion. Linking motivation to cognition is central to this process; as noted by Holland and Gallagher (38),

conventional associative learning paradigms have been adapted to allow systematic study of expectancy and action in a range of species, including humans … "expectancy" refers to the associative activation of such reinforcer representations by the events that predict them, before the delivery of the reinforcer itself (p. 148).

Other Complex Psychological Processes

Expectancy also has been related to a wide variety of psychological processes including the development of language [65], expert sport performance (e.g., [66]), empathy [67], and a broad sampling of the clinical domain (see [68]). In that area, it has been related to mood dysfunction, fear, pain reduction, sexual dysfunction, asthma, drug abuse, alcohol abuse and alcoholism, smoking, psychotherapy, hypnosis, and therapeutic effects of psychotropic [69] and analgesic medications [70]. Expectancy concepts have dominated explanations of placebo [71–74] and nocebo effects (psychologically induced illness or even death); these theories emphasize that expectancy effects are responsible for fundamentally similar, if not identical, brain responses to those induced by active medications.

11.3 Alcohol Expectancy as an Intermediate Phenotype: A Probe of Decision Processing Immediately Before Drinking

It is, of course, a complex cascade of the multiple influences described above that influences the decision to drink; as with all decisions, neural processing pathways use various kinds of information collected in the past, and interwoven with current contextual circumstances, to set up a competition between "go" signals, "stop" signals, and "do something else" signals. Three decades of alcohol expectancy research have suggested that it is possible to probe at least some aspects of this cascade with conventional psychological measures. That is, alcohol expectancies, to some extent, can be seen as the coming together of the above expectancy processes as they bear upon the control of drinking decisions. While it is possible that not all of these decisional influences are effectively probed by the measures that have been developed for assessing alcohol expectancies, many studies suggest that they do represent significant levels of systematic influence. Literally hundreds (possibly thousands) of studies have been published that show that explicit responses to expectancy questionnaires relate to drinking concurrently and prospectively, even over lag periods of many years. That expectancies can be measured in children before drinking ever begins is most critical to the expectancy logic model because these findings show that explicitly measured expectancies are not just a by-product (an epiphenomenon) of already accumulated drinking patterns. We shall now review some of this literature.

11.3.1 Self-Report Measures
Conventional paper-and-pencil self-report measures have yielded a great deal of information on drinking behavior and are described in recent reviews [8, 75, 76]. As noted above, alcohol expectancies correlate with drinking, predicting up to 50% of drinking variance in error attenuated models. Both cross-sectional and longitudinal developmental designs have shown alcohol expectancies to predict the onset of drinking in children. When placed in models with other known antecedents of drinking (e.g., family, peer, or cultural factors), alcohol expectancies statistically mediate a portion of the relationship between these antecedents and drinking. Finally, when activated or diminished in controlled experiments, alcohol expectancies have caused increases and decreases in drinking, respectively (see [75] for a more detailed summary of this research).

11.3.2 Beyond Psychometric Models

To more comprehensively study expectancy theory and drinking behavior, our research group has embarked on multiple lines of research to advance alcohol expectancy research beyond static psychometric (predictive) models into theoretically driven models of how expectancies might actually operate. These advances were undertaken in two stages. First, the foundation for operational models of alcohol expectancies was provided by construction of memory network models of the kind used in cognitive science. From the perspective of expectancy theory, activation of associative memory networks provides an internal context ("mind-set"), much of which is not accessible to awareness, that guides prospective decision making [55]. This theoretical framework is now shared across many domains of cognitive/affective science (see (13)).

In one kind of modeling using multidimensional scaling and clustering, expectancy memory networks have been well-described with two orthogonal dimensions: (1) valence (positive/negative) and (2) activation (arousal/sedation). These models have shown that people tend to hold alcohol expectancies across the full range of these dimensions. That is, for any given individual, alcohol's anticipated effects may simultaneously include such outcomes as, for example, sleepiness, euphoria, relaxation, and vomiting. In other words, the same person might hold what may appear to be competing expectations, even to the point that the same drinking episode could lead to both happiness and sickness. That individuals hold all these conflicting expectancies simultaneously turns out to be less important, however, than the relative emphasis (weight) they place on each expectancy and dimension [77–79]. These and subsequent studies have repeatedly shown that heavier drinkers place a greater emphasis on positive and arousing alcohol expectancies than do lighter drinkers.

More recent evolutions of network (graph) theory propose that links (connections) between nodes (concepts) in a network are not evenly distributed. Whether the network be comprised of connections between people, Internet sites, airports, or authors, specific key nodes have a disproportionately large number of links to other nodes. These key nodes, called hubs, allow for the average number of links necessary to move from one node in the network to another to be exponentially small (relative to the size of the network). The result of network hubs is the "small world" phenomenon which allows people to travel by air to any city on the globe with fewer than five stops and accounts for the six degrees of separation demonstrated by the psychologist Stanley Milgram [80] decades ago.

Recently, we began to apply these newer models by studying one potential hub in the alcohol expectancy memory network: "drunk" [81]. "Drunk" is an important expectancy because it is widely held [79], the affective meaning can vary and therefore carries links to many other expectancies [78], and "drunk" refers to effects associated specifically with ingesting high doses of alcohol [82]. In this study [81], 647 18- to 19-year-olds completed three different measures of memory network associations: a free associates task, a paired similarities task, and the Alcohol Expectancy Multiaxial Assessment [83]. Consistently, in all of these measures, results demonstrated that individuals who linked positive expectancies with "drunk" consumed more than those who linked more negative expectancies. This outcome is important because it demonstrates that the actual behavior of drinking is related to memory network relationships with this potential "hub."

These models of alcohol expectancy network structure provided the framework for the second line of research which experimentally investigated expectancy operation; that is, that alcohol expectancy memory networks depend on context to activate anticipatory cognitions/affect that, in turn, influence behavioral outputs (both consumption and alcohol-related behaviors once consumption takes place). Because alcohol expectancy memory networks arise in intimate connection with contexts, they may, for example, differ at a football game compared to dinner at a nice restaurant, and consequent behavior should (and does) differ accordingly. A series of studies were conducted in our laboratory using tasks developed by cognitive psychologists to test memory activation following implicit primes. The term implicit simply means that participants were unaware of (not conscious of) the prime that influenced their performance on a subsequent task. Because initiation and pursuit of goals may be largely nonconscious [84], approaches that tap these nonconscious processes are quite applicable to the investigation of alcohol-related goals. In addition, the use of implicit methods reduces demand characteristics (by indirectly encouraging participants to provide the results that the experimenter "wants" to see), thereby increasing the validity of studies employing these approaches.

A final advantage of using implicit measures is their sensitivity to very subtle contextual shifts. For example, priming effects vastly differ with either slight or dramatic changes in stimuli; Reich, Noll, and Goldman [85] demonstrated this sensitivity to slight modifications in alcohol expectancy task stimuli by simply varying the first word of a list to be later recalled (beer vs. milk). Other implicit priming studies in our lab have used alcohol words followed by the Stroop task [86] or a false memory task [87]. Each of these studies demonstrated that alcohol primes, when compared to alcohol-neutral primes, led to enhanced memory activation of positive and/or arousing alcohol expectancies. Not surprisingly, in each study activation was contingent on the typical drinking level of the participants. The greatest activation was found in the heaviest drinkers—those with the most drinking experience were the most sensitive to alcohol-specific contexts.

Studies that showed memory network activation to be sensitive to context, although important, only provided support for a hypothetical mechanism that might influence drinking. To complete the picture, several studies in our and related laboratories also have shown that contextual priming can influence *actual* drinking. To this end, participants in these studies were exposed to primes in a procedure that was seemingly unrelated to a later "taste test" of an alcoholic beverage. These "unrelated experiment" studies have shown that exposure to primes, including television clips [88], alcohol expectancy words [89], or a simulated bar [90], can lead to increases in actual drinking during the later taste test.

Beyond our immediate research group, many other investigators have used widely different implicit memory network measures (e.g., reaction time, recall, free associates) to study alcohol expectancies and related cognitions. A prevailing view that drove the use of implicit measures in these studies was that implicit measures somehow gained access to psychological processes that were in some way more central to decision making than those processes that could be accessed by explicit measurement approaches. To summarize this body of studies and to compare the predictive power of explicit and implicit measures, Reich, Below, and Goldman [91] conducted a meta-analysis of studies that included both kinds of measures in one study. Contrary to the prevailing view, however, in 13 of the 16 studies that met this criterion, explicit measures accounted for more drinking variance than implicit measures (in 12 out of 13 studies accounting for unique variance). Meta-analysis of the relative mean effect sizes confirmed the observed differences. Hence, these empirical findings indicated that implicit measurement was not somehow more central or fundamental to the prediction of behavior; that is, implicit measures were not supported as accessing "deeper" or more critical decision-making processes. At the same time, however, implicit measures did account for significant (incremental) drinking variance beyond that accounted for by explicit measures. This pattern of results supported the use of both explicit and implicit measures of alcohol expectancy and similar associative processes to predict drinking.

A very recent line of research in our laboratory has added one form of neuroimaging to open a new window on alcohol expectancy operation. To this end, Fishman, Goldman, and Donchin [92] measured the P300 component of the event-related potential as an index of alcohol expectancy violation. Elevations in this component usually have been understood as indexing violations of preconceived expectancies; that is, individuals who strongly expected a particular outcome showed large P300s if a different outcome appeared. After alcohol expectancies were assessed in college students using conventional paper-and-pencil measures, participants who were monitored using a dense array electroencephalogram (EEG) were presented with the sentence stem "On a Friday night, alcohol makes me ...," which was completed with an alcohol expectancy word (e.g., happy). The stem was repeated multiple times with different alcohol expectancy completions representing both positive and negative dimensions. For each of these presentations, participants either agreed or disagreed with the statement using a computer keyboard. Participants who more strongly endorsed positive alcohol expectancies on the paper-and-pencil measures had larger P300 EEG waves when presented with negative alcohol expectancy sentences. That is, expectancy violation was greatest for participants presented with expectancies incongruent with their own. These P300 results occurred too quickly (within 300 ms) to be contingent on conscious verbal if–then propositions and suggested that decisions about alcohol outcomes were in process well before individuals might be aware of these decisions. This lack of awareness supports the idea that implicit processes are at least partially guiding drinking-related cognition.

11.4 Alcohol Reward: A Psychosocial Enhancer?

It is evident from years of research that alcohol is reinforcing, at least in part, through its chemical effects in the brain; that is, it is known to influence neurobiological pathways involved in reward. However, research in the psychosocial domain has suggested additional reward pathways. Earlier in this chapter, we described how humans are tuned into biologically meaningful rewards and that these rewards include information that is social in nature. Alcohol consumption may be understood as one pathway to such rewards [76]. Put another way, the question can be raised of whether drinking is fully reinforced by it effects in the brain or whether its use may at times serve as means of achieving ends further downstream, such as social effects. What has yet to be explicitly tested using any type of alcohol expectancy measure is whether social alcohol expectancies are accorded "privileged" cognitive status [21]. Although studies that directly test this hypothesis are as yet unavailable, several lines of evidence point to this possibility.

First, in addition to the many anticipated effects already mentioned that seem to reflect pharmacological effects of ethanol, social and sexual effects were often identified as highly salient alcohol expectancies. Among these effects, expectancies commonly reported in our research included sociability, talkativeness, attractiveness, and "horniness" [79]. Rarer, but repeatedly observed, expectancies included enhanced dancing and fighting. Second, we have found that social and sexual expectancies were always highly positively related to actual drinking. Reich and Goldman [75] isolated social and sexual expectancies from other expectancies following a free associate task where participants could give up to five responses to complete the phrase "Alcohol makes me____." The more that each individual produced social and sexual responses, the heavier their typical self-reported drinking tended to be.

Third, among other types of alcohol expectancies on standardized measures, social and sexual scales typically accounted for the greatest statistical prediction of drinking behavior (e.g., [83]). And fourth, effective expectancy challenge studies have had at their core the manipulation of social and sexual activities (e.g., [93, 94]). During these procedures participants were randomly assigned to receiving drinks containing alcohol or placebo alcohol. Participants then interacted in social games with (rating the attractiveness of models in pictures) or without (Win, Lose or Draw) sexual content to elicit expectancy-consistent behavior. Later, when asked to guess who had consumed alcohol based on their behaviors during these activities, participants were correct only at about the level of a coin flip (chance). This level of accuracy, or more precisely inaccuracy, occurred even when they were guessing about their own personal consumption. The expectancy challenge manipulation, which showed participants that many social and sexual effects of alcohol could be the result of placebo effects, has led to observable decreases in drinking in some college student samples [95].

11.5   Conclusion

The expectancy/anticipatory model can have transdisciplinary application as a means of tying together concepts developed in molecular and behavioral genetics with the manifestations of genetic processes in neurobiological systems and, ultimately, with behavioral decision making as it influences amounts and patterns of alcohol consumption. As measured using most current psychosocial assessment methods, alcohol expectancies can legitimately be considered an intermediate phenotype, but this level of assessment must be understood as representing complex multilevel processing. These multilevel processes can also be understood within the framework of epigenesis. Because these many complex and interplaying process can be indexed with alcohol expectancy measurement tools, these tools may offer a less cumbersome array of targets for prevention research.

Acknowledgment

This chapter was supported in part by National Institute on Alcohol and Alcoholism Grant R01 AA08333.

References

1. Crabbe, J. C. (2011). Exploring the E in GXE: How can rodents help? In C. Martin (Ed.), *Conceptualization and measurement of the environment when studying the gene–environment interplay in alcohol research* (Atlanta, GA, Symposium conducted at the meeting of the Research Society on Alcoholism).

2. Crabbe, J. C., Phillips, T. J., Harris, R. A., Arends, M. A., & Koob, G. F. (2006). Alcohol-related genes: contributions from studies with genetically engineered mice. *Addiction Biology, 11*, 195–269.

3. Guo, G., Wilhelmsen, K., & Hamilton, N. (2007). Gene–lifecourse interaction for alcohol consumption in adolescence and young adulthood: five monoamine genes. *American Journal of Medical Genetics. Part B, Neuropsychiatric Genetics, 144B*, 417–423.

4. Dick, D. M., Bierut, L., Hinrichs, A., et al. (2006). The role of GABRA2 in risk for conduct disorder and alcohol and drug dependence across developmental stages. *Behavior Genetics, 36,* 577–590.

5. Heath, A. C., Whitfield, J. B., Martin, N. G., et al. (2011). A quantitative-trait genome-wide association study of alcoholism risk in the community: findings and implications. *Biological Psychiatry, 70,* 513–518.

6. Meyer-Lindenberg, A., & Weinberger, D. R. (2006). Intermediate phenotypes and genetic mechanisms of psychiatric disorders. *Nature Reviews. Neuroscience, 7,* 818–827.

7. Sher, K. J., Dick, D. M., Crabbe, J. C., et al. (2010). Consilient research approaches in studying gene × environment interactions in alcohol research. *Addiction Biology, 15,* 200–216.

8. Goldman, M. S., Darkes, J., Reich, R. R., & Brandon, K. O. (2010). From DNA to conscious thought: anticipatory processing as a transdisciplinary bridge in addiction. In D. Ross, H. Kincaid, D. Spurrett, & P. Collins (Eds.), *What Is Addiction?* (pp. 291–334). Cambridge, MA: MIT Press.

9. Kendler, K. S., Chen, X., Dick, D., et al. (2012). Recent advances in the genetic epidemiology and molecular genetics of substance use disorders. *Nature Neuroscience, 15,* 181–189.

10. Clark, A. (2012). Prediction and the brain: a response. (http://opinionator.blogs.nytimes.com/2012/01/18/prediction-and-the-brain-a-response/).

11. Tolman, E. C. (1932). *Purposive Behavior in Animals and Man.* New York: Appleton-Century-Crofts.

12. Dennett, D. C. (1991). *Consciousness Explained.* New York: Little, Brown.

13. Bar, M. (2011). *Predictions and the Brain.* New York: Oxford University Press.

14. Redish, A. D., Jensen, S., & Johnson, A. (2008). A unified framework for addiction: vulnerabilities in the decision process. *Behavioral and Brain Sciences, 31,* 415–437, discussion 437–487.

15. Bolles, R. C. (1967). *Theory of Motivation.* New York: Harper and Row.

16. Bolles, R. C. (1972). Reinforcement, expectancy and learning. *Psychological Review, 79,* 394–409.

17. Kupfermann, I., Kandel, E. R., & Iversen, S. (2000). *Motivational and Addictive States.* New York: McGraw-Hill.

18. Kirsch, I. (Ed.). (1999). *How Expectancies Shape Experience.* Washington, DC: American Psychological Association.

19. Calvin, W. H., & Bickerton, D. (2000). *Lingua ex Machina: Reconciling Darwin and Chomsky with the Human Brain.* Cambridge, MA: MIT Press.

20. Glimcher, P. (2002). Decisions, decisions, decisions: choosing a biological science of choice. *Neuron, 36,* 323–332.

21. Mitchell, J. P. (2008). Contributions of functional neuroimaging to the study of social cognition. *Current Directions in Psychological Science, 17,* 142–146.

22. Jablonka, E., & Lamb, M. J. (2002). The changing concept of epigenetics. *Annals of the New York Academy of Sciences*, *981*, 82–96.

23. Gottesman, I. I., & Hanson, D. R. (2005). Human development: biological and genetic processes. *Annual Review of Psychology*, *56*, 263–286.

24. Ross, D., Kincaid, H., Spurrett, D., & Collins, P. (Eds.), *What Is Addiction?* Cambridge, MA: MIT Press.

25. Katner, S. N., Kerr, T. M., & Weiss, F. (1996). Ethanol anticipation enhances dopamine efflux in the nucleus accumbens of alcohol-preferring (P) but not Wistar rats. *Behavioural Pharmacology*, *7*, 669–674.

26. McCarthy, D. M., Wall, T. L., Brown, S. A., & Carr, L. G. (2000). Integrating biological and behavioral factors in alcohol use risk: the role of ALDH2 status and alcohol expectancies in a sample of Asian Americans. *Experimental and Clinical Psychopharmacology*, *8*, 168–175.

27. McCarthy, D. M., Brown, S. A., Carr, L. G., & Wall, T. L. (2001). ALDH2 status, alcohol expectancies, and alcohol response: preliminary evidence for a mediation model. *Alcoholism, Clinical and Experimental Research*, *25*, 1558–1563.

28. Breiter, H. C., Aharon, I., Kahneman, D., Dale, A., & Shizgal, P. (2001). Functional imaging of neural responses to expectancy and experience of monetary gains and losses. *Neuron*, *30*, 619–639.

29. McClure, S. M., Laibson, D. I., Loewenstein, G., & Cohen, J. D. (2004). Separate neural systems value immediate and delayed monetary rewards. *Science*, *306*, 503–507.

30. Schultz, W. (2004). Neural coding of basic reward terms of animal learning theory, game theory, microeconomics and behavioural ecology. *Current Opinion in Neurobiology*, *14*, 139–147.

31. Schultz, W., Dayan, P., & Montague, P. R. (1997). A neural substrate of prediction and reward. *Science*, *275*, 1593–1599.

32. Glimcher, P. W., & Lau, B. (2005). Rethinking the thalamus. *Nature Neuroscience*, *8*, 983–984.

33. Minamimoto, T., Hori, Y., & Kimura, M. (2005). Complementary process to response bias in the centromedian nucleus of the thalamus. *Science*, *308*, 1798–1801.

34. Tobler, P. N., Fiorillo, C. D., & Schultz, W. (2005). Adaptive coding of reward value by dopamine neurons. *Science*, *307*, 1642–1645.

35. Alexander, W. H., & Brown, J. W. (2011). Medial prefrontal cortex as an action-outcome predictor. *Nature Neuroscience*, *14*, 1338–1344.

36. Berridge, K. C., & Robinson, T. E. (1998). What is the role of dopamine in reward: hedonic impact, reward learning, or incentive salience? *Brain Research. Brain Research Reviews*, *28*, 309–369.

37. Montague, P. R., & Berns, G. S. (2002). Neural economics and the biological substrates of valuation. *Neuron, 36,* 265–284.

38. Holland, P. C., & Gallagher, M. (2004). Amygdala–frontal interactions and reward expectancy. *Current Opinion in Neurobiology, 14,* 148–155.

39. Iversen, S., Kupfermann, I., & Kandel, E. R. (2000). Emotional states and feelings. In E. R. Kandel, J. H. Schwartz, & T. M. Jessell (Eds.), *Principles of Neural Science* (pp. 982–997). New York: McGraw-Hill.

40. Phelps, E. A. (2004). Human emotion and memory: interactions of the amygdala and hippocampal complex. *Current Opinion in Neurobiology, 14,* 198–202.

41. Cardinal, R. N., & Everitt, B. J. (2004). Neural and psychological mechanisms underlying appetitive learning: links to drug addiction. *Current Opinion in Neurobiology, 14,* 156–162.

42. Heinrichs, S. C., & Koob, G. F. (2004). Corticotropin-releasing factor in brain: a role in activation, arousal, and affect regulation. *Journal of Pharmacology and Experimental Therapeutics, 311,* 427–440.

43. Matsumoto, K., & Tanaka, K. (2004). The role of the medial prefrontal cortex in achieving goals. *Current Opinion in Neurobiology, 14,* 178–185.

44. Norretranders, T. (1999). *The User Illusion: Cutting Consciousness Down to Size.* New York: Penguin Group.

45. Maunsell, J. H. (1995). The brain's visual world: representation of visual targets in cerebral cortex. *Science, 270,* 764–769.

46. Grossberg, S. (1995). The attentive brain. *American Scientist, 83,* 438–449.

47. Singer, W. (1995). Development and plasticity of cortical processing architectures. *Science, 270,* 758–764.

48. Tsodyks, M., & Gilbert, C. (2004). Neural networks and perceptual learning. *Nature, 431,* 775–781.

49. Miyashita, Y. (2004). Cognitive memory: cellular and network machineries and their top-down control. *Science, 306,* 435–440.

50. Kerzel, D. (2005). Representational momentum beyond internalized physics: embodied mechanisms of anticipation cause errors in visual short-term memory. *Current Directions in Psychological Science, 14,* 180–184.

51. Correa, A., Lupianez, J., & Tudela, P. (2005). Attentional preparation based on temporal expectancy modulates processing at the perceptual level. *Psychonomic Bulletin & Review, 12,* 328–334.

52. Haith, M. M., Hazan, C., & Goodman, G. S. (1988). Expectation and anticipation of dynamic visual events by 3.5-month-old babies. *Child Development, 59,* 467–479.

53. Huron, D. (2006). *Sweet Anticipation: Music and the Psychology of Expectation*. Cambridge, MA: MIT Press.

54. Nelson, D. L., McEvoy, C. L., & Pointer, L. (2003). Spreading activation or spooky action at a distance? *Journal of Experimental Psychology. Learning, Memory, and Cognition, 29*, 42–52.

55. Bar, M. (2011). The proactive brain. In M. Bar (Ed.), *Predictions in the Brain: Using Our Past to Generate a Future* .(pp. 13–26). New York: Oxford University Press.

56. Dragoi, V., & Staddon, J. E. (1999). The dynamics of operant conditioning. *Psychological Review, 106*, 20–61.

57. Kirsch, I., Lynn, S. J., Vigorito, M., & Miller, R. R. (2004). The role of cognition in classical and operant conditioning. *Journal of Clinical Psychology, 60*, 369–392.

58. Van Hamme, L. J., & Wasserman, E. A. (1994). Cue competition in causality judgments: the role of nonpresentation of compound stimulus elements. *Learning and Motivation, 25*, 127–151.

59. Heekeren, H. R., Marrett, S., Bandettini, P. A., & Ungerleider, L. G. (2004). A general mechanism for perceptual decision-making in the human brain. *Nature, 431*, 859–862.

60. Ritov, I. (2000). The role of expectations in comparisons. *Psychological Review, 107*, 345–357.

61. Bower, G. (2000). A brief history of memory research. In E. Tulving & F. I. M. Craik (Eds.), *The Oxford Handbook of Memory*. (pp. 3–32). New York: Oxford University Press.

62. Wulf, G., & Prinz, W. (2001). Directing attention to movement effects enhances learning: a review. *Psychonomic Bulletin & Review, 8*, 648–660.

63. Barrett, L. F., & Bar, M. (2011). See it with feeling: Affective predictions during object perception. In M. Bar (Ed.), Predictions in the Brain: Using Our Past to Generate a Future. (pp. 107–121). New York: Oxford University Press.

64. Bechara, A. (2011). The somatic marker hypothesis and its neural basis: Using past experiences to forecast the future in decision making. In M. Bar (Ed.), *Predictions and the Brain* (pp. 122–133). New York: Oxford University Press.

65. Colunga, E., & Smith, L. B. (2005). From the lexicon to expectations about kinds: a role for associative learning. *Psychological Review, 112*, 347–382.

66. Gray, R., Beilock, S. L., & Carr, T. H. (2007). "As soon as the bat met the ball, I knew it was gone": outcome prediction, hindsight bias, and the representation and control of action in expert and novice baseball players. *Psychonomic Bulletin & Review, 14*, 669–675.

67. Baron-Cohen, S., Knickmeyer, R. C., & Belmonte, M. K. (2005). Sex differences in the brain: implications for explaining autism. *Science, 310*, 819–823.

68. Kirsch, I. (1999). *How Expectancies Shape Experience*. Washington, DC: American Psychological Association.

69. Kirsch, I., & Scoboria, A. (2001). Apples, oranges, and placebos: heterogeneity in a meta-analysis of placebo effects. *Advances in Mind–Body Medicine, 17,* 307–309, discussion 312–318.

70. Waber, R. L., Shiv, B., Carmon, Z., & Ariely, D. (2008). Commercial features of placebo and therapeutic efficacy. *Journal of the American Medical Association, 299,* 1016–1017.

71. Fiske, S. T., Schacter, D. L., & Sternberg, R. (Eds.), *Annual Review of Psychology.* Palo Alto: Annual Reviews.

72. Wager, T. D. (2005). The neural bases of placebo effects in anticipation and pain. *Seminars in Pain Medicine, 3,* 22–30.

73. Wager, T. D., Rilling, J. K., Smith, E. E., et al. (2004). Placebo-induced changes in fMRI in the anticipation and experience of pain. *Science, 303,* 1162–1167.

74. Petrovic, P., Dietrich, T., Fransson, P., et al. (2005). Placebo in emotional processing-induced expectations of anxiety relief activate a generalized modulatory network. *Neuron, 46,* 957–969.

75. Goldman, M. S., Reich, R. R., & Darkes, J. (2006). Expectancy as a unifying construct in alcohol related cognition. In R. W. Wiers & A. W. Stacy (Eds.), *Handbook of Implicit Cognition and Addiction* (pp. 105–121). Thousand Oaks, CA: Sage.

76. Reich, R. R., & Goldman, M. S. (2012). Drinking in college students and their age peers: The role of anticipatory processes. In H. L. White, & D. L. Rabiner (Eds.), *College Drinking and Drug Use* (pp. 105–120). New York: Guilford.

77. Rather, B. C., Goldman, M. S., Roehrich, L., & Brannick, M. (1992). Empirical modeling of an alcohol expectancy memory network using multidimensional scaling. *Journal of Abnormal Psychology, 101,* 174–183.

78. Rather, B. C., & Goldman, M. S. (1994). Drinking-related differences in the memory organization of alcohol expectancies. *Experimental and Clinical Psychopharmacology, 2,* 167–183.

79. Reich, R. R., & Goldman, M. S. (2005). Exploring the alcohol expectancy memory network: the utility of free associates. *Psychology of Addictive Behaviors, 19,* 317–325.

80. Milgram, S. (1967). The small world problem. *Psychology Today, 2,* 60–67.

81. Reich, R. R., Ariel, I., Darkes, J., & Goldman, M. S. (2012). What do you mean "drunk"? Convergent validation of multiple methods of mapping alcohol expectancy memory networks. *Psychology of Addictive Behaviors, 26,* 406–413.

82. Kerr, W. C., Greenfield, T. K., & Midanik, L. T. (2006). How many drinks does it take you to feel drunk? Trends and predictors for subjective drunkenness. *Addiction, 101,* 1428–1437.

83. Goldman, M. S., & Darkes, J. (2004). Alcohol expectancy multiaxial assessment: a memory network-based approach. *Psychological Assessment, 16,* 4–15.

84. Custers, R., & Aarts, H. (2010). The unconscious will: how the pursuit of goals operates outside of conscious awareness. *Science, 329,* 47–50.

85. Reich, R. R., Noll, J. A., & Goldman, M. S. (2005). Cue patterns and alcohol expectancies: how slight differences in stimuli can measurably change cognition. *Experimental and Clinical Psychopharmacology, 13*, 65–71.

86. Kramer, D. A., & Goldman, M. S. (2003). Using a modified Stroop task to implicitly discern the cognitive organization of alcohol expectancies. *Journal of Abnormal Psychology, 112*, 171–175.

87. Reich, R. R., Goldman, M. S., & Noll, J. A. (2004). Using the false memory paradigm to test two key elements of alcohol expectancy theory. *Experimental and Clinical Psychopharmacology, 12*, 102–110.

88. Roehrich, L., & Goldman, M. S. (1995). Implicit priming of alcohol expectancy memory processes and subsequent drinking behavior. *Experimental and Clinical Psychopharmacology, 3*, 402–410.

89. Stein, K. D., Goldman, M. S., & Del Boca, F. K. (2000). The influence of alcohol expectancy priming and mood manipulation on subsequent alcohol consumption. *Journal of Abnormal Psychology, 109*, 106–115.

90. Lau-Barraco, C., & Dunn, M. E. (2008). Evaluation of a single-session expectancy challenge intervention to reduce alcohol use among college students. *Psychology of Addictive Behaviors, 22*, 168–175.

91. Reich, R. R., Below, M. C., & Goldman, M. S. (2010). Explicit and implicit measures of expectancy and related alcohol cognitions: a meta-analytic comparison. *Psychology of Addictive Behaviors, 24*, 13–25.

92. Fishman, I., Goldman, M. S., & Donchin, E. (2008). The P300 as an electrophysiological probe of alcohol expectancy. *Experimental and Clinical Psychopharmacology, 16*, 341–356.

93. Darkes, J., & Goldman, M. S. (1998). Expectancy challenge and drinking reduction: process and structure in the alcohol expectancy network. *Experimental and Clinical Psychopharmacology, 6*, 64–76.

94. Darkes, J., & Goldman, M. S. (1993). Expectancy challenge and drinking reduction: experimental evidence for a mediational process. *Journal of Consulting and Clinical Psychology, 61*, 344–353.

95. Dunn, M. E., Lau, H. C., & Cruz, I. Y. (2000). Changes in activation of alcohol expectancies in memory in relation to changes in alcohol use after participation in an expectancy challenge program. *Experimental and Clinical Psychopharmacology, 8*, 566–575.

12 Intermediate Phenotypes for Alcohol Use and Alcohol Dependence: Empirical Findings and Conceptual Issues

Michael Windle

Over the past ten years or so, intermediate phenotypes and candidate endophenotypes have become prominent foci in the study of complex disorders, including alcohol and other substance use disorders and mental health disorders [1, 2]. A major impetus for the focus on intermediate phenotypes has been the lack of precision of psychiatric categorical diagnostic systems, combined with rapid advances in new technologies (e.g., high throughput molecular genetics, neuroimaging techniques) that facilitate the objective measurement of some intermediate phenotypes and associated pathway models [3]. It has been suggested that continuously distributed intermediate phenotypes may more closely approximate underlying biological processes, including pathophysiology, associated with psychiatric disorders than binary clinical diagnostic categories [4, 5]. Intermediate phenotypes are also assumed to be more directly influenced by genes relevant to underlying psychiatric disorders than are the manifest symptoms of disorders [6]. Furthermore, to the extent that such intermediate phenotypes may be identified early in development (i.e., preceding the onset of the disorder), they may serve as biomarkers or risk factors that would enhance opportunities for early identification and preventive intervention activities. For example, genetic markers or neural signatures that signified substantial risk for the subsequent development of problem drinking or alcohol dependence would provide clearer targets for pharmacological or behavioral interventions.

This chapter focuses on four areas related to intermediate phenotypes for alcohol use and alcohol dependence. First, a brief historical perspective on alcohol disorders is provided to impart a conceptual lens for focusing on specific intermediate phenotypes for alcohol-related phenotypes. Second, alcohol disorders have been identified as heterogeneous (i.e., as multiple disorders rather than a single disorder) [7, 8], and conceptual approaches to the heterogeneity issue in alcohol studies are presented and related to the selection of intermediate phenotypes. Third, a brief selective review of intermediate phenotypes in alcohol studies is provided, including alcohol-specific (e.g., ALDH; intensity of response to alcohol) and alcohol-nonspecific (e.g., behavioral undercontrol or disinhibition) intermediate phenotypes. The identification of both specific and nonspecific intermediate phenotypes is important both in actuarial-prediction models (i.e., statistical models for optimal prediction) and causal-explanatory models (i.e., statistical models that examine hypothesized causal relations among variables). Fourth, some current conceptual and methodological limitations of intermediate phenotype research for alcohol phenotypes are discussed and future directions suggested that incorporate developmental-temporal dynamics and the gene–environment (GE) and gene–gene (GG) interplay as part of a more comprehensive conceptual model.

12.1   Brief Historical Perspective on Alcohol Disorders

The National Institute on Alcohol Abuse and Alcoholism was established in 1970. In the early 1970s alcoholism was conceptualized as a disease driven by physiological determinants and as primarily occurring among middle-aged men. However, the current perspective on alcoholism is that it is a multidetermined, GE disorder (or disorders) with multilevel causal determinants ranging from genes to social-historical factors [8–10]. Furthermore, and importantly, it is recognized as affecting both men and women and as reflecting *developmental processes* beginning with childhood non–alcohol specific influences (e.g., aggression, attentional difficulties, early attitudes and beliefs about alcohol) and progressing across time to initiation, escalation, problem drinking, and alcohol dependence [11]. Epidemiology data presented in figure 12.1 using a nationally representative sample demonstrates the peak ages for alcohol dependence are ages 18–24 rather than middle age [12]. Similarly, for the alcohol phenotype of heavy episodic drinking (i.e., HED, drinking five or more drinks in a single setting), findings from the National Survey on Drug Use and Health [13] indicated that HED peaks in the age range of ages 18–24 (see figure 12.2).

The significance of these major changes in the conceptualization of alcohol disorders from that of a middle-aged, male-dominated phenomenon to a developmental, late adolescent/early young adulthood peak period phenomenon has had implications for the selection and prioritization of intermediate phenotypes. For example, proposed intermediate phenotypes based on middle-aged samples may reflect neural processes of neuroadaptation, neurodegeneration, or neuroplasticity in response to the chronic use of alcohol (i.e., they may be a consequence or end point of prolonged use) rather than being a biomarker or intermediate phenotype closely associated with an alcohol disorder in terms of the underlying genetic architecture or the proximal causal mechanisms. By contrast, a focus on precursors and correlates of childhood, adolescent, and young adult alcohol use and alcohol disorders has spawned the investigation of intermediate phenotypes for risk and protective factors that occur during these portions of the life span.

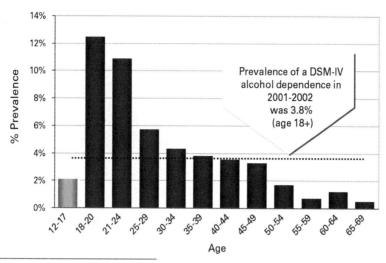

18 + yrs. -NIAAA NESARC (Grant, et al., (2004) Drug and Alcohol Dependence, 74:223-234)
12-17 yrs -U.S. Substance Abuse and Mental Health Services Administration 2003 National
Survey on Drug Use and Health (NSDUH)

Figure 12.1

Prevalence of past-year *Diagnostic and Statistical Manual of Mental Disorders* (4th ed.; *DSM–IV*) alcohol dependence by age—United States.

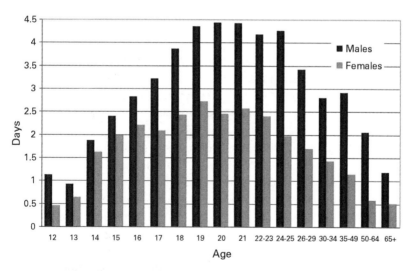

U.S. Substance Abuse and Mental Health Services Administration 2003 National
Survey on Drug Use and Health (NSDUH)

Figure 12.2

Number of days in the past 30 days on which survey respondents drank five or more drinks.

The focus on these earlier onset intermediate phenotypes holds promise for the earlier identification of these factors, their time course in relation to the disorder, the study of mediators and moderators of intermediate phenotypes and their association with alcohol disorders, and the study of relevant GE and GG processes. Collectively, this more comprehensive approach should provide a stronger basis for the testing of explanatory models that may serve to guide targeted prevention and intervention programs. Parenthetically, this focus on earlier onset is not intended to minimize significant intermediate phenotype research conducted on adults regarding, for example, vulnerability to end-organ damage (e.g., alcoholic liver cirrhosis) or individual variation in medication response (e.g., pharmacogenomics)—the study of intermediate phenotypes across the life course is essential for alcohol use and alcohol disorders [14].

12.2 Approaches to the Heterogeneity Issue and the Selection of Intermediate Phenotypes

Based both on population-based studies of comorbidity [12, 15] and clinical treatment studies [16, 17], there has been a recognition that alcohol disorders are multifactorial and heterogeneous rather than a single, homogenous disorder [14, 18]. Hence, different conceptual and analytic strategies have been adopted to try to model the multifactorial influences on alcohol disorders. These conceptual and analytic approaches have been significant in identifying important intermediate phenotypes and are briefly described now.

One of the most influential investigations of alcoholic subtypes was the Swedish Adoption Study that identified two distinct types of alcoholism, referred to as type I ("milieu limited") and type II ("male limited") [17, 19]. The first subtype—*type I alcoholics*, represented the largest portion (76%) of alcoholics in the Swedish sample and were characterized as having a genetic diathesis toward alcoholism that was exacerbated by environmental (milieu) stressors. These type I alcoholics had a later onset of alcoholism than type II alcoholics, displayed psychological rather than physical dependence on alcohol, and did not manifest high levels of early onset and persistent antisocial behavior. By contrast, *type II alcoholics* were characterized as having a genetic diathesis for alcoholism and a symptom pattern that included an early onset pattern of alcohol abuse and antisocial behaviors that persisted across time. Subsequent to the Swedish Adoption Study, Cloninger [20] delineated a biogenetic temperament theory to describe the type I and type II alcoholics. According to Cloninger's theory, type I alcoholics have a temperament profile characterized by low novelty seeking, high harm avoidance, and high reward dependence. Type II alcoholics have a temperament profile characterized by high novelty seeking, low harm avoidance, and low dependence. Cloninger proposed that specific neurobiological structures and neuromodulators were associated with each of the temperament traits and that the inherited biochemical nature of these traits, and their impact on learning, contributed to the underlying vulnerability to different forms (or subtypes) of alcoholism.

In a separate set of studies that included alcohol treatment samples, Babor and his colleagues (e.g., [16, 21]) identified two subtypes based on the use of cluster analytic techniques with data assessed across several dimensions. These dimensions included premorbid risk factors (e.g., family history of alcoholism, childhood behavior problem symptoms), severity of alcohol dependence and alcohol-related consequences, chronicity and adverse social consequences, and comorbid psychopathology (e.g., levels of general anxiety, antisocial behaviors). The two subtypes of alcoholics were identified as type A and type B. *Type A alcoholics* were characterized as lower on premorbid risk factors (e.g., childhood behavior problems), with a later onset of alcoholism and less severity with regard to dependence and adverse social consequences, with a less chronic course, and with lower levels of comorbidity. By contrast, *type B alcoholics* were characterized as higher on premorbid risk factors, with an earlier onset of alcoholism and more severity with regard to dependence and adverse social consequences, with a more chronic course, and with more comorbid psychopathology. While other typological models have proposed additional alcoholic subtypes, for example, a Negative Affect alcoholic subtype in which the alcoholics are also high on depressive or anxiety symptoms [22–24], they have largely retained the subtypes identified separately by Cloninger and Babor [16, 19]. There is, as yet, no consensus on the exact number of alcoholic subtypes, but the notion that there is more than one form of alcoholism is widely accepted in the field of alcohol studies.

The significance of these typological approaches for intermediate phenotypes is that they serve as guideposts to identify and further study salient factors (or intermediate phenotypes) that characterize alcoholics. For example, critical to the type 2 (Cloninger) or type B (Babor) antisocial alcoholic is a pattern of early onset (childhood) symptoms of disruptive behavior disorders (e.g., symptoms of conduct problems, oppositional behavior) and associated temperament characteristics (e.g., impulsivity) and alcohol time course features such as early age of alcohol onset and more adverse social consequences as imparting greater risk. Some of the more prominent intermediate phenotypes under investigation for alcohol disorders that have been influenced by this typology research are temperament and personality characteristics (e.g., impulsivity, high activity level, sensation seeking, negative mood), disinhibition or behavioral undercontrol, negative affect (e.g., dysfunctional affective circuitry), stress-dampening, and neurobehavioral regulation [8, 25–27].

Yet another approach to the study of the heterogeneity of alcohol disorders has been that of *trajectory analyses*. Trajectory analyses, or more formally latent growth mixture models, focus on heterogeneity associated with different time-ordered trajectories of intraindividual change that may be masked by a single, aggregated sample analysis of change across time. That is, time-related changes may be better characterized by a number of distinct subpopulation trajectories (e.g., increasing, decreasing, stable trajectories) rather than a single, aggregated shape (e.g., linear) across the entire population. For example, in a longitudinal panel design, aggregated (total sample) statistics may suggest that the average change (i.e., mean level) of alcohol use may remain approximately equal over three waves of measurement, leading a researcher to conclude that the behavior under study, alcohol use, has manifested a high level of continuity across time. However, it is possible (hypothetically) that 50% of the sample, characterized by a family history of alcoholism, conduct disorder, and positive alcohol expectancies, actually manifested a pattern of significantly increasing levels of alcohol use across time whereas the remainder of the sample actually manifested a pattern of significantly decreasing levels of alcohol use across time. The two significant trends (increasing for one group and decreasing for the other) could, in effect, cancel each other out, and the conclusion drawn (based on aggregated sample data) would be that there was continuity across time, even though there was significant change that would have been revealed if groups had been identified according to trajectory profiles.

A number of studies exist regarding the use of trajectory analysis with different alcohol phenotypes (e.g., alcohol use, binge drinking, alcohol dependence symptoms) and have typically identified four to five trajectory groups. For example, as illustrated in figure 12.3, Schulenberg et al. [28] used large-sample data (over 9,000 cases) to

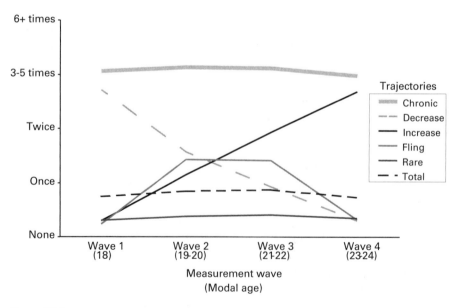

Figure 12.3
Mean score for five or more drinks in a row in the past two weeks by binge drinking trajectory.
Source: Schulenberg et al. (1996) [28].

identify five trajectory groups for binge drinking across the age interval of 18 to 24 years. The aggregate data, represented as the dotted line in figure 12.3, suggests a linear, flat trajectory across this age interval. However, based on trajectory analyses, using cluster analysis in this application, five quite distinct patterns of change were indicated that differed in level and shape of change across time. The Chronic binge drinking group reported binge drinking between three and five times in the past two weeks across this six-year interval whereas the Decrease group started at approximately the same level as the Chronic group at age 18 but decreased to less than once a week by ages 23 to 24. This study demonstrates the heterogeneity in binge drinking trajectories that may be captured by focusing on subgroup variation rather than assuming that a single growth pattern adequately represents the data across time.

Some have proposed that trajectories may serve as useful phenotypes to study the developmental course of alcohol and other substance use disorders [29]. In addition to identifying developmental course trajectories of alcohol and other substance use phenotypes, there has also been research activity in specifying the predictors that distinguish the trajectory groups [28, 30]. This is where the study of intermediate phenotypes may be applied in a significant way. Different alcohol-specific and non-specific intermediate phenotypes have been identified across different levels of assessment ranging from metabolic pathways and neural signals to neuropsychological functioning and disinhibited behaviors. The combination of intermediate phenotypes such as these, along with more traditionally defined risk (e.g., stressful events) and protective (e.g., parental monitoring) factors, may provide powerful predictive and explanatory models for alternative developmental courses of alcohol phenotypes. The specificity of this information may prove valuable for targeting particular variables (intermediate phenotypes, risk, or protective factors) at particular times based on developmental course; furthermore, one may be able to capitalize on this developmental course information if it varies by other sociodemographic variables such as sex, ethnicity, or socioeconomic group.

In summary, efforts to address the heterogeneity issue of alcohol disorders have spawned conceptual and analytic approaches in the form of alcohol typologies (subtypes) and trajectory analyses. These approaches have been valuable both as guideposts for the identification of intermediate phenotypes and for the further development and refinement of intermediate phenotypes that may distinguish different multifactorial patterns of change representing subtypes or subgroups of alcoholics.

12.3   Selective Brief Review of Intermediate Phenotypes in Alcohol Studies

On the basis of a confluence of prior research findings, including those stemming from approaches to the heterogeneity of alcohol disorders, as well as traditional risk and protective research, intermediate phenotypes have been identified across multiple levels of analysis, including electrocortical responses (e.g., P3 and EEG patterns), neuroimaging responses (e.g., neural activation patterns associated with craving), pharmacological response patterns, neuropsychological factors (e.g., executive cognitive functioning; cognitive-affective dysregulation), temperament (e.g., impulsivity, sensation seeking), and childhood externalizing (e.g., attention-deficit/hyperactivity disorder [ADHD], antisocial behavior) and internalizing (e.g., depression and anxiety) problems. In critically discussing intermediate phenotypes for alcohol phenotypes, it is important to distinguish between *alcohol-specific* and *alcohol-nonspecific* intermediate phenotypes. Based on the extant literature, alcohol-specific intermediate phenotypes refer to factors that influence alcohol phenotypes only and not other mental health or substance-related phenotypes. By contrast, alcohol nonspecific intermediate phenotypes refer to factors that may influence multiple substance use and mental health phenotypes, including alcohol phenotypes. Selected, specific examples of specific and nonspecific intermediate phenotypes are provided subsequently. It is important to identify the nature of intermediate phenotypes along this dimension of specific and nonspecific for theories of etiology, prevention, and treatment.

12.3.1 Alcohol-Specific Intermediate Phenotypes in Alcohol Studies

Perhaps the most widely known intermediate phenotype for alcohol dependence is the alcohol metabolizing enzyme of *aldehyde dehydrogenase*. Ethanol is converted to acetaldehyde by alcohol dehydrogenase and then to acetic acid for elimination by aldehyde dehydrogenase (ALDH). A mutation of the ALDH*2 gene on chromosome 12 results in an enzyme that cannot convert and eliminate acetaldehyde at the usual levels and rate. Acetaldehyde is toxic and at high levels induces a range of autonomic responses including a "facial flushing response" that is reported to be unpleasant and uncomfortable. The homozygote mutation of the ALDH*2 gene occurs in about 10% of Asians and results in intense responses of facial flushing, nausea, heart palpitations, and sweaty palms. For the 10% of Asians who are homozygotes for the ALDH*2 gene, there is a near zero prevalence of alcohol disorders.

These intense aversive responses are specific to alcohol (and the associated metabolic processes) and do not occur in response to other substances. There is also no association among Asians who are homozygotes for the ALDH*2 gene and the prevalence of other substance disorders. Hence, the ALDH*2 gene is viewed as an alcohol-specific intermediate phenotype that serves a "protective" function by increasing the aversion of consuming alcohol via unpleasant physiological responses (e.g., facial flushing) and thereby decreases risk for the development of alcohol dependence.

A second widely known intermediate phenotype for alcohol dependence is the *intensity of response to alcohol* investigated by Schuckit and his colleagues [31, 32]. Intensity of response to alcohol is conceptualized as an individual-difference variable with regard to sensitivity to the pharmacological effects of alcohol. A low level of response to alcohol (LR) is postulated by Schuckit to be an intermediate phenotype for alcohol dependence because low sensitivity to alcohol may contribute to higher levels of intake to achieve the same desired end state (e.g., euphoria, escape). That is, individuals with low intensity of response to alcohol may need to drink substantially more alcohol than those with moderate or high levels of response to alcohol to obtain similar goal states (e.g., euphoria). Therefore, individual variation in intensity of response to alcohol may impact drinking patterns that in turn influence the expression and severity of symptoms of alcohol dependence.

The basis of the intensity of response to alcohol was originally determined in laboratory alcohol challenge studies and has generated substantial corroboration in a number of other cross-sectional and long-term prospective studies conducted by Schuckit and colleagues [32, 33]. For example, in a 20-year follow-up study, LR still predicted recent alcohol-related problems while controlling for prior levels of alcohol related problems and other significant predictors such as family history of alcoholism, alcohol expectancies, peer drinking, and coping motives for drinking [31]. Furthermore, these robust associations were also supported with structural equation models with other samples [33] that enabled the specification of models to test both the direct and indirect effects of LR on alcohol-related outcomes. This more comprehensive, longitudinal modeling approach facilitated the testing of the multiple ways, or pathways, that an intermediate phenotype like LR may assume in impacting alcohol phenotypes. Hence, there has been consistent support for the potential role of LR as an intermediate phenotype for alcohol dependence. Furthermore, this influence appears to be alcohol specific because LR has not been significantly associated with heavy use, problems, or disorders of other substances (e.g., marijuana, cocaine).

12.3.2 Alcohol Nonspecific Intermediate Phenotypes in Alcohol Studies

There are a larger number of nonspecific intermediate phenotypes for alcohol disorders than specific intermediate phenotypes. Part of this may stem from common, underlying genetic or environmental influences. For example, confirmatory factor analytic models have indicated that a large number of psychiatric disorders can be adequately accounted for by two factors—internalizing and externalizing disorders [34]. Biometric models with data on twins have also indicated that large portions of variance for these larger common factors are attributable primarily to common genetic influences, though there are also significant portions of gene-specific and environmental sources of variation [35]. Consistent with these findings, Plomin, Haworth, and Davis [36] proposed the notion of "generalist genes and specialist environments." According to this notion, pleiotropic genetic factors tend to promote correlations among phenotypes whereas environmental factors tend to promote their differentiation. As such, there may be a larger portion of common genetic factors across a number of disorders that become distinguishable based on environmental influences. Thus, hypothetically, a common genetic diathesis could significantly contribute either to alcohol or cocaine disorders contingent on environmental factors (e.g., exposure to and availability of these substances in the environment). Whatever the causal explanation, behavioral undercontrol or disinhibition is now presented to illustrate a nonspecific factor associated with alcohol use and alcohol disorders.

A nonspecific (or general) intermediate phenotype for alcohol use and alcohol disorders is *behavioral undercontrol or disinhibition*. The scope of this intermediate phenotype is broad, including dimensions (symptoms) associated with disruptive behavior disorders (e.g., ADHD, conduct disorder, oppositional defiant disorder), temperament and personality dimensions (e.g., sensation seeking, impulsivity), and dimensional scores on neuropsychological tests measuring different aspects of impulsivity, decision making, and response inhibition. While a comprehensive presentation of the behavioral undercontrol literature is beyond the scope of this chapter (for reviews, see [37, 38, 39]), there are several points critical to appreciating its role as a nonspecific intermediate phenotype.

First, different features of behavioral undercontrol in childhood have been consistently associated both concurrently and prospectively with a broad range of adverse outcomes in adolescence and adulthood [25, 40]. These include an earlier onset of alcohol and other drug use, heavier alcohol use, binge drinking, alcohol-related problems during adolescence and adulthood, heavier use of other illicit substances in adolescence and adulthood, and in adulthood with alcohol and other substance-related disorders, antisocial personality disorder, and criminality [8, 27, 40, 41]. Second, confirmatory factor analytic studies have generally supported a higher-order factor structure for externalizing (behavioral undercontrol) disorders for disruptive behaviors and for alcohol and other substance abuse disorders and antisocial personality disorder [42] with twin studies suggesting a large heritable component for this common factor [35]. Third, a number of studies have begun to identify genes and gene variants associated with this spectrum of behavioral undercontrol intermediate phenotypes, including MAO-A, GABRA2, and CHRM2, among others [43].

Collectively, the findings on behavioral undercontrol provide a strong working model of pathways in which genes may influence important cognitive, affective, and behavioral features, or domains, of intermediate phenotypes that, in turn, influence both broad (e.g., psychiatric disorders) and narrow (e.g., craving) substance use and mental health phenotypes. However, these influences are not alcohol specific but rather influence a range of substance abuse and mental health outcomes, possibly due to common genetic and environmental influences. The pervasiveness of the influence of behavioral undercontrol across a range of behaviors and disorders across the life span contributes to its recognition as a high-priority area of study that could be gainfully targeted for prevention and treatment. Furthermore, it is a useful intermediate phenotype (or set of intermediate phenotypes) for integrating preclinical (animal), clinical, and population-based studies.

12.4 Conceptual and Methodological Limitations of Intermediate Phenotype Research for Alcohol Phenotypes and Future Directions

Although there has been substantial progress on identifying intermediate phenotypes for alcohol use and alcohol disorders, as well as other complex psychiatric phenotypes, much of this research has been conceptualized and conducted with a focus on the pathways from genes → neurotransmitters or metabolic processes → intermediate phenotypes (e.g., reward response system, arousal systems, localized neural activation patterns, response inhibition) → outcome (e.g., substance abuse or mental health disorder). Although this approach is systematic and complex, and is advancing the study of intermediate phenotypes, there remain some significant features that need to be included in a more comprehensive model to maximize our appreciation of the impact of intermediate phenotypes on selected outcomes, as well as to provide direction for prevention and treatment activities and for public health policy. Two of these issues are discussed now.

First, while the pathway notion described above, from gene to intermediate phenotype to outcome, has provided a general conceptual frame of reference, there has been an insufficient appreciation of the inclusion of environmental variables and contextual influences in more comprehensive models. Pathway models conceptualized within a systems biology or systems neuroscience framework have been proposed [44] that carefully delineate genetic and neurobiological pathways, but environmental influences are typically viewed as largely extraneous to these system dynamics with the exception that alcohol is viewed as an environmental exposure. Although no reconceptualization or technological advancement exists to provide a comprehensive taxonomy of environmental influences, there are a number of replicated findings that illustrate the importance of environmental influences in understanding intermediate phenotypes and their associations with substance use and other psychiatric disorders. For example, there is extensive research on the role of stressful events, particularly severe events such as child sexual abuse, that interact with the serotonin promoter gene variant (5-HTTLPR) to influence fear circuitry in the brain (an intermediate phenotype) and the occurrence of major depressive disorders in adulthood [45].

Similarly, there has been extensive research conducted on the role of social environmental influences (e.g., parents, siblings, peers) that influence and are influenced by intermediate phenotypes (e.g., [8, 46]), and influence alcohol and other substance use phenotypes via exposure and learning factors. For example, Tarter et al. [27] reported that neurobehavioral disinhibition (an intermediate phenotype) was associated with involvement with more deviant peers to predict early onset substance use.

Peer alcohol use is often identified as the most potent (i.e., highest magnitude) proximal predictor of adolescent alcohol use [46]. Studies focusing on more macrolevel environmental variables such regional variation [47], density of bars in neighborhoods [48], and density of substance use in schools [49] have all identified significant environmental influences on patterns of alcohol use. Hence, there is abundant evidence that environmental factors are part of the essential matrix for understanding complex phenotypes such as alcohol use and alcohol disorders and other complex disorders. More comprehensive, integrative conceptualizations of alcohol use and alcohol disorders that include the broader matrix of factors that influence both intermediate phenotypes and outcomes would benefit the field and provide a larger corpus of research findings to facilitate the translation of scientific findings for clinical practice and public health policy.

A second issue is the need for the greater inclusion of conditional and nonlinear influences such as GE and GG interactions, in more comprehensive model formulations, as well as the modeling of mediators. There have been an increasing number of studies supporting GE interactions with regard alcohol and other substance use across childhood and adolescence. For example, Brody [50] reported a significant interaction between parenting and the *5-HTTLPR* polymorphism (*s* allele) on increases in adolescent substance use. Those adolescents with low supportive parenting and the risk allele (*s* allele) reported higher levels and greater increases in substance use across time relative to adolescents with low supportive parenting and the nonrisk allele (*l* allele). Under conditions of high supportive parenting, the risk allele did not distinguish subgroups in terms of increases in level of substance use across time. Similarly, Chen et al. [51] reported a significant interaction between parental monitoring during middle childhood and CHRNA5 (rs16969968) on subsequent smoking dependence for a sample that ranged in age from 25 to 44. Under conditions of low parental monitoring, the risk single nucleotide polymorphism was associated with higher smoking dependence, and under conditions of high parental monitoring, risk for smoking dependence was significantly lower. In addition to these examples of GE interactions, there have also been increases in the number of GG interactions [52, 53] that highlight the need for a broader and more integrated perspective on intermediate phenotypes associated with complex variables such as alcohol disorders.

There are a number of challenging issues raised in the investigation of GE relations that merit thoughtful, ongoing proposals, strategies, evaluations, and modifications as needed. For example, many extant GE studies are statistically underpowered to evaluate more than a few candidate genes or gene variants, and replication of significant GE interactions has often been difficult, though not impossible to achieve [54]. This small-sample-size issue is further compounded in longitudinal-developmental studies where critical research design parameters (e.g., age span of assessment, number and spacing of time intervals) and measures may vary in significant ways that make pooling across studies challenging. Three strategies are currently available to address these issues. First, replication of findings is still a powerful scientific strategy to garner support for selected genes or gene variants interacting with identified environmental risk or protective factors. Without doubt, the seminal research of Caspi [45] triggered a landslide of studies on GE interactions between the serotonin transporter gene (5HTTPLR) and childhood maltreatment on major depressive disorders [54, 55], but there have been a quite limited set of GE studies on other specific genes and specific environmental factors. Guided by GE findings in animal and human studies on other specific genes and specific environmental exposures, theoretically and empirically informed candidate GE studies remain a viable strategy for pursuing GE relations.

Second, meta-analysis provides a set of quantitative methods to evaluate average effect sizes across GE studies and predictors (e.g., general population vs. clinical sample; age, sex, and ethnic distribution of samples; method of measurement for environmental variables) of variation in effect sizes. Third, there has been substantial interest and progress in developing harmonized data sets that permit the pooling of samples to address issues related to statistical power, corrections to alpha levels for multiple testing (i.e., reduce the false discovery rate), and the testing of adequately powered GE hypotheses.

In addition to these three strategies, the National Genome Research Institute of the National Institutes of Health is developing an online resource of domains and measures of phenotypes, including some intermediate phenotypes, and exposures (environmental factors), via expert consensus panels (https://www.phenx.org). The goal of this resource, or tool kit, is to assist researchers in identifying and using state-of-the-art measures across studies to facilitate cross-study harmonization. This resource was designed to be of particular value in large-scale genome-wide association studies (GWASs) that typically have not included any significant environmental measures and are therefore ill equipped to address substantive issues and statistical models related to GE relations. However, this resource could be valuable far beyond GWAS and could, for example, also assist informed candidate GE models in that it could facilitate data harmonization and increase statistical power to test specific GE hypotheses. Hence, while there are challenges to investigating GE relations, there are also some solutions already available or in progress.

Addressing these two major issues (and others) will facilitate the development of more comprehensive working models to increase our understanding of how and when (under what conditions) intermediate phenotypes are influential in predicting phenotypically complex outcomes. In addition, more comprehensive working models may serve as a useful vehicle to develop, implement, and evaluate prevention and treatment programs and to guide public health policy.

12.5   Conclusions and Future Directions

The investigation of intermediate phenotypes for alcohol use and alcohol disorders has been a productive area of research that has drawn upon both prior research findings on risk and protective factors in general-population and clinical samples and advances in technologies associated with molecular genetics and neuroimaging. The selection of intermediate phenotypes has also been guided to some extent by epidemiological findings on the life course developmental nature of alcohol use and alcohol disorders, with the highest prevalence of alcohol dependence occurring between the ages of 18–24. This has intensified interest in early onset intermediate phenotypes such as behavioral undercontrol, dysregulation of neural affective circuitry, brain markers (e.g., evoked potential responses, structural and functional imaging neural signatures), individual differences in pharmacological responses, and ethanol-specific genetic variation. Many of the current intermediate phenotypes were foreshadowed by alcohol subtype research that identified critical systems (e.g., neurotransmitters) and behaviors (e.g., early onset antisocial behaviors) that remain important mechanisms to enrich with current technologies and advances in conceptual thinking (e.g., systems biology, developmental systems theory). There remains a need to further develop conceptual models and methodological procedures to better integrate information across multiple levels of organismic functioning within variable environments that are reacted to and created across the developmental life course. For alcohol use and alcohol disorders, there is also a need to integrate both alcohol-specific and alcohol nonspecific intermediate phenotypes into pathway models that enable the study of pleiotropic effects, GE, and GG relations.

Acknowledgment

This research was supported by National Institute on Alcoholism and Alcohol Abuse Grant No. K05AA021143 and National Institute on Drug Abuse Grant No. P30DA027827. The contents are solely the responsibility of the author and do not necessarily represent the official views of the National Institutes of Health.

References

1. Ducci, F., & Goldman, D. (2008). Genetic approaches to addiction: genes and alcohol. *Addiction (Abingdon, England), 103*, 1414–1428.

2. Meyer-Lindenberg, A., & Weinberger, D. R. (2006). Intermediate phenotypes and genetic mechanisms of psychiatric disorders. *Nature Reviews. Neuroscience, 7*, 818–827.

3. Bearden, C. E., & Freimer, N. B. (2006). Endophenotypes for psychiatric disorders: ready for primetime? *Trends in Genetics, 22*, 306–313.

4. Gottesman, I. I., & Gould, T. D. (2003). The endophenotype concept in psychiatry: etymology and strategic intentions. *American Journal of Psychiatry*, *160*, 636–645.

5. Insel, T. R., & Cuthbert, B. N. (2009). Endophenotypes: bridging genomic complexity and disorder heterogeneity. *Biological Psychiatry*, *66*, 988–989.

6. Waldman, I. D. (2005). Statistical approaches to complex phenotypes: evaluating neuropsychological endophenotypes for attention-deficit/hyperactivity disorder. *Biological Psychiatry*, *57*, 1347–1356.

7. Goldman, D. (1995). Identifying alcoholism vulnerability alleles. *Alcoholism, Clinical and Experimental Research*, *19*, 824–831.

8. Zucker, R. A. (2006). Alcohol use and the alcohol use disorders: a developmental-biopsychosocial systems formulation covering the life course. In D. Cicchetti & D. J. Cohen (Eds.), *Developmental Psychopathology: Vol. 3. Risk, Disorder and Adaptation* (2nd ed., pp. 620–656). New York: Wiley.

9. Windle, M. & Zucker, R.A. (2010). Reducing underage and young adult drinking: how to address critical drinking problems during this developmental period. *Alcohol Research & Health*, *33*, 29–44.

10. Li, T. K. (2004). Alcohol abuse increases, dependence declines across decade young adult minorities emerge as high-risk subgroups. NIH News.

11. Zucker, R. A., Donovan, J. E., Masten, A. S., Mattson, M. E., & Moss, H. B. (2008). Early developmental processes and the continuity of risk for underage drinking and problem drinking. *Pediatrics*, *121*(Suppl 4), S252–S272.

12. Grant, B. F., Dawson, D. A., Stinson, F. S., et al. (2004). The 12-month prevalence and trends in DSM–IV alcohol abuse and dependence: United States, 1991–1992 and 2001–2002. *Drug and Alcohol Dependence*, *74*, 223–234.

13. Substance Abuse and Mental Health Services Administration. (2006). National Survey on Drug Use and Health (Washington, DC).

14. Gunzerath, L., Hewitt, B. G., Li, T. K., & Warren, K. R. (2011). Alcohol research: past, present, and future. *Annals of the New York Academy of Sciences*, *1216*, 1–23.

15. Kessler, R. C., Chiu, W. T., Demler, O., Merikangas, K. R., & Walters, E. E. (2005). Prevalence, severity, and comorbidity of 12-month DSM–IV disorders in the National Comorbidity Survey Replication. *Archives of General Psychiatry*, *62*, 617–627.

16. Babor, T. F., Hofmann, M., DelBoca, F. K., et al. (1992). Types of alcoholics: I. evidence for an empirically derived typology based on indicators of vulnerability and severity. *Archives of General Psychiatry*, *49*, 599–608.

17. Bohman, M., Sigvardsson, S., & Cloninger, C. R. (1981). Maternal inheritance of alcohol abuse: Cross-fostering analysis of adopted women. *Archives of General Psychiatry*, *38*, 965–969.

18. Leggio, L., Kenna, G. A., Fenton, M., Bonenfant, E., & Swift, R. M. (2009). Typologies of alcohol dependence: from Jellinek to genetics and beyond. *Neuropsychology Review*, *19*, 115–129.

19. Cloninger, C. R., Bohman, M., & Sigvardsson, S. (1981). Inheritance of alcohol abuse: cross-fostering analysis of adopted men. *Archives of General Psychiatry, 38,* 861–868.

20. Cloninger, C. R. (1987). Neurogenetic adaptive mechanisms in alcoholism. *Science, 236,* 410–416.

21. Litt, M. D., Babor, T. F., DelBoca, F. K., Kadden, R. M., & Cooney, N. L. (1992). Types of alcoholics: II. application of an empirically derived typology to treatment matching. *Archives of General Psychiatry, 49,* 609–614.

22. Windle, M., & Scheidt, D. M. (2004). Alcoholic subtypes: are two sufficient? *Addiction, 99,* 1508–1519.

23. Zucker, R. A. (1994). Pathways to alcohol problems and alcoholism: a developmental account of the evidence for multiple alcoholisms and for contextual contributions to risk. In R. Zucker, G. Boyd, & J. Howard (Eds.), *The Development of Alcohol Problems: Exploring the Biopsychosocial Matrix of Risk* (NIH Publication No. 94–3495, pp. 255–289). Rockville, MD: National Institutes of Health.

24. Del Boca, F. K., & Hesselbrock, M. N. (1996). Gender and alcoholic subtypes. *Alcohol Health and Research World, 20,* 56–62.

25. Schulenberg, J. E., & Maggs, J. L. (2008). Destiny matters: distal developmental influences on adult alcohol use and abuse. *Addiction (Abingdon, England), 103*(Suppl 1), 1–6.

26. Sher, K. J., Grekin, E. R., & Williams, N. A. (2005). The development of alcohol use disorders. *Annual Review of Clinical Psychology, 1,* 493–523.

27. Tarter, R. E., Kirisci, L., Mezzich, A., et al. (2003). Neurobehavioral disinhibition in childhood predicts early age at onset of substance use disorder. *American Journal of Psychiatry, 160,* 1078–1085.

28. Schulenberg, J., O'Malley, P. M., Bachman, J. G., Wadsworth, K. N., & Johnston, L. D. (1996). Getting drunk and growing up: trajectories of frequent binge drinking during the transition to young adulthood. *Journal of Studies on Alcohol, 57,* 289–304.

29. Jackson, K. M., Sher, K. J., Rose, R. J., & Kaprio, J. (2009). Trajectories of tobacco use from adolescence to adulthood: Are the most informative phenotypes tobacco specific? In G. E. Swan, T. B. Baker, L. Chassin, D. V. Conti, C. Lerman, & K. A. Perkins (Eds.), *Phenotypes and Endophenotypes: Foundation for Genetic Studies of Nicotine Use and Dependence* (pp. 289–335). Bethesda, MD: National Institutes of Health.

30. Windle, M., Mun, E. Y., & Windle, R. C. (2005). Adolescent-to-young adulthood heavy drinking trajectories and their prospective predictors. *Journal of Studies on Alcohol, 66,* 313–322.

31. Schuckit, M. A., Smith, T. L., Anderson, K. G., & Brown, S. A. (2004). Testing the level of response to alcohol: social information processing model of alcoholism risk—a 20-year prospective study. *Alcoholism, Clinical and Experimental Research, 28,* 1881–1889.

32. Trim, R. S., Schuckit, M. A., & Smith, T. L. (2008). Level of response to alcohol within the context of alcohol-related domains: an examination of longitudinal approaches assessing changes over time. *Alcoholism, Clinical and Experimental Research, 32,* 472–480.

33. Schuckit, M. A., Smith, T. L., Danko, G. P., et al. (2009). An evaluation of the full level of response to alcohol model of heavy drinking and problems in COGA offspring. *Journal of Studies on Alcohol and Drugs, 70,* 436–445.

34. Krueger, R. F. (1999). The structure of common mental disorders. *Archives of General Psychiatry, 56,* 921–926.

35. Kendler, K. S., Prescott, C. A., Myers, J., & Neale, M. C. (2003). The structure of genetic and environmental risk factors for common psychiatric and substance use disorders in men and women. *Archives of General Psychiatry, 60,* 929–937.

36. Plomin, R., Haworth, C. M., & Davis, O. S. (2009). Common disorders are quantitative traits. *Nature Reviews. Genetics, 10,* 872–878.

37. de Wit, H. (2009). Impulsivity as a determinant and consequence of drug use: a review of underlying processes. *Addiction Biology, 14,* 22–31.

38. Gorenstein, E. E., & Newman, J. P. (1980). Disinhibitory psychopathology: a new perspective and a model for research. *Psychological Review, 87,* 301–315.

39. Windle, M. (in press). Behavioral undercontrol: A multifaceted concept and its relationship to alcohol and substance use and abuse. In R. A. Zucker & S. Brown (Eds.). *Oxford Handbook of Adolescent Substance Abuse.* New York: Oxford University Press.

40. Iacono, W. G., Malone, S. M., & McGue, M. (2008). Behavioral disinhibition and the development of early-onset addiction: common and specific influences. *Annual Review of Clinical Psychology, 4,* 325–348.

41. Hussong, A. M., Wirth, R. J., Edwards, M. C., et al. (2007). Externalizing symptoms among children of alcoholic parents: entry points for an antisocial pathway to alcoholism. *Journal of Abnormal Psychology, 116,* 529–542.

42. Krueger, R. F., Markon, K. E., Patrick, C. J., & Iacono, W. G. (2005). Externalizing psychopathology in adulthood: a dimensional-spectrum conceptualization and its implications for DSM–V. *Journal of Abnormal Psychology, 114,* 537–550.

43. Dick, D. M. (2007). Identification of genes influencing a spectrum of externalizing psychopathology. *Current Directions in Psychological Science, 16,* 331–335.

44. Spanagel, R. (2009). Alcoholism: a systems approach from molecular physiology to addictive behavior. *Physiological Reviews, 89,* 649–705.

45. Caspi, A., Sugden, K., Moffitt, T. E., et al. (2003). Influence of life stress on depression: moderation by a polymorphism in the 5-HTT gene. *Science, 301,* 386–389.

46. Windle, M. (2000). Parental, sibling, and peer influences on adolescent alcohol use and alcohol problems. *Applied Developmental Science, 4,* 98–110.

47. Rose, R. J., Dick, D. M., Viken, R. J. & Kaprio, J. (2001). Gene–environment interaction in patterns of adolescent drinking: regional residency moderates longitudinal influences on alcohol use. *Alcoholism, Clinical and Experimental Research, 25,* 637–643.

48. Weitzman, E. R., Folkman, A., Folkman, M. P., & Wechsler, H. (2003). The relationship of alcohol outlet density to heavy and frequent drinking and drinking-related problems among college students at eight universities. *Health & Place, 9*, 1–6.

49. Mrug, S., Gaines, J., Su, W., & Windle, M. (2010). School-level substance use: effects on early adolescents' alcohol, tobacco, and marijuana use. *Journal of Studies on Alcohol and Drugs, 71*, 488–495.

50. Brody, G. H., Beach, S. R., Philibert, R. A., et al. (2009). Parenting moderates a genetic vulnerability factor in longitudinal increases in youths' substance use. *Journal of Consulting and Clinical Psychology, 77*, 1–11.

51. Chen, L. S., Johnson, E. O., Breslau, N., et al. (2009). Interplay of genetic risk factors and parent monitoring in risk for nicotine dependence. *Addiction, 104*, 1731–1740.

52. Ribbe, K., Ackermann, V., Schwitulla, J., et al. (2011). Prediction of the risk of comorbid alcoholism in schizophrenia by interaction of common genetic variants in the corticotropin-releasing factor system. *Archives of General Psychiatry, 68*, 1247–1256.

53. Ressler, K. J., Bradley, B., Mercer, K. B., et al. (2010). Polymorphisms in CRHR1 and the serotonin transporter loci: gene × gene × environment interactions on depressive symptoms. *American Journal of Medical Genetics. Part B, Neuropsychiatric Genetics, 153B*, 812–824.

54. Karg, K., Burmeister, M., Shedden, K., & Sen, S. (2011). The serotonin transporter promoter variant (5-HTTLPR), stress, and depression meta-analysis revisited: evidence of genetic moderation. *Archives of General Psychiatry, 68*, 444–454.

55. Risch, N., Herrell, R., Lehner, T., et al. (2009). Interaction between the serotonin transporter gene (5-HTTLPR), stressful life events, and risk of depression: a meta-analysis. *Journal of the American Medical Association, 301*, 2462–2471.

13 Epigenetic Effects and Intermediate Phenotypes

Steven R. H. Beach, Meg Gerrard, Gene H. Brody, Ronald L. Simons,
Steven M. Kogan, Frederick X. Gibbons, and Robert A. Philibert

An intermediate phenotype approach to understanding genetic causation of substance use disorders (SUDs) has many advantages [1], including helping to clarify mechanisms intervening between genes and disorders, substantially clarifying mediational pathways, and opening the way to the use of potentially powerful experimental approaches to better probe environmental influences on genetic effects (e.g., [2, 3]). In the current chapter we suggest, however, that an intermediate phenotype approach is incomplete unless it is coupled with attention to the role of epigenetics. Epigenetic modification is a ubiquitous biological process that involves attaching or removing chemical groups at key genetic locations, literally "above the genome," a process we describe in more detail below. Epigenetic modifications of the genome are presumed to combine with genetics, learning history, and current stimulus and reinforcement environments to produce the broad behavioral intermediate phenotypes and substance use patterns of interest to intervention and prevention researchers. Accordingly, greater exploration of epigenetic processes is likely to be essential in order to develop more complete mechanistic models and to stimulate research on mechanisms of change.

We begin by providing a general, very brief overview of several common epigenetic change processes. We note the likely relevance of coregulation of intermediate phenotypes and provide an initial model, followed by a discussion of the evolutionary plausibility of epigenetic influence on intermediate phenotypes. We conclude with a brief discussion of recent findings, an overall organizing model for future research, and a brief discussion of methodological issues. Throughout, we note the rapid development and expansion of this area of research and focus particularly on the epigenetic change mechanism for which measurement is best developed at the current time: gene methylation. Given the fast-paced nature of research in epigenetics, enhanced measurement of several mechanisms of epigenetic regulation will become more available in the near future, and so it is important to note that the broad epigenetic framework we present is generalizable to all forms of epigenetic change. Central to the argument we develop below is the possibility that epigenetic change may result from several types of environmental exposures, including variation in parenting, and that stressful events in childhood, such as those deriving from actions of an attachment figure, are a likely source of epigenetic change. Finally, central to the model presented is the expectation that epigenetic processes exert their effect by influencing gene transcription and ultimately gene products, including changes in morphology and function, that is, we expect epigenetic influences to utilize the same mechanisms by which genotypic variation is assumed to influence intermediate phenotypes, creating a "biological interaction" between the impact of epigenetic change processes and the impact of genotypic variation on intermediate phenotypes and substance use outcomes.

13.1 Types of Epigenetic Change

Epigenetic change refers to alterations in DNA that can be transmitted from the parent cell to the daughter cell but that do not involve changes in the base pairs that make up the genetic code. In some, but not all, cases these alternations may be transmitted to subsequent generations. However, intergenerational transmission seems more likely from mothers to offspring than from fathers to offspring for several reasons, including the fact that DNA contributed by the paternal line is systematically demethylated during early embryogenesis in many organisms [4]. In mammals, molecular-level epigenetic effects appear to be the result of just a few basic mechanisms including histone modifications, processes influencing RNA and RNA cross talk, and DNA methylation or acetylation . Each of these basic mechanisms is present for humans and has the potential to influence gene expression or activity. In turn, changes in gene expression or activity are thought to give rise to downstream outcomes of interest such as changes in developmental trajectories, structural changes in organ systems with biological significance, or changes in cellular environments that alter posttranslational products of genes.

13.1.1 Histone Modification
Histones provide the scaffold (or infrastructure) around which DNA can wind, either more tightly or loosely, allowing DNA to form into structures such as nucleosomes, which themselves bundle together to form larger structures such as chromatin fibers, which in turn make up chromosomes. An important aspect of histones, and the characteristic that makes histone modifications a vehicle of cellular information, is that they have a flexible ("N-terminus") tail that protrudes and that can be modified. Modifications of the tail has a substantial effect on the shape of the "infrastructure" of the DNA, making information in some areas more or less available. In particular, histone tail modifications are related to alterations in the structure of the DNA molecule, and this, in turn, is substantially associated with gene expression levels. Histone modifications influence structure, either loosening (opening up for greater transcrip-

tion) or tightening (reducing transcription), by altering the electrostatic charge of the histones, leading to shifts in chromatin structure and so the ability of molecules involved in gene expression to bind to key regions and exert regulatory influence over the many biological processes leading to gene expression. These modifications may be transient and highly reversible in some cases. Moreover, it appears that a relatively small number of histone modifications may accurately predict gene expression [5].

It should be noted that many histone modifications (and methylation changes and RNA interference processes as well) are normative and essential for normal cell development and organ differentiation. As epigenetic changes are copied in daughter cells within a particular cell line, this allows them to retain the altered function acquired by the parent cell. So, silenced areas remain silenced as organs differentiate. Therefore, histone modifications (and methylation and acetylation changes) are important in understanding the way that different cell types are formed and the way that developmentally specific and cell-specific patterns can be executed from the same general genetic blueprint. Thus, it is shifts in pattern of histone modification over time, relative to the background of development, that are of greatest interest to behavioral researchers. It is also noteworthy that histone modifications and DNA methylation are correlated processes.

In brief, histone modifications and changes in methylation exert their influence by changing the information contained in the chromatin structure that "shapes" the genome, rendering protein encoding instructions in DNA more or less available for transcription. In this way, the information carried by an individual's DNA is supplemented by modifiable information from epigenetic mechanisms. These supplemental effects may occur at the transcription regulatory phase. Epigenetic changes are transmitted to daughter cells, conferring a lasting effect on the individual that continues beyond the life of the initially affected cell. These processes help explain the impact of the environment on gene expression or activity, provide a means for individual organisms to respond to changing environments, and may also help explain rapid evolutionary changes in heritable phenotypes [6].

13.1.2 Processes Influencing RNA and RNA Cross Talk

Only one fifth of gene transcription across the human genome is directly associated with coding protein products [7], suggesting that there are a range of important regulatory functions mediated by RNAs. Epigenetic changes influencing RNA cross talk are of particular interest because the effects can occur rapidly in response to the environment and produce profound effects. Because RNA-mediated gene regulation is widespread, examination of this source of phenotypic variation may prove very interesting to behavioral researchers in the future, providing a biological basis for positing and explaining some epistatic (gene × gene) interactions involving widely separated genes.

In addition, a number of different mechanisms for control of cell function by small RNAs have been discovered. In particular, messenger RNA (mRNA) can be targeted for cleavage by short interfering RNA (siRNA)-protein complexes, or translation can be prevented by microRNAs (miRNAs). Either way, the mRNA is eventually destroyed by the cell, and so gene function is regulated and cell function is changed. If this occurs over time or at key developmental stages, there could be implications for phenotypic structures that influence later behavior. Of particular interest in the context of stress are short interspersed nuclear Alu elements that are often transcribed in response to environmental stresses [8] and potentially inhibit preinitiation complexes, creating broad and rapid repression of gene expression in response to certain stressors [9], making them of great interest as measurement technologies become increasingly accessible to behavioral scientists. Likewise, RNA editing, a posttranslation process that may influence miRNAs and may selectively alter protein products in response to the environment [10], is an area of growing interest. In brief, epigenetic changes influencing RNA cross talk alter gene expression but more often affect the regulation of mRNA that has already been transcribed, providing another level of epigenetic control over the activity of cells, potentially up- or downregulating cellular processes.

13.1.3 DNA Methylation

Despite the considerable promise of greater attention to histone modification and epigenetic modification of the activity of noncoding RNAs for future theorizing and research, we restrict our attention in the remainder of the chapter to DNA methylation. This reflects the revolution that has occurred in the technology for DNA methylation assessment and analysis over the past decade. There has been considerably more work done in model organisms and humans related to methylation than to other forms of epigenetic control, creating a substantial foundation of biologically plausible mechanisms for behavioral scientists to explore. In addition, analyses that previously could be performed only at specific CpG sites can now be performed on a genome-wide basis, and entire methylomes can be characterized at single-base-pair resolution, creating excitement about the potential precision of epigenetic characterization that is now possible. Likewise, costs, particularly estimated future costs, now make contemplating inclusion of such analyses in the context of behavioral investigation feasible and potentially attractive. Finally, DNA methylation can be characterized as patterns, potentially allowing behavioral researchers to examine similarity in patterns across tissues. These characteristics make DNA methlation particularly attractive to behavioral researchers interested in characterizing biological change that may predict future stable intermediate behavioral phenotypes. Notwithstanding this excitement, it should be noted that the opposing process of acetylation is also likely to be of considerable interest to behavioral scientists as measurement strategies become more tractable.

One factor contributing to interest in DNA methylation is that it occurs across the entire genome. However, a sobering corollary of this observation is that a large proportion of global DNA methylation occurs in noncoding regions such intragenic regions that are rich in DNA of viral origin. In this context it is likely that DNA methylation is meant to protect the organism from viral intrusions rather than to fine-tune regulatory processes. This should caution behavioral scientists as we explore behavioral connections to the epigenome. Broad patterns of DNA methylation are established during early embryonic and fetal life, and, as noted above, are essential to normal cellular development and differentiation. However, several lines of research demonstrate that methylation also can be influenced by environmental factors throughout the life span, perhaps especially during some periods of rapid developmental change. In humans, methylation is mediated by three methyltransferases that facilitate the transfer of a methyl group from a methyl donor to the carbon-5 position of a cytosine–guanosine (CpG) dinucleotide pair, resulting in a methylated residue. It is this methylated residue that is assessed when methylation at specific CpG sites are examined. Typically, methylation is reported as the fraction of molecules in the sample that are methylated at a specific CpG site, creating a possible range of values from 0% of the cells in the sample exhibiting methylation at the specified CpG site to 100% exhibiting methylation.

Methylation at a particular location typically has the effect of reducing the expression of gene products associated with that particular location. DNA methylation adds a methyl group into the major groove of the DNA, creating a physical barrier to transcription factor binding proteins, thus inhibiting transcription. At the same time, methylated DNA is more readily bound by proteins that initiate "chromatin remodeling," which ultimately generates a different shape for the methylated DNA (as described above under histone modification). In this way methylation of a region can lead to an entirely inactive structure, that is, one that is tightly wound, permitting no interaction with the cellular environment. This highlights several ways that CpG methylation could have an impact on phenotypes or elements of phenotypes by altering gene expression, and it demonstrates that this impact could occur in both an all-or-none or a graduated manner.

It has long been believed that methylation patterns are responsible for stably maintained differences in gene expression across tissue types [11]. Likewise, the relationship between promoter DNA methylation and changes in gene activity are well established [12, 13], with hypermethylation in promoter CpG islands typically resulting in decreased transcription of downstream genes [14] whereas experimentally induced hypomethylation of promoter regions leads to an increase in gene transcription (e.g., [15]). More recent research has shown that DNA methylation patterns are also predictive of individual differences in gene expression [16, 17] and cross-species differences in gene expression [18] and has also confirmed the particular importance of methylation of promoter regions in predicting gene expression [16, 17]. Accordingly, methylation of regions associated with gene regulation, that is, gene promoter regions, is likely to be of particular interest to behavioral scientists.

CpG Island Methylation

Interestingly, CpG pairs, which are the targets of methylation, are not randomly distributed across the human genome. In most regions of the genome, the frequency of CpG dinucleotide pairs in the human genome is much less than would be expected by chance. In those regions where CpG pairs do exist they are typically concentrated in areas known as "CpG islands." In these areas there is a high density of CpG residues, and these CpG islands are often found in close proximity to gene promoters, that is, regions known to be important in regulating gene expression. Because dense arrays of "CpGs" frequently occur in areas affecting gene transcription, this makes the regions of DNA responsible for regulation of gene transcription more likely to be responsive to environmental input due to increased susceptibility to epigenetic regulation. From the standpoint of understanding intermediate phenotypes, therefore, this suggests that the genome may be prepared to respond to the environment and to modify various genetic programs based on experience. Because methylation in these areas typically alters the gene expression pattern in cells in a relatively stable manner, it suggests that CpG island methylation occurring as a result of experiences or exposures during development might result in broad, stable alterations in patterns of RNA transcription. Accordingly, it is possible to provide a priori, probabilistic predictions regarding the biological effect of differing levels of methylation that may result from developmental exposures. Two caveats of potential importance are (1) some changes in methylation may be rapid and/or transitory [19], suggesting dynamic or cumulative processes that are as yet poorly understood, and (2) changes in methylation may vary according to proximity to gene transcription start site [16], adding some potential complexity to the measurement of level of overall promoter region methylation.

13.2   Coregulation of Intermediate Phenotypes by Genetics and Epigenetics

As the brief overview of epigenetic change processes suggests, one way in which the environment may work in concert with genotype to influence intermediate phenotypes is by acting on gene expression via epigenetic change. Just as functional genetic polymorphisms exert their influence by altering gene expression or gene activity, so too can gene methylation regulate gene expression by altering rate of transcription or by altering the nature or activity of the resulting transcripts. Accordingly, environmental influences alter patterns of gene methylation, which, in turn, influence gene expression or activity. Thus, gene methylation will influence the same biological pathways as those thought to link genotype to intermediate and distal SUD pheno-

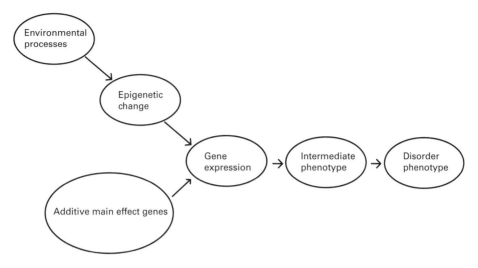

Figure 13.1
Epigenetic and additive genetic effects share a common pathway to intermediate phenotypes.

types, allowing for joint and interactive transactions between genotype and epigenetic change through a shared biological pathway. This understanding of the link between the impact of epigenetic changes and the additive main effects of genotypes through their joint impact on gene expression is summarized in figure 13.1. We explore the theoretical and practical implications of a model of joint environmental and genetic control over intermediate phenotypes below, but first we briefly consider the evolutionary plausibility of epigenetic reprogramming by environments, addressing the question of how a mechanism of epigenetic control of intermediate phenotypes could have been selected.

13.3 Evolutionary and Biological Plausibility of Epigenetic Influence

An evolutionary foundation for understanding the regulatory significance of epigenetic change has the potential to add conceptual clarity. In particular, to the extent that the regulatory effects of epigenetic processes have evolutionary significance, one might anticipate finding a recognizable set of environmental stressors that occur with some frequency, that regularly require an adaptive solution, and that produce coordinated patterns of response, that is, that influence intermediate phenotypes. Thus, evolutionary plausibility would provide considerable encouragement for the inclusion of epigenetic processes in accounts of the development of intermediate phenotypes. In particular, if the requirements for high- versus low-stress environments are typically sufficiently different that differential phenotype development might confer an adaptive advantage in preparing for future environmental stress, one might expect evolutionary pressure in favor of flexibility in the development of phenotypes. This would provide a foundation for the hypothesis that some coherent intermediate phenotypes emerge, in part, as a response to epigenetic change processes. In our examples we focus primarily on current parental behavior because it may sometimes create stress, and it may also convey information about external pressures and stress on the parents, creating the opportunity for parenting behavior to signal a need for epigenetic change and guide differential development. However, similar arguments could be made for other sources of environmental stress and support that confer information about the future environment with sufficiently long-term implications.

How could parental behavior function as the conduit of complex information about likely future stressful circumstances for which the developing child must be prepared? Unfortunately, high levels of stress on caregivers produce reliable negative changes in parenting behavior as well as patterns of differential parental investment [20]. Therefore, among species, like humans, who care for their relatively helpless young for an extended period of time, protracted poor-quality parenting could provide a reliable signal for the developing child about level of stress in the external environment and the likely implications of that stress for future demands on the self. If so, to the extent that poor parenting predicts future environmental constraints and threats, poor parenting could be a useful and informative stimulus for epigenetic reprogramming of intermediate phenotypes. This reprogramming could allow individuals to "make the best of a bad situation" in the perhaps brief and stressful time available to them while discounting potential future costs that may be less relevant than short-term survival (e.g., [21]). Accordingly, when parents are under severe environmental stress, the poor parenting that results might be hypothesized to entrain developmental pathways and prepare the individual for a life of deprivation or hardship in the future. Likewise, other factors that directly impinge upon the child, such as malnutrition, community violence, or exposure to racism, might be hypothesized to similarly entrain developmental processes.

Further accentuating the predictive value of negative parenting and other stressful contextual processes for a child's future environment is the potential for low levels of parental involvement to be associated with differential parental investment in offspring, a process to which siblings are known to be attentive [22] and that may be exacerbated by external stress such as discrimination [23]. Differential parental investment may be particularly useful in predicting future parental investment at key transitions, such as the transition to independent adult status, suggesting substantial predictive power for a likely stressful transitional period in the youth's future. Given the substantial implications of parental investment for life course, particularly in a stressful environment, the amount of parental support that is likely during the transition to adulthood could signal the value of alternative intermediate phenotypes. These insights and hypotheses can be incorporated into life history theory (LHT) [24], a well-developed evolutionary perspective on flexible intermediate phenotypes, yielding additional implications for phenotypic consequences of early family environment.

As described above, methylation of CpG residues in response to particular family environments is one mechanism that would allow for responsiveness to the environment. Patterned change in the methylation across key CpG sites could, in turn, create broad shifts in morphology or function during development or might alter gene expression into adulthood. Such changes, perhaps leading to a particular intermediate phenotype, might prepare the individual to be better adapted to future stressful environmental conditions. This would be particularly important in preparing for the transition to adulthood. If the transition to adulthood is likely to occur in the midst of an adverse environment, epigenetic influences on intermediate phenotypes that shifted responses in a way that increased probably of survival or reproductive success could pass the test of evolutionary plausibility.

13.3.1 Comment on Potential Genetic Moderation of Environmental Effects

There may be genetically influenced variability in the extent to which individuals are open to entrainment by the environment [25]. This could lead to individual differences in sensitivity [26] or susceptibility [27] to the impact of contextual effects, rather than uniform epigenetic responsiveness to adversity or to the presence of enriched environments. Some individuals may respond to environmental conditions in a "for better or for worse" manner, with genotype moderating the impact of environments on methylation or other downstream biological effects, ultimately influencing behavioral outcomes. If such effects of genotype on environmental susceptibility are common and replicable, as they currently appear to be [28], they could introduce important moderators of epigenetic effects as illustrated in figure 13.2. In particular, susceptibility and sensitivity effects raise questions about whether epigenetic reprogramming effects are the result of interactions between genes and environment (pathway 1 in figure 13.2). This could happen if certain genes amplified the effect of environment on circulating stress hormones or other biological consequences, eventually leading to an amplified environmental impact on epigenetic changes within susceptible individuals. Conversely, epigenetic change and susceptibility genes could interact to influence gene expression (pathway 2 in figure 13.2). This could happen if low-expressing or low-activity alleles were more susceptible to being blocked by methylation of promoter regions or if threshold effects resulted in greater phenotypic change for those with low-expressing genotypes. Finally, susceptibility could result from an interaction between gene expression at a distal location and variation in susceptibility genes (pathway 3 in figure 13.2). This could happen if the epigenetic effects influenced key signaling or regulatory pathways that were then differentially amplified or dampened by the susceptibility allele. Each of the potential pathways of influence would suggest

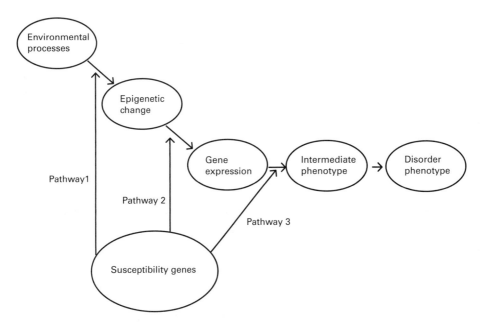

Figure 13.2
Epigenetic and susceptibility alleles.

a different mechanism and so lead to a different understanding of the processes driving the susceptibility effect. Given the current lack of data to distinguish among these different mechanisms of effect, below we focus on main effects of epigenetic response to adversity without precluding the possibility that there may be sensitivity or susceptibility genes that further amplify negative environmental influence or that are associated with greater responsiveness to supportive, nurturing, enriched environments.

13.3.2   Biological Plausibility of Epigenetic Effects on Intermediate Phenotypes

The proposal that environmental stressors or other aspects of the environment may entrain child development, leading to epigenetic reprogramming of developing individuals, and ultimately contributing to intermediate phenotypes, requires good evidence that environmental exposures can produce change in methylation. Evidence for this proposition comes from several sources: dietary manipulation, toxic exposure, and recent preliminary work on family/parenting stress which we review separately below. For example, in the Agouti mouse, dietary manipulations (restricting folate) can create changes in methylation that, in turn, produce dramatically different phenotypes for genetically identical mice. Lack of methylation of specific CpG sites within the Agouti gene, due to a maternal diet low in folate during gestation, results in a mouse with yellow fur, high body weight, and a propensity for illness. Conversely, a genetically vulnerable fetus whose mother is fed a diet high in folate during gestation will have well-methylated CpG sites within the Agouti gene and will appear brown, slim, and healthy into adulthood. With regard to toxic exposures, arsenic exposure is particularly informative regarding impact on methylation. Recent work has established that the toxic effect of arsenic is mediated through its impact on widespread DNA hypomethylation, apparently leading to many of its harmful long-term effects. A similar set of processes appears to account for the toxic effects of benzene, which also induces global DNA hypomethylation. Although illustrative of the potential for environmental exposures to influence methylation, and the potential for these changes to influence phenotypes, these examples do not directly address the question of whether methylation patterns are influenced by variation in social circumstances; that is, whether rearing environments, environmental deprivation, or other psychosocial stressors linked to adverse outcomes may lead to methylation changes and, in turn, lead to noticeable phenotypic effects. Fortunately, recent research with animal models and with humans has begun to address this issue as well.

13.4   Parenting Stress as a Source of Epigenetic Change

How could changes in maternal (or other parental) behavior have an impact on epigenetic change such as DNA methylation? The mechanism highlighted in research to date is that maternal behavior may influence stress hormone levels, changing the cellular environment in key areas of the brain [29], producing methylation changes and so leading to impact on stress response later in life [30–33]. More broadly, the effect of early adverse rearing environments on methylation patterns may be initiated by prolonged hormonal changes. As noted by Gunnar and Quevedo [34], although stress responses are necessary for survival, prolonged stress has been linked to an increased probability of future physical and mental health problems [35]. In particular, chronic stressors occurring during development appear to produce sustained hormonal changes with the potential to alter function of many tissues [36]. Such changes seem counterproductive from an evolutionary perspective unless considered in a broader conceptual framework such as LHT [37], which highlights the relative advantages of different phenotypes as adaptations to likely future challenges and threats. From a life history perspective, it may make evolutionary sense in some contexts for the developing individual to optimize his or her phenotype for a shorter, faster life plan. This provides some opportunity for development, and perhaps reproduction, at a younger age, or perhaps the opportunity to accrue other short-term survival benefits, even if these come at some cost in other respects.

13.4.1   Life History Theory as an Organizing Framework
LHT describes strategic patterns of differential resource allocation depending on the likely payoff for future growth, maintenance, and reproduction. Depending on likely

future environmental pressure, investments in the development of different phenotypes can be expected to result in different relative payoffs, with calculations being
organized by a "speed" of life course calculation [24]. Slow life strategies and associated
phenotypes are favored in contexts that suggest strong payoff of delayed reproduction.
Such contexts encourage a stronger focus on the future, optimizing growth, high
parental investment, delayed parenthood, and commitment in reproductive relationships. Slow strategies and phenotypes should be favored in less stressful environments
because these suggest longer life expectancies, allowing for greater average reproductive success through higher quality offspring who will have greater reproductive
success on average. In contrast, fast strategies are favored in contexts that suggest a
relatively high probability of an early death or an abbreviated reproductive career. In
stressful, dangerous contexts with poor long-term prospects, evolutionary pressure
should favor reproductive strategies that involve less commitment, greater likelihood
of early sexual behavior with multiple partners, and less aversion to long-term risks
[38, 39]. In more stressful or dangerous environments, fast strategies provide a mechanism allowing individuals to adopt adult roles early, before they are killed or incapaci-
tated, thereby enhancing reproductive fitness in the context of poor long-term survival
prospects, albeit at a probable cost to longevity or health later in the life course.

13.4.2   Integration of Life History Theory and Parenting Stress

Because poor parenting and differential (worse) treatment relative to siblings is one
signal of a potentially negative environment for the self in the future, one would
anticipate that poor parental care or other contexts directly affecting the child during
key developmental periods could reliably signal the need for a faster life history strategy. In addition, to the extent that parents are a source of stress by way of their poor
parenting behavior, they signal the need for faster development among offspring in
a manner that is consistent with their own reproductive interests in stressful contexts,
that is, minimizing per-offspring investment and increasing degree of differential
treatment across siblings to maximize strategic parental investment. From the perspective of LHT, poor parenting and other stressors that directly affect the developing child
can be seen as conveying information about level of chronic stress, danger, and disadvantage, in the current as well as the likely future environment, triggering the
development of epigenetic intermediate phenotypes associated with "fast" life strategies. Further, effects on epigenetic change and intermediate phenotypes could, in
principle, be initiated by activity of the hypothalamic–pituitary–adrenal (HPA) axis or
other biological responses to chronic stress (i.e., allostatic load variables; [36, 40]).

13.4.3   Likely Relevance of HPA-Axis Activity

The HPA axis is an integrated system of great interest for behavioral researchers because
it is implicated in a variety of behavioral tendencies. It is also of particular interest

from the standpoint of induction of methylation changes because HPA system response results in the production of glucocorticoids (i.e., cortisol) which can cross the blood–brain barrier and so has the potential to impact a range of different tissues throughout the body and brain, potentially initiating a range of changes in a systematic and coordinated manner. In addition, many of its effects appear to occur by inducing changes in gene expression [41]. Finally, because the hormonal cascade produced by the HPA axis in response to stress has the potential to alter cellular environments in peripheral tissue and in the central nervous system (CNS), sustained HPA activation has ideal characteristics to produce epigenetic reprogramming that could be detected in peripheral tissue and not just in tissue taken from the CNS, simplifying research strategies for behavioral researchers. Conversely, because tissue-specific silencers, enhancers, and transcription factors produce differences in gene expression between tissue types even when methylation patterns are similar, gene expression patterns may show greater variability than gene methylation patterns across tissue types.

13.4.4 Evidence from Animal Models

Illustrating the potential of LHT to organize patterns of results, pups born to mothers who exhibited high levels of grooming were less anxious in a novel environment and showed a reduced steroid response to stress compared with offspring of low-grooming mothers [42, 43], a pattern that would be adaptive if maternal grooming predicted future threat in the environment. In addition, cross-fostering studies confirmed that the creation of differing phenotypes was mediated by variation in maternal care [32, 44], confirming the environmental source of the effect. Likewise, there was differential CNS expression of neuron-specific glucocorticoid receptors (GRs) [29] and differential methylation of a relevant gene promoter region [43, 45, 46], confirming epigenetic change. Consistent with an intermediate phenotype approach, the differences remained after weaning and into adulthood. That is, there was a constellation of changes in hormones (more corticosterone), brain morphology (i.e., shorter dendritic branches and changes in density of hippocampal neurons), and complex, functional behavioral dispositions including changes in behavior under stress [47], changes in open-field exploration, early sexual behavior, and parenting [48, 49] that might be expected to work together to enhance survival and reproduction under adverse circumstances. Consistent with the model presented in figure 13.1, low-quality maternal care (i.e., low levels of maternal grooming of the pup) had a substantial, long-term impact on pups' stress reactivity, with effects apparently mediated by epigenetic change [50]. The range of the changes associated with poor maternal care is striking and suggestive of just the sort of change in "intermediate phenotype" that might have been selected through evolutionary pressure to help rat pups "make the best of a bad situation." Specifically, the complex pattern of impacts seems to form a coherent intermediate phenotype designed for enhanced performance under high-threat conditions.

13.4.5   Research with Humans

Supporting the potential for rearing environments and early stressful experiences to exert an influence on epigenetic change in human development, methylation differences have been demonstrated in the postmortem hippocampus obtained from suicide victims with a history of childhood abuse relative to those from either suicide victims with no childhood abuse or nonsuicide controls [51, 52]. Likewise, the GR gene shows increased methylation in cord-blood DNA drawn from newborns of depressed or anxious mothers [53], and the IGF2 gene shows reduced methylation in blood DNA of individuals exposed to famine prenatally [54]. In each of these cases, it has been speculated that environments may stimulate changes in stress hormones or other systemic biological effects that, in turn, produce specific patterns of change in the epigenome, regulating growth, metabolism, immune responsiveness, developmental pace, and behavior.

Despite the compelling nature of the heuristic models already available, and good evidence that childhood events producing chronic HPA axis arousal among children and youth are associated with identifiable and reliable negative long-term outcomes (e.g., [36, 55]), it is yet to be established that such events are associated with reliable changes in DNA methylation signatures or other broad epigenetic patterns or that changes in methylation patterns are a mediator of such impacts. Likewise, it is unknown if different types of threatening or arousing events are capable of producing different patterns of change or whether level of threat may produce different epigenetic change depending on developmental stage or the presence of "safety signals" such as parents to whom one has a strong, positive attachment. As research progresses it will be useful to focus attention on possible shifts toward hyper- or hypomethylation of specific areas of the genome, as well as broader impacts across gene networks or the entire genome. An additional pattern of likely future interest will be global DNA hypomethylation secondary to particular patterns of substance use. Hypermethylation [56, 57] or hypomethylation in coding areas, in promoter regions, or in areas around transcription start sites will also be of particular interest. Likewise, genome-wide patterns [58] are of considerable interest, as well. Accordingly, to illustrate emerging work in this area, we describe two recent studies in more detail, setting the stage for a discussion of methodological issues likely to be important as behavioral scientists incorporate assessments of methylation into their research.

13.5   Empirical Investigations of Family Processes and Epigenetic Change

13.5.1   Gene-Specific Epigenetic Effects

As a preliminary examination of the plausibility of methylation change as a function of family stressors, Beach et al. [57] conducted an investigation of the impact of child abuse on changes in methylation in the promoter region of *5HTT* [56, 57]. We examined *5HTT* because this is a highly conserved and phylogenetically old component of the serotonergic regulatory system, with a range of implications for behavioral dispositions. Regulation of this gene, and the gene network within which it is embedded, has implications for a range of behaviors relevant to survival in stressful circumstances. In addition, primates exposed to a stressful environment during early development show altered CNS serotonin system functioning that produces long-term effects on behavior [59]. These effects include heightened aggressiveness, impaired impulse control [60], and problematic social behavior [61]. Similar effects involving decreased responsiveness of the serotonin system have been observed in humans [62]. Given these observations, we proposed that hypermethylation of the promoter region of *5HTT* might be one of the mechanisms by which the experience of child sex abuse contributes to an intermediate phenotype characterized by antisocial and other risky health-related behavior that may increase risk for substance abuse.

DNA was prepared from lymphoblast cell lines derived from 155 female participants in the Iowa Adoptee Study. Methylation at 71 CpG residues in the promoter region of *5HTT* was determined by quantitative mass spectroscopy, and the resulting values were averaged to produce an average CpG ratio for each participant. Simple associations and path analyses within an Mplus framework were examined to characterize the relationships among childhood sex abuse by a family member, overall level of methylation, and subsequent antisocial behavior in adulthood. Because the sample was drawn from an adoption study, we were also able to control the effect of biological parent psychopathology as well as *5HTT* genotype, and so control for additive gene effects. Replicating prior work [56], a significant effect of childhood sex abuse on methylation of the *5HTT* promoter region emerged for women. In addition, a significant effect of methylation at *5HTT* on symptoms of antisocial personality disorder (ASPD) emerged for women. This pattern of association suggested the possibility of mediation in a manner consistent with figure 13.1, that is, mediation of the effect of family context on a behavioral intermediate phenotype by epigenetic change. To examine mediation, we examined the effect of constraining the direct path from child sex abuse to symptoms of ASPD to be zero (see figure 13.3). Using nested model comparisons, we found that constraining the direct pathway to be zero yielded a nonsignificant change in model fit, indicating that the direct pathway was not needed and so supporting the hypothesis that the direct effect of child sex abuse on symptoms of ASPD was fully mediated through methylation of *5HTT*.

We concluded that child sex abuse may create long-lasting changes in a behavioral phenotype by producing changes in methylation of the promoter region of *5HTT* in women. Further, we concluded that hypermethylation in key gene promoters, such as the promoter region for *5HTT*, may be one mechanism linking childhood sex abuse to changes in risk for adult antisocial behavior in women. Accordingly, this study suggests potential for environmental effects on methylation to contribute to an important

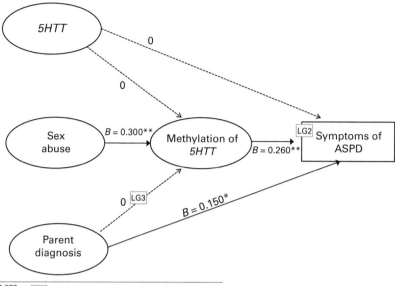

Figure 13.3
The main effect model in which the direct path from sex abuse to antisocial personality disorder (ASPD) is constrained to be zero, indicating full mediation through methylation at *5HTT*.

intermediate phenotype known to be associated with elevated substance use. An important caveat is that comparable effects of child abuse on methylation of *5HTT* were not observed in men in the replication sample.

Genome-wide Patterns

It is also of interest to examine whether environmental exposures during early childhood might result in coordinated patterns of methylation across the genome. A recent study by Essex et al. [58] indicates both the challenge and promise of consideration of the full genome. They examined a sample of 109 adolescents drawn from the Wisconsin Study of Families and Work for whom considerable longitudinal data were available. Measures of parental stress were obtained from mothers and fathers in several domains: (a) depression symptoms, (b) family expressed anger, (c) parenting stress, (d) role overload, and (e) financial stress, with early assessments at 1, 4, and 12 months postdelivery of the target child, and later, preschool period assessments when children were 3.5 and 4.5 years of age. To their existing longitudinal data set Essex et al. [58] added an assessment of DNA methylation using buccal cells obtained from the youth when they were in midadolescence. Microarray technology was used to examine methylation at 28,000 CpG residues across the genome, with a focus on promoter regions and first exons of coding regions. Following identification of CpG sites for which methylation was associated with maternal or paternal experience of stress, results were interpreted using the web-based "DAVID" platform (http://david.abcc.ncifcrf.gov/) to identify functional clusters of affected genes.

Using overall stress across measures as an index, it was found that maternal stressors in infancy and paternal stressors in the preschool years were most strongly predictive of methylation, with patterns varying somewhat by children's gender. Differences in methylation at 139 CpG sites were associated with greater maternal stress during the target's infancy even after correcting for multiple comparisons. The "functional cluster" that was identified as significantly enriched for differential methylation suggested that the mother's experience had an impact on biosynthetic and metabolic processes of the target. Paternal stress during infancy was not associated with differential methylation for targets, but paternal stress during the preschool period was associated with increased DNA methylation at 31 CpG sites, again after controlling for multiple comparisons.

The Essex et al. [58] findings offer initial evidence that there may be genome-wide epigenetic consequences of parenting stress, with developmental and gendered effects consistent with our understanding of the somewhat different impact of maternal and paternal care across different developmental periods. As a consequence, in keeping with the broad expectations for social causation of epigenetic change, it appears that parental stress may "get under the skin" of children and become part of the biological legacy imparted to each child during development.

13.6   An Organizing Model

The Beach et al. [57] and the Essex et al. [58] studies extend prior research in animal models and postmortum tissue to suggest the potential importance of epigenetic reprogramming of intermediate phenotypes. Epigentic processes initiated by environmental stress for the child or the parent may influence development and influence the development of intermediate phenotypes. Accordingly, efforts to incorporate epigenetic change into models of intermediate phenotypes are timely. In line with the expectation that there will be a period of expanding exploration of stress, epigenetic processes, and effects on intermediate phenotypes, we offer figure 13.4 as an overall heuristic model to help guide the integration of methylation patterns, or other epigenetic change processes, into the investigation of intermediate phenotypes or other outcomes by behavioral researchers.

As can be seen in figure 13.4, we propose that some environmental circumstances, including adverse childhood events and forms of problematic parenting, may lead to reliable epigenetic change reflected in stable alterations of DNA methylation at one or more behaviorally relevant CpG sites. The epigenetic changes, in turn, are expected

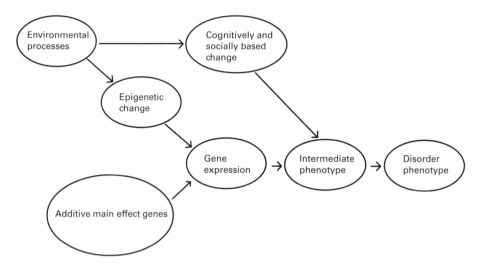

Figure 13.4
Full environmental model.

to be associated with shifts in gene expression or gene activity, influencing developmental change and accounting for some or all of the impact of the environment on
later changes in behavioral dispositions or key aspects of the intermediate phenotype.
In turn, the intermediate behavioral phenotype is expected to set the stage for the
emergence of later disorder such as the development of substance use or abuse.

Although not emphasized elsewhere in the current chapter due to our focus on
biological pathways, the heuristic model in figure 13.4 also highlights the potential
for psychological and environmental pathways from childhood experiences and exposures to problematic outcomes and intermediate phenotypes, making figure 13.4 a
more conceptually complete model. In particular, environmental impacts may be
mediated by their effect on opportunity and constraint vis-à-vis particular behaviors,
highlighting the potential importance of effects on monitoring and direct reinforcement or punishment of particular alternatives, as well as effects on cuing of behavioral
responses and behavioral intentions. These are well-explored pathways for environmental effects on intermediate phenotypes and so should be included and examined
as research on intermediate phenotypes moves forward. Evidence from behavioral
genetics further suggests the relevance of nonbiological pathways of influence on
intermediate phenotypes, with results from several genetically informed designs supporting the notion that social environments have the potential to interact with genotype, including documentation of the impact of family environments [63], key
environmental exposures [64], peer groups [63], and lower social integration [65, 66].

Despite the potential for considerable excitement regarding the exploration of epigenetic pathways and their association with the development of intermediate phenotypes, there is still considerable uncertainty about some aspects of epigenetic change processes. Therefore, tempering the excitement about the potential of epigenetic research is the need to stay alert to a number of methodological issues, several of which we discuss briefly below.

13.7 Methodological Issues

13.7.1 Tissue Type

It seems likely that behavioral scientists increasingly will begin to include methylation and other epigenetic markers in their assessment strategies. As they do so, they will confront choices about what types of tissue to use and whether to conduct analyses that are focused on a particular gene or instead to focus on gene networks or to focus even more broadly on genome-wide patterns. The two studies reviewed above provide a contrast regarding these choices. With regard to tissue type examined, in the Beach et al. [57] study methylation was examined in lymphoblast cell lines, that is, using immortalized cells derived from blood samples. This approach has several advantages, for example, that the cells can reproduce indefinitely, creating an unlimited supply of biological material for analyses, and that they supply material for examination of gene expression and other cellular processes. Likewise, the use of a single cell type has advantages in the exploration of individual differences because it reduces the potential for artifacts that can arise, for example, in the context of using whole blood for analyses. However, these advantages come at the price of potential artifacts in methylation patterns that may be introduced by the process of converting the cells into an immortalized cell line. In addition, as with all cells of peripheral origin, there is also the disadvantage that the cell is not drawn directly from the CNS or from the specific tissue type most directly related to behavioral change. This may introduce interpretive challenges for behavioral scientists to the extent that there are differences in the factors influencing methylation of lymphocytes compared to those influencing CNS tissue. At a minimum, for genes that are not expressed in peripheral tissue but are expressed centrally, it is unlikely that peripheral tissue will provide a useful index of central methylation.

Conversely, in the Essex et al. [58] study, methylation was examined in buccal cells, that is, epithelial cells, drawn from the lining of the cheek. This tissue has the advantage of being minimally invasive to collect, and it is also a tissue that is directly exposed to many potential factors of potential interest to behavioral researchers including tobacco and alcohol consumption. On the negative side, however, differences in patterns of exposure and gene expression for this tissue relative to CNS tissue may be greater than for tissue taken from blood or the immune system, potentially further accentuating interpretive problems and differences in epigenetic patterns observed. Likewise, because it may be exposed to biological processes to a different degree and in different ways than CNS cells, its use as a window on epigenetic change in the CNS may be limited. Conversely, direct exposure to some substances may increase their impact, increasing sensitivity to such influences. At present the relative costs and benefits of different tissue choices are still being debated.

In addition to potential choices between buccal cells and lymphoblasts, behavioral researchers working with human populations may also consider several other commonly used sources of tissue including lymphocytes, leukocytes, whole blood, tissue biopsies, T cells, and macrophages, among others, with each tissue type having its own potential strengths and weaknesses. Animal researchers may directly examine CNS tissue as well. Accordingly, it will be important for behavioral researchers to consider the purpose of their investigation and whether the strengths of the particular tissue selected map onto the purpose of their study. For example, in some cases choice of tissue type will involve examination of correspondence between gene expression in peripheral tissue with that in CNS regions of interest. Although environmental exposures may trigger widespread chromatin remodeling, or more selective methylation in different organ systems, it should be expected that effects will be similar across tissues only to the extent that tissues are similarly exposed to the changed intracellular environment and that the genes being examined are similarly expressed across the type of tissue being compared.

Preliminary evidence from gene expression patterns suggests that there may be considerable cross-tissue similarity in degree of methylation and gene expression for many tissues and for many genomic regions. For example, Vawter and colleagues [67] compared human genome-wide gene expression in three tissues: lymphoblasts, lymphocytes, and cerebellar tissue [67]. After filtering for genes known to be affected by the transformation of the lymphoblasts, they showed that lymphocyte mRNA expression accurately predicted cerebellar mRNA expression ($r = 0.981$) for the ~6,800 genes expressed at the same level in each cell [68]. Such results suggest the potential utility of careful extrapolation from epigenetic examination of cells drawn from peripheral sources and the potential relevance of patterns identified in this manner for understanding effects for gene regions that are similarly expressed. However, this should not obscure well-known, stable difference in gene expression and patterns of methylation across tissues [16].

13.7.2 Promoters versus Other Genomic Regions
Another broad set of issues will arise as behavioral researchers consider genome-wide data sets. First, it is important to consider whether to treat methylation changes in promoter regions differently than methylation in introns and noncoding regions of the genome. Current levels of information about the impact of methylation on func-

tional change limit our ability to predict with confidence exactly which regions should be most consequential and which patterns of methylation should influence intermediate phenotypes to the greatest degree. As noted in our overview above, there are good reasons to focus on promoter regions because of their potential regulatory significance, but ample reason as well to anticipate potentially important regulatory impact from methylation in other regions, including potential impact on gene products such as different splice variants or interfering RNAs. In addition, methylation may vary across the genome, with level of methylation typically lower in regions adjacent to transcription start sites. This normal variation may complicate the characterization of change in methylation across a gene or gene network.

Second, behavioral researchers will need to consider whether there are theoretically linked sets of genes that would be anticipated to interact and change together. In particular, it will be useful to identify gene networks and look for coordinated patterns of change in methylation across networks of related genes. Likewise, it may prove useful to examine particular regulatory motifs across the genome, or to characterize epigenetic changes across all regions of the genome. The two studies reviewed above illustrate something of the range of choices in this regard. In the Beach et al. [57] study, the authors focused on a specific gene region of interest and used a labor-intensive approach involving mass spectrometry to assess all potential residues (76 CpG sites) that could be assessed in that single promoter region. This approach is currently impractical for profiling large regions of the genome, but it provided highly detailed information about a specific area of interest. In contrast, Essex et al. [58] used microarray technology to assess 28,000 CpG sites that were dispersed across the genome. This approach focused on a particular motif, promoter regions, but did so genome-wide. Currently, there are also platforms that allow assessment of more than 485,000 CpG sites distributed across the genome, and it is likely that future platforms, such as those being developed by Pacific Biological, will allow for complete sequencing of all methylation data, providing an exhaustive account of methylation patterns across the genome. Accordingly, the range of choices available to behavioral researchers is likely to grow over the next several years. However, choosing a methodological approach is likely to remain complex for the foreseeable future, with no one approach enjoying universal approbation. Cost is likely to be one factor influencing decision making about which approach to use, but as costs continue to decline over the next several years, other factors related to the adequacy of sampling in each approach are likely to become more salient. In particular, approaches in which only a few samples are drawn to cover a region of interest may be less useful if there is substantial variability in methylation level across the region of interest or if certain regions need to be characterized in greater detail to accurately predict gene expression or impact on outcomes. Likewise, known patterns of systematic change in methylation across various gene motifs may require more intensive sampling of CpG sites in some areas, suggesting that in some cases, optimal sampling may require use of more than one assessment platform.

Third, behavioral researchers will need to remain alert to ongoing examination of assumptions about the association of methylation and gene expression. In both studies reviewed above an assumption was made that hypermethylation in a given region of the genome would be associated with downregulation of gene expression in that region, and this is a relatively safe assumption, on average [16, 17]. However, it is likely that there are complexities in the regulatory patterns associated with methylation that are not yet fully appreciated. Accordingly, behavioral researchers will need to cultivate ongoing partnerships with basic researchers to remain abreast of developments in epigenetic regulatory processes. Likewise, they will need to remain vigilant for patterns that do not conform to preconceptions. One of the challenges for research in this area will be to better understand the patterning of methylation across the genome and the way these general patterns are modified to yield different intermediate phenotypes. To the extent that methylation represents a coordinated response to environmental input that extends across many different organ systems, one might anticipate broad, nonrandom patterns of methylation that extend beyond a single region or system, as well as broad, coordinated changes in these patterns that extend beyond a single gene. In the study of such coordinated changes, behavioral researchers can play an important role.

Not addressed in either study reviewed above is the strong likelihood that substance use will also function as an exposure, resulting in patterns of differential methylation across the genome. However, as work continues in this area, it will become increasingly important to account for such effects. In particular, it will be important to determine whether substance use obscures or adds to the epigenetic signatures of early adversity. As alluded to above, the salience of effects of recent substance use may vary by tissue type, making choice of tissue type a particularly important issue for substance use researchers. It will also be important to examine situations in which adversity has the potential for substantially different impact on the optimal survival and reproductive strategies of males and females. In such cases we may expect to see differential impact on methylation patterns as a function of gender. Likewise, to the extent that the adversity carries different implications by developmental stage, we might anticipate important moderation by developmental stage. Hints of such effects are already present in the studies reviewed above. In addition, there is strong potential for development of epigenetic tests of exposure to substances and strong potential for epigenetic effects of substance use to mediate downstream health impacts. As interest develops in these areas, it is likely that behavioral researchers will have an important role to play.

13.8   Final Remarks

Despite several cautionary notes, epigenetic profiling appears likely to add important information to the study of genetic influence on behavior. There are several potential mechanisms of epigenetic influence, but for the foreseeable future, most research is likely to remain focused on methylation, with considerable attention to methylation of promoter regions. The influence of adverse childhood events and parenting problems on methylation or other epigenetic change processes seems plausible from an evolutionary perspective, and the biological mechanisms continue to be explicated. Epigenetic profiling and the identification of epigenetic patterns has the potential to identify an additional mechanism linking the environment to intermediate phenotypes. As a result, epigenetic profiles may identify environmental factors influencing the health and well-being of groups or subpopulations as well as helping to better define the developmental specificity of key epigenetic influences. Finally, given the potential for epigenetic marks to be inherited in subsequent generations under some circumstances, particularly through maternal transmission, the extent to which particular epigenetic signatures can be transmitted across generations and are responsible for variance in outcomes previously attributed to genotype alone remains an important issue in need of further study. As is reflected in the models presented above, we expect that the study of methylation and other epigenetic processes will have a profoundly stimulating effect on theories connecting family environments to health and resilience and that the examination of epigenetic intermediate phenotypes will enhance the intermediate phenotype approach in the study of substance use and abuse.

Acknowledgment

Preparation of this chapter was supported by the Center for Contextual Genetics and Prevention Studies (1P30DA027827) funded by the National Institute on Drug Abuse. The content of this report is solely the responsibility of the authors and does not necessarily represent the official views of the National Institute on Drug Abuse or the National Institutes of Health.

References

1. MacKillop, J., & Munafò, M. R. (Eds.). (this volume). *Genetic Influences on Addiction: An Intermediate Phenotype Approach*. Cambridge, MA: MIT Press.

2. Howe, G. W., Beach, S. R., & Brody, G. H. (2010). Microtrial methods for translating gene–environment dynamics into preventive interventions. *Prevention Science, 11*, 343–354.

3. Dodge, K. A., & Rutter, M. (Eds.), *Gene–Environment Interactions in Developmental Psychopathology*. New York: Guilford.

4. Fulka, H., Mrazek, M., Tepla, O., & Fulka, J., Jr. (2004). DNA methylation pattern in human zygotes and developing embryos. *Reproduction, 128*, 703–708.

5. Karlic, R., Chung, H. R., Lasserre, J., Vlahovicek, K., & Vingron, M. (2010). Histone modification levels are predictive for gene expression. *Proceedings of the National Academy of Sciences of the United States of America, 107*, 2926–2931.

6. Meagher, R. B. (2010). The evolution of epitype. *Plant Cell, 22*, 1658–1666.

7. Kapranov, P., Willingham, A. T., & Gingeras, T. R. (2007). Genome-wide transcription and the implications for genomic organization. *Nature Reviews. Genetics, 8*, 413–423.

8. Liu, W. M., Chu, W. M., Choudary, P. V., & Schmid, C. W. (1995). Cell stress and translational inhibitors transiently increase the abundance of mammalian SINE transcripts. *Nucleic Acids Research, 23*, 1758–1765.

9. Mariner, P. D., Walters, R. D., Espinoza, C. A., et al. (2008). Human Alu RNA is a modular transacting repressor of mRNA transcription during heat shock. *Molecular Cell, 29*, 499–509.

10. Englander, M. T., Dulawa, S. C., Bhansali, P., & Schmauss, C. (2005). How stress and fluoxetine modulate serotonin 2C receptor pre-mRNA editing. *Journal of Neuroscience, 25*, 648–651.

11. Urnov, F. D., & Wolffe, A. P. (2001). Above and within the genome: epigenetics past and present. *Journal of Mammary Gland Biology and Neoplasia, 6*, 153–167.

12. Jaenisch, R., & Bird, A. (2003). Epigenetic regulation of gene expression: how the genome integrates intrinsic and environmental signals. *Nature Genetics, 33*(Suppl), 245–254.

13. Murrell, A., Rakyan, V. K., & Beck, S. (2005). From genome to epigenome. *Human Molecular Genetics, 14*(Spec No 1), R3–R10.

14. Stein, R., Razin, A., & Cedar, H. (1982). In vitro methylation of the hamster adenine phosphoribosyltransferase gene inhibits its expression in mouse L cells. *Proceedings of the National Academy of Sciences of the United States of America, 79*, 3418–3422.

15. Hansen, R. S., & Gartler, S. M. (1990). 5-Azacytidine-induced reactivation of the human X chromosome-linked PGK1 gene is associated with a large region of cytosine demethylation in the 5′ CpG island. *Proceedings of the National Academy of Sciences of the United States of America, 87*, 4174–4178.

16. Bell, C. G., Wilson, G. A., Butcher, L. M., et al. (2012). Human-specific CpG "beacons" identify loci associated with human-specific traits and disease. *Epigenetics, 7*, 1188–1199.

17. Plume, J. M., Beach, S. R., Brody, G. H., & Philibert, R. A. (2012). A cross-platform genome-wide comparison of the relationship of promoter DNA methylation to gene expression. *Frontiers in Genetics, 3*, 12.

18. Pai, A. A., Bell, J. T., Marioni, J. C., Pritchard, J. K., & Gilad, Y. (2011). A genome-wide study of DNA methylation patterns and gene expression levels in multiple human and chimpanzee tissues. *PLOS Genetics, 7*, e1001316.

19. Unternaehrer, E., Luers, P., Mill, J., et al. (2012). Dynamic changes in DNA methylation of stress-associated genes (OXTR, BDNF) after acute psychosocial stress. *Transcultural Psychiatry, 2*, e150.

20. Sturge-Apple, M. L., Skibo, M. A., Rogosch, F. A., Ignjatovic, Z., & Heinzelman, W. (2011). The impact of allostatic load on maternal sympathovagal functioning in stressful child contexts: implications for problematic parenting. *Development and Psychopathology, 23*, 831–844.

21. Shonkoff, J. P., Boyce, W. T., & McEwen, B. S. (2009). Neuroscience, molecular biology, and the childhood roots of health disparities: building a new framework for health promotion and disease prevention. *Journal of the American Medical Association, 301*, 2252–2259.

22. Brody, G. H. (2004). Siblings' direct and indirect contributions to child development. *Current Directions in Psychological Science, 13*, 124–126.

23. Brody, G. H., Chen, Y. F., Kogan, S. M., Murry, V. M., Logan, P., & Luo, Z. (2008). Linking perceived discrimination to changes in African American mothers' parenting practices. *Journal of Marriage and the Family, 70*, 319–331.

24. Charnov, E. L. (1993). *Life History Invariants.* Oxford, UK: Oxford University Press.

25. MacKillop, J., & Munafò, M. R. (Eds.). (this volume). *Genetic Influences on Addiction: An Intermediate Phenotype Approach.* Cambridge, MA: MIT Press.

26. Boyce, W. T., & Ellis, B. J. (2005). Biological sensitivity to context: I. an evolutionary-developmental theory of the origins and functions of stress reactivity. *Development and Psychopathology, 17*, 271–301.

27. Belsky, J., & Pluess, M. (2009). Beyond diathesis stress: differential susceptibility to environmental influences. *Psychological Bulletin, 135*, 885–908.

28. Ellis, B. J., Boyce, W. T., Belsky, J., Bakermans-Kranenburg, M. J., & van Ijzendoorn, M. H. (2011). Differential susceptibility to the environment: an evolutionary-neurodevelopmental theory. *Development and Psychopathology, 23*, 7–28.

29. Champagne, F. A., Weaver, I. C., Diorio, J., et al. (2006). Maternal care associated with methylation of the estrogen receptor-alpha1b promoter and estrogen receptor-alpha expression in the medial preoptic area of female offspring. *Endocrinology, 147*, 2909–2915.

30. Barha, C. K., Pawluski, J. L., & Galea, L. A. (2007). Maternal care affects male and female offspring working memory and stress reactivity. *Physiology & Behavior, 92*, 939–950.

31. Trollope, A. F., Gutierrez-Mecinas, M., Mifsud, K. R., et al. (2012). Stress, epigenetic control of gene expression and memory formation. *Experimental Neurology, 233*, 3–11.

32. Champagne, F. A., & Meaney, M. J. (2007). Transgenerational effects of social environment on variations in maternal care and behavioral response to novelty. *Behavioral Neuroscience, 121,* 1353–1363.

33. Menard, J. L., & Hakvoort, R. M. (2007). Variations of maternal care alter offspring levels of behavioural defensiveness in adulthood: evidence for a threshold model. *Behavioural Brain Research, 176,* 302–313.

34. Gunnar, M., & Quevedo, K. (2007). The neurobiology of stress and development. *Annual Review of Psychology, 58,* 145–173.

35. Lupien, S. J., McEwen, B. S., Gunnar, M. R., & Heim, C. (2009). Effects of stress throughout the lifespan on the brain, behaviour and cognition. *Nature Reviews. Neuroscience, 10,* 434–445.

36. Miller, G. E., Chen, E., & Parker, K. J. (2011). Psychological stress in childhood and susceptibility to the chronic diseases of aging: moving toward a model of behavioral and biological mechanisms. *Psychological Bulletin, 137,* 959–997.

37. Gadgil, M., & Bossert, W. H. (1970). Life historical consequences of natural selection. *American Naturalist, 104,* 1–24.

38. Belsky, J., Steinberg, L., & Draper, P. (1991). Childhood experience, interpersonal development, and reproductive strategy: and evolutionary theory of socialization. *Child Development, 62,* 647–670.

39. Figueredo, A. J., Vásquez, G., Brumbach, B. H., Sefcek, J. A., Kirsner, B. R., & Jacobs, W. J. (2005). The K factor: individual differences in life history strategy. *Personality and Individual Differences, 39,* 1349–1360.

40. McEwen, B. S. (2002). Sex, stress and the hippocampus: allostasis, allostatic load and the aging process. *Neurobiology of Aging, 23,* 921–939.

41. de Kloet, E. R., Rots, N. Y., & Cools, A. R. (1996). Brain–corticosteroid hormone dialogue: slow and persistent. *Cellular and Molecular Neurobiology, 16,* 345–356.

42. Liu, L., Li, Y., & Tollefsbol, T. O. (2008). Gene–environment interactions and epigenetic basis of human diseases. *Current Issues in Molecular Biology, 10,* 25–36.

43. McGowan, P. O., Sasaki, A., D'Alessio, A. C., et al. (2009). Epigenetic regulation of the glucocorticoid receptor in human brain associates with childhood abuse. *Nature Neuroscience, 12,* 342–348.

44. Barros, V. G., Rodriguez, P., Martijena, I. D., et al. (2006). Prenatal stress and early adoption effects on benzodiazepine receptors and anxiogenic behavior in the adult rat brain. *Synapse, 60,* 609–618.

45. Szyf, M., Weaver, I. C., Champagne, F. A., Diorio, J., & Meaney, M. J. (2005). Maternal programming of steroid receptor expression and phenotype through DNA methylation in the rat. *Frontiers in Neuroendocrinology, 26,* 139–162.

46. Weaver, I. C. (2007). Epigenetic programming by maternal behavior and pharmacological intervention. Nature versus nurture: let's call the whole thing off. *Epigenetics, 2,* 22–28.

47. Champagne, D. L., Bagot, R. C., van Hasselt, F., et al. (2008). Maternal care and hippocampal plasticity: evidence for experience-dependent structural plasticity, altered synaptic functioning, and differential responsiveness to glucocorticoids and stress. *Journal of Neuroscience, 28,* 6037–6045.

48. Cameron, N. M., Champagne, F. A., Parent, C., Fish, E. W., Ozaki-Kuroda, K., & Meaney, M. J. (2005). The programming of individual differences in defensive responses and reproductive strategies in the rat through variations in maternal care. *Neuroscience and Biobehavioral Reviews, 29,* 843–865.

49. Cameron, N., Del Corpo, A., Diorio, J., et al. (2008). Maternal programming of sexual behavior and hypothalamic–pituitary–gonadal function in the female rat. *PLoS ONE, 3,* e2210.

50. Kaffman, A., & Meaney, M. J. (2007). Neurodevelopmental sequelae of postnatal maternal care in rodents: clinical and research implications of molecular insights. *Journal of Child Psychology and Psychiatry, and Allied Disciplines, 48,* 224–244.

51. McGowan, P. O., Sasaki, A., Huang, T. C. T., et al. (2008). Promoter-wide hypermethylation of ribosomal RNA gene promoter in the suicide brain [electronic article]. *PLoS ONE, 3,* e2085.

52. McGowan, P. O., Sasaki, A., D'Alessio, A. C., et al. (2009). Epigenetic regulation of the glucocorticoid receptor in human brain associates with childhood abuse. *Nature Neuroscience, 12,* 342–348.

53. Oberlander, T. F., Weinberg, J., Papsdorf, M., et al. (2008). Prenatal exposure to maternal depression, neonatal methylation of human glucocorticoid receptor gene (NR3C1) and infant cortisol stress responses. *Epigenetics, 3,* 97–106.

54. Heijmans, B. T., Tobi, E. W., Stein, A. D., et al. (2008). Persistent epigenetic differences associated with prenatal exposure to famine in humans. *Proceedings of the National Academy of Sciences of the United States of America, 105,* 17046–17049.

55. Felitti, V. J., & Anda, R. F. (2010). The relationship of adverse childhood experiences to adult medical disease, psychiatric disorders and sexual behavior: implications for healthcare. In R. A. Lanius, E, Vermetten, & C. Pain (Eds.), *The Impact of Early Life Trauma on Health and Disease: The Hidden Epidemic* (pp. 77–86). Cambridge, UK: Cambridge University Press.

56. Beach, S. R., Brody, G. H., Todorov, A. A., Gunter, T. D., & Philibert, R. A. (2010). Methylation at SLC6A4 is linked to family history of child abuse: an examination of the Iowa Adoptee sample. *American Journal of Medical Genetics. Part B, Neuropsychiatric Genetics, 153B,* 710–713.

57. Beach, S. R., Brody, G. H., Todorov, A. A., Gunter, T. D., & Philibert, R. A. (2011). Methylation at 5HTT mediates the impact of child sex abuse on women's antisocial behavior: an examination of the Iowa adoptee sample. *Psychosomatic Medicine, 73,* 83–87.

58. Essex, M. J., Thomas Boyce, W., Hertzman, C., et al. (2013). Epigenetic vestiges of early developmental adversity: childhood stress exposure and DNA methylation in adolescence. *Child Development*, 84, 58–75.

59. Shannon, C., Schwandt, M. L., Champoux, M., et al. (2005). Maternal absence and stability of individual differences in CSF 5-HIAA concentrations in rhesus monkey infants. *American Journal of Psychiatry*, 162, 1658–1664.

60. Ichise, M., Vines, D. C., Gura, T., et al. (2006). Effects of early life stress on [11C]DASB positron emission tomography imaging of serotonin transporters in adolescent peer- and mother-reared rhesus monkeys. *Journal of Neuroscience*, 26, 4638–4643.

61. Mehlman, P. T., Higley, J. D., Faucher, I., et al. (1995). Correlation of CSF 5-HIAA concentration with sociality and the timing of emigration in free-ranging primates. *American Journal of Psychiatry*, 152, 907–913.

62. Carver, C. S., Johnson, S. L., & Joormann, J. (2008). Serotonergic function, two-mode models of self-regulation, and vulnerability to depression: what depression has in common with impulsive aggression. *Psychological Bulletin*, 134, 912–943.

63. Dick, D. M., Viken, R., Purcell, S., et al. (2007). Parental monitoring moderates the importance of genetic and environmental influences on adolescent smoking. *Journal of Abnormal Psychology*, 116, 213–218.

64. Kendler K. S., Gardner, C., & Dick, D. M. (2010). Predicting alcohol consumption in adolescence from alcohol-specific and general externalizing genetic risk factors, key environmental exposures and their interaction. *Psychological Medicine*, 14, 1–10.

65. Dick, D. M., Rose, R. J., Viken, R. J., Kaprio, J., & Koskenvuo, M. (2001). Exploring gene–environment interactions: socioregional moderation of alcohol use. *Journal of Abnormal Psychology*, 110, 625–632.

66. Dick, D. M., Bernard, M., Aliev, F., Viken, R., Pulkkinen, L., Kaprio, J., et al. (2009). The role of socioregional factors in moderating genetic influences on early adolescent behavior problems and alcohol use. *Alcoholism, Clinical and Experimental Research*, 33, 1739–1748.

67. Rollins, B., Martin, M. V., Morgan, L., & Vawter, M. P. (2010). Analysis of whole genome biomarker expression in blood and brain. *American Journal of Medical Genetics. Part B, Neuropsychiatric Genetics*, 153B, 919–936.

68. Yuferov, V., Nielsen, D. A., Levran, O., et al. (2011). Tissue-specific DNA methylation of the human prodynorphin gene in post-mortem brain tissues and PBMCs. *Pharmacogenetics and Genomics*, 21, 185–196.

14 Differential Sensitivity to Context: *GABRG1* Enhances the Acquisition of Prototypes That Serve as Intermediate Phenotypes for Substance Use

Ronald L. Simons, Man Kit Lei, Steven R. H. Beach, Gene H. Brody, Robert A. Philibert, Frederick X. Gibbons, and Meg Gerrard

In the past 15 years, a profusion of molecular genetic studies have investigated the association between various genetic polymorphisms and particular substances use disorders. Another approach to investigating the impact of genes on substance abuse involves testing models of gene × environment (G × E) interactions [1]. A G × E framework can complement traditional main effect approaches by highlighting different types of intermediate phenotypes. Whereas main effect studies focus upon underlying biological features in the gene-to-behavior pathways, G × E research often emphasize the mediating role of cognitive schemas, attitudes, and beliefs that facilitate or discourage the use of substances. This is particularly true of the differential susceptibility model of G × E that is tested in this chapter. In the following section, we contrast the diathesis–stress model that informs most G × E research, including substance abuse studies, with the differential susceptibility perspective. The latter framework posits that some people are predisposed by their genes to be more sensitive than others to environmental context [2, 3]. Thus those persons most vulnerable to adverse social environments are the same ones who reap the most benefit from environmental support. We go on to argue that *GABRG1*, a gene that has been linked to substance use, likely operates as a differential susceptibility gene. Based upon this idea, we test the idea that *GABRG1* haplotypes interact with social environmental factors to foster cognitive schemas (intermediate phenotypes) that increase the use of alcohol and marijuana. Further, we expect that the pattern of this G × E interaction is consistent with the differential susceptibility perspective and mediates much of the effect of G × E on use of alcohol and marijuana.

14.1 The Differential Susceptibility Hypothesis

Genetically informed behavioral science requires models of the manner in which genetic variables combine with environmental context to influence behavioral outcomes [4–6]. The diathesis–stress perspective is the model utilized in the vast majority of G × E studies. This approach asserts that some individuals are by nature more vulnerable than others as they possess dysfunctional "risk alleles." These alleles amplify the probability that exposure to some adverse social condition (e.g., abusive parenting, stressful life events) will lead to a problem behavior (e.g., depression, substance abuse). This assumption is contradicted, however, by the fact that over the past several thousand years evolution seems to have conserved the various alleles that have been identified as risk factors [3]. While truly dysfunctional genetic variants should largely disappear over time, most of the so-called risk alleles studied by behavioral science researchers are highly prevalent, often being present in 40% to 50% of the members of the populations being investigated [3]. Thus, contrary to the negative view usually taken of these alleles, this suggests that, at least in certain contexts, these genetic variants must provide advantages over other genotypes. This idea is an essential component of the alternative model of G × E proposed by Jay Belsky and his colleagues [2, 3, 7].

After reviewing scores of studies that purported to show evidence of a diathesis–stress effect, Belsky and company [2, 7] concluded that a careful inspection of the data actually points to a different interpretation. Rather than showing that some individuals are more vulnerable to stress than others, they asserted that the data support the idea that some people are simply genetically predisposed to be more susceptible to environment influence than others. This suggests that those persons most vulnerable to adverse social environments are the same ones who reap the most benefit from environmental support. In other words, some people are programmed by their genes to be more sensitive to environmental context, for better or worse [7].

Support for the differential susceptibility or plasticity argument is evident when the slopes for a gene by environment interaction show a crossover effect with the susceptibility group showing worse outcomes than the comparison group when the environment is negative but demonstrating better outcomes than the comparison group when the environment is positive [2, 3]. Recently, Belsky and Pleuess [2] reviewed scores of published G × E studies, including Caspi's widely cited article in *Science*, and found widespread evidence of significant crossover effects. In most of these studies, however, this pattern was not recognized or discussed because the authors were operating out of the diathesis–stress paradigm.

How would genes cause some individuals to be more sensitive than others to their environment? Belsky and Pleuss [2] observe that the genes included in the studies that they reviewed involved the dopaminergic system, which has been implicated in reward sensitivity and sensation seeking, and the serotonergic system, which has been linked to sensitivity to punishment and displeasure [8, 9]. They therefore posit that some individuals may be more responsive to their environment than others because they have different thresholds for experiencing pleasure or displeasure. That is, because of their genetic endowment, the behavior of some individuals may be more readily shaped by salient environmental rewards and punishments.

Recently, Simons and colleagues extended the differential susceptibility perspective to include intermediate phenotypes [10, 11]. Generally, the behavioral sciences assume that social experiences give rise to schemas and sentiments that, in turn, influence one's interpretation of and response to subsequent situations [12]. Based on this assumption, Simons and associates argued that individuals who are genetically predisposed to be more sensitive to their environment than others would be expected to manifest more of an emotional response to environmental conditions and to learn the lessons inherent in recurrent environmental events more readily than less environmentally sensitive individuals. In support of this idea, Simons et al. [10, 11] reported results showing that young adults with various putative plasticity alleles (l-*DRD4*, s-*MAOA*, s-*5HTTLPR*) are more likely than other genotypes to adopt the code of the street, a hostile view of relationships, and a cynical view of conventional norms when they grow up in a dangerous social environment but are less likely than other genotypes to adopt these deviant schemas when they are raised in a more favorable social milieu. Further, this interaction effect mediated the impact of social environment × genetic plasticity on violence. In other words, the results supported a mediated moderation model where social schemas served as intermediate phenotypes.

The present study extends this general model to the etiology of substance use. We investigate the extent to which *GABRG1* haplotypes interact with the social environment in a manner suggested by the differential susceptibility perspective to influence the learning of cognitive schemas relating to the use of substances. Further, we examine the degree to which these schemas serve as intermediate phenotypes that mediate the impact of genes and the environment on the use of substances. These ideas are elaborated in the following section.

14.1.1 *GABRG1*, Differential Susceptibility, and Substance Use

Several decades of research has investigated prototypes, or social images of substance users, as possible explanations for the use of substances such as alcohol and marijuana [13, 14]. This line of investigation assumes that adolescents and young adults have clear and salient images of the types of people their age who engage in substance use and that favorable prototypes encourage participation in such behavior [15, 16]. A wide variety of studies have provided support for this view. In studies focusing on the use of substances such alcohol, marijuana, and tobacco, positive prototypes encourage whereas negative prototypes discourage utilization of the substance [14, 17–20]. The substance use prototype utilized in the present study focused upon the extent to which people who use alcohol or marijuana are viewed as popular, smart, cool, attractive, and dull.

Past research has shown that a variety of child and adolescent experiences influence substance use. As noted earlier, a strong test of the differential susceptibility perspective requires that one include the full range of the social environment, from conditions promoting to those discouraging the phenotype being investigated. Thus in the present study we focused upon factors that might be expected to encourage substance use (viz., peer use, prevalence of use in the community, parents' favorable attitudes toward use, childhood trauma) as well as factors thought to discourage use (viz., school involvement, supportive parenting, prosocial friends, informal community control) . We expect that in large measure these factors exert their effect through their impact on substance use prototypes. In other words, we assume a general model where social environmental conditions during adolescence foster a substance use prototype that, in turn, influences use of alcohol and marijuana during early adulthood. The next step was to elaborate this model to include genetic effects. We selected a gene— *GABRG1*—that has been shown to be related to substance use but that might also be viewed as contributing to differential susceptibility.

Considerable effort has been made to identify genes that increase the risk for substance use disorders. Genome-wide linkage scans have implicated a region on chromosome 4p12 that accommodates a cluster of four genes (*GABRA2*, *GABRA4*, *GABRB1*, and *GABRG1*) that encode GABA$_A$ subunits [21–23]. This is significant as several preclinical studies have linked GABA$_A$ receptors to tolerance, dependence, and withdrawal from substances, especially alcohol [24, 25]. Of this four-gene cluster, it is *GABRA2* that has received the most attention. Several studies have reported haplotype and single nucleotide polymorphism (SNP) associations between *GABRA2* and alcoholism [26–31]. Other studies, however, have failed to find this association [32–34]. For example, neither of the two published studies of African Americans found an association between *GABRA2* and substance use.

Recently, some researchers have focused on the gene *GABRG1*. SNPs and haloptypes from this gene have shown rather robust associations with alcohol dependence [24, 32]. This research also investigated *GABRA2* and concluded that prior findings of an association of the *GABRA2* gene with alcohol dependence were probably due to moderate linkage disequilibrium between *GABRA2* and *GABRG1*. These studies were based on samples of European Americans, Finnish Caucasians, and Plains Indians. The current study focuses upon the role of *GABRG1* in substance use among African Americans.

Animal research indicates that *GABRG1* is expressed primarily in the amygdala and areas receiving innervation from the striatum such as the substantia nigra [35, 36]. The latter regions are implicated in reward and addiction. This would argue for a main effect of *GABRG1* on substance use. However, *GABRG1* is also likely to operate as a plasticity gene that increases sensitivity to context. The fact that the gene is expressed in the amygdala as well as the subtantia nigra suggests that it influences thresholds for sensitivity to both punishments and rewards. Polymorphisms in such a gene might be expected to produce differential susceptibility to context with individuals harboring minor alleles learning lessons provided by the environment regarding substances more quickly than other genotypes [10]. Therefore, we posit that in addition to any main effect on substance use, *GABRG1* will interact with the environment to influence the development of both substance use prototypes and substance use and that it will do so in a manner consonant with the differential susceptibility perspective.

More specifically, we hypothesize that individuals with the minor version of the haplotypes will report a more positive view of alcohol and marijuana users than other genotypes when the environment encourages substance use but a less positive view of alcohol and marijuana users than other genotypes when the environment discourages substance use. Similarly, we posit that those with the minor version of the haplotypes will report more use of alcohol and marijuana than other genotypes when the environment encourages use but less use of alcohol and marijuana than other genotypes when the environment discourages use. Finally, given our assumption that substance use prototypes operates as intermediate phenotypes, we expect that controlling for the impact of $G \times E$ on prototype will eliminate the effect of $G \times E$ on substance use. In other words, we expect to find support for a mediated moderator model.

14.2   Research Design

14.2.1   Sample

We tested our hypotheses using data from waves 1 to 5 of the Family and Community Health Study (FACHS), a multisite investigation of neighborhood and family effects on health and development [11]. FACHS was designed to identify neighborhood and family processes that contribute to school-age African American children's development in families living in a wide variety of community settings outside the inner-city core. Each family included a child who was in fifth grade at the time of recruitment. The response rate for the contacted families was 84%.

At the first wave, the FACHS sample consists of 889 African American children (411 boys and 478 girls) and their primary caregivers. At study inception, about half of the sample resided in Georgia and the other half in Iowa. The target children were 10–11 years of age at wave 1, 12–13 at wave 2, 14–15 at wave 3, 17–18 at wave 4, and 20–21 at wave 5. Of the original sample, 689 (78%) continued to participate at Wave 5. As part of wave 5 data collection, targets were asked to provide DNA for genotype analysis. Of the 689 participants, 549 (80%) agreed to DNA collection. Successful genotyping for *GABRG1* SNPs was achieved for 542 individuals. Finally, SNPs with a minor-allele frequency ≤0.05 were dropped, and listwise deletion of missing cases was used. The final sample size used in our analysis was 474 individuals (202 boys and 272 girls).

Analyses indicated that those individuals who did not participate at various waves did not differ significantly from those who participated with regard to youths' age, sex, delinquency, substance use, primary caregivers' education, household income, or neighborhood characteristics. To further assess attrition bias, we used Heckman's [37] two-step procedure to estimate sample selection bias. Including Heckman's Lambda in our models did not change the findings.

14.2.2 Procedures

The questionnaires were administered in the respondent's home and took on average about two hours to complete. In waves 1–4, the instruments were presented on laptop computers. The researcher read each question aloud, and the participant entered an anonymous response using a separate keypad. Many of the instruments administered in wave 5 included questions regarding illegal or potentially embarrassing sexual activities. Hence, in an effort to further enhance anonymity, audio-enhanced, computer-assisted, self-administered interviews were utilized. Using this procedure, the respondent sat in front of a computer and responded to questions as they were presented both visually on the screen and auditorily via earphones.

Participants' were also asked to contribute DNA at wave 5 using Oragene DNA kits (Genotek; Calgary, Alberta, Canada). Those who chose to participate rinsed their mouths with tap water and then deposited 4 ml of saliva in the Oragene sample vial. The vial was sealed, inverted, and shipped via courier to a laboratory at the University of Iowa, where samples were prepared according to the manufacturer's specifications.

14.2.3 Variable Measurement

Substance Use

At wave 5, respondents were asked how often in the past year they had drunk alcohol (i.e., beer, wine, wine coolers, whiskey, gin, or other liquor), had three or more drinks of alcohol at one time, smoked marijuana [38, 39]. Response categories ranged from 1 "never" to 6 "more than once a week." Cronbach's alpha for this three-item scale was 0.75. Only the last two items were available in the wave 4 instrument. Cronbach's alpha for this two-item measure was 0.51.

Substance Use Prototypes

At waves 4 and 5, respondents were asked to rate the extent (1 = "not at all" to 4 = "very") to which they perceive persons their age who use alcohol to be popular, smart, cool, attractive, and dull/boring (reverse coded). They were then asked to perform the same ratings for individuals their age who use drugs. Cronbach's alpha for both the alcohol and drug prototypes was 0.84. Based upon prior research [19], these two sub-scales were summed to create a composite measure of substance use prototype.

Substance Use Risk Factors

We assessed environment risk using four variables that past research has shown to enhance the probability of adolescence substance use: lack of parental communication, community norms regarding substance use, neighborhood victimization, and friends' substance use. *Lack of parental communication about substance use* was assessed at waves 1–4 using two items [18] that asked respondents to report how often during the past year their primary caregiver talked to them about drinking alcohol and about using illegal drugs. The response format for the items ranged from 1 ("many times") to 4 ("never"). Coefficient alpha for the measure was approximately 0.90 at each wave. Scores were standardized and averaged across waves to form a composite measure of low parental communication about substance use. *Community norms* were assessed using four items. The first two items asked the primary caregiver to report what people in their neighborhood think (1 = "they do not think it is okay," 5 = "they think it is okay") about teens using drugs or drinking alcohol and getting drunk. The other two items involved caregiver ratings (1 = "not at all a problem" to 3 = "a big problem") of alcohol and drug use in their neighborhood. The four items were standardized and summed. Coefficient alpha was roughly 0.75 at each wave. A composite measure of community norms was created by standardizing and averaging the scores across waves. *Neighborhood victimization* was measured at waves 1–4 by asking respondents to report how often during the past year they or their friends had been the victim of mugging, physical attack, or sexual assault [40]. At each wave, 9% to 17% of the respondents reported that they or their friends had been the victim of at least one violent act during the preceding year. Scores were standardized and summed across waves to form a composite measure of exposure to violent victimization. *Friends' substance use* was assessed using a three-item scale [41]. The items asked respondents to report how many (1 = "none," 5 = "all") of their close friends had used alcohol or illicit drugs (i.e., marijuana, hashish, inhalants, LSD, or cocaine) in the past year. Cronbach's alpha was approximately 0.55 at each wave. Scores were standardized and averaged across waves.

Substance Use Protective Factors

We assessed environmental protection against substance use using four variables that past research has shown to decrease the probability of adolescence substance use: supportive parenting, school involvement, neighborhood informal social control, and prosocial friends. *Supportive parenting* was assessed at waves 1–4 using target responses to 11 items concerned with caregiver warmth, monitoring, problem solving, and help with homework. Response categories ranged from 1 ("never") to 4 ("always"). Coefficient alpha for this scale was roughly 0.80 at each wave. A composite measure of supportive parenting was created by standardizing and averaging the scores across waves. *School involvement* was assessed at waves 1–4 using nine items [42] that asked targets to indicate how much they agreed (1 = "strongly disagree," 4 = "strongly agree") that various statements describe themselves. The statements referred to having positive relationships with teachers, completing homework, interest in school, involvement in school activities, and academic performance. Coefficient alpha for the scale was approximately 0.70 at each wave. A composite measure of school involvement was obtained by standardizing and averaging scores across waves. *Neighborhood informal social control* was assessed with a revised version of neighborhood monitoring from the Project on Human Development in Chicago Neighborhoods [43]. Respondents were asked to rate (1 = "very unlikely," 4 = "very likely") the probability that neighborhood residents would intervene (issue a verbal reprimand, call the police) if teens spray-painted graffiti on a building, showed disrespect to an adult, or skipped school and were hanging out on a street corner and were asked to rate (1 = "very difficult," 4= "very easy") the degree to which neighborhood residents are easy to pick out who are outsiders. Coefficient alpha for this four-item scale was roughly 0.65 at each wave. Scores were standardized and then averaged across waves to form a composite measure of neighborhood informal social control. *Prosocial friends* was measured at wave 4 using the following general question: How important is avoiding problems of drug and alcohol use to your closet friends? The response format for this item ranged from 1 ("not at all important") to 5 ("extremely important"). At wave 4, 31% of the respondents reported that their friends think avoiding drug and alcohol use is extremely important. In order to obtain a composite measure of the social environment that ranged from very protective to very risky (*substance use protective/risk factors*), we reverse coded the measures of protective factors and then standardized and summed the eight measures of protective and risk factors. Using Nunnally's reliability formula for composite variables [44], the reliability of this measure was 0.856.

Genotyping

All participants were genotyped using Taqman MGB assays (Applied Biosystems, Foster City, California) and the Fluidigm Biomark Genetic Analysis System (Fluidigm, South San Francisco, California). The *GABRG1* (GABA$_A$ receptor subunits gamma-1) were genotyped using a panel of seven SNPs: rs1497565, rs1497571, rs2350438, rs1497576, rs1391166, rs10938426, and rs7654165. Genotype distributions were consistent with Hardy–Weinberg equilibrium for all seven SNPs.

14.2.4 Analytic Strategy

Haploblock analysis was preformed with Haploview and PHASE software [45–47]. Hierarchal regression models were employed to test the differential susceptibility hypothesis using the statistical software Mplus 6.1 [48]. Parameters in the hierarchical regression models were examined using maximum likelihood estimation with robust standard errors and utilized bootstrapping to adjust model fit for nonnormal distributions [49].

We used multiple imputation techniques for missing data at the item level. All independent variables were standardized in order to make coefficients easier to interpret, to reduce multicollinearity, and to make the simple slope easier to test [50]. When gene–environment interaction effects were present, post hoc analyses of significant interaction terms were conducted using the Johnson–Neyman technique [51, 52]. This procedure identifies regions of significance for interactions between continuous (social environment) and categorical variables (genotypes).

Lastly, we employed the mediated-moderation model [53] to examine the extent to which substance use prototype mediates the main effect of substance use risk/ protective environment, and, more importantly, the interaction effect of environment and genotype on substance use. The logic of the mediated-moderation model is similar to traditional mediation models, but this model focuses only on the relationship among an interaction term, a mediator, and an outcome rather than considering the effects of multiple independent variables [54]. Unlike traditional mediating theory using three steps to test for mediating effects [55], the bootstrapping option in Mplus determines the significance of mediation effects [56] and enables one to examine all direct and indirect effects [53].

To assess the goodness of fit of our models, standardized root mean squared residual [57], the comparative fit index (CFI) [58], and the chi-square divided by its degrees of freedom (fit ratio) were used. The CFI is truncated to the range of 0 to 1, and values close to 1 indicate a very good fit [58]. A root mean square error of the approximation (RMSEA) smaller than 0.05 indicates a close fit whereas a RMSEA between 0.05 and 0.08 suggests a reasonable fit [57].

14.3   Results

14.3.1   Single-SNP Analysis of *GABRG1*

For each of the seven SNPs, the risk allele was identified by examining associations with substance use. For each SNP, the allele showing the highest mean score on substance use was selected as a risk allele. This procedure resulted in the T allele of rs1497565 and rs1391166, the C allele of rs1497571, rs2350438, and rs7654165, and the A allele of rs1497576 and rs10938426 being defined as risk alleles.

14.3.2   Haplotype Analysis of *GABRG1*

As shown in figure 14.1, haplotype analyses using Haploview software identified two haplotype blocks: block 1 (18 kb in length; rs1497565–rs1497571) and block 2 (21 kb in length; rs1391166–rs7654165). PHASE software was used to estimate and reconstruct haplotypes; haplotypes with a frequency ≥5% were selected. Block 1 included two haplotypes (TC; GT), and block 2 consisted of two haplotypes (TAC; AGT). Association analysis indicated that the risk haplotype was TC in block 1 and was TAC in block 2. Table 14.1 shows the mean differences of substance use by the haplotype groups. Among the 474 respondents in the sample, 16.2% possess two copies of the risk haplotype (TC:TC) at block 1 and 7.2% are homozygous for the risk haplotype

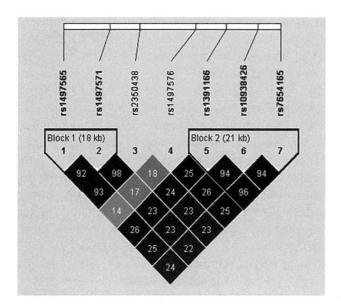

Figure 14.1
GABRG1 haplotype block structure.

Table 14.1
GABRG1 haplotype frequencies (A1 and A2 alleles) and substance use

				ANOVA test	
GABRG1	Haplotypes	%	Mean of substance use	$F_{(1, 472)}$	*p*
Block 1	TC: TC	16.2	4.403	1.933	.165
	Others	83.8	4.019		
Block 2	TAC: TAC	7.2	4.875	1.024	.312
	Others	92.8	4.021		

Note: Haplotypes with minor allele frequency ≥ 0.05 are used; n = 474. ANOVA, analysis of variance.

(TAC:TAC) at block 2. Analysis indicated that 18.1% of respondents were either TC:TC and/or TAC:TAC. The observed distribution of these two blocks did not differ significantly from that predicted by the Hardy–Weinberg equilibrium law.

14.3.3 Initial Findings

Response categories for each of the three items in our substance use measure ranged from 0 ("never") to 5 ("several times per week"). Frequency analysis indicated that 38% of respondents reported using marijuana in the past year, and 15% indicated that they used it several times per week. Eighty percent of respondents had drunk alcohol in the past year, and 8% reported doing so more than once a week. Sixty-four percent of respondents reported binge drinking in the past year, and 5% indicated that they did so several times per week.

Table 14.2 presents the means, standard deviations, and zero-order correlations for the study variables. As expected, substance use protective/risk factors is not significantly associated with GABRG1, indicating an absence of gene–environment correlation. In addition, GABRG1 is not significantly related to substance use prototype or substance use. This finding is consistent with previous molecular studies indicating that so-called risk alleles generally have little main effect on problem behavior [59, 60]. Finally, consistent with prior research, table 14.2 also shows that substance use protective/risk factors, substance use prototype, and substance use are interrelated.

14.3.4 The Effect of G × E on Substance Use Prototype and Substance Use

Table 14.3 shows the results of using hierarchical regression to examine the effects of substance use protective/risk factors (E) and GABRG1 on both substance use prototype and substance use, controlling for various demographic variables. To simplify the analysis, the two blocks of GABRG1 were combined into a single dichotomous variable: those with at least one risk (or plasticity) haplotype (18.1%) and those with no copies of a risk (or plasticity) haplotype (81.9%). Model 1a shows that E has a significant effect on substance use prototype whereas GABRG1 does not. Model 2b adds the interaction of E by GABRG1 to the regression equation. This interaction term is significant, and neither E nor G has a significant main effect when this moderating effect is included in the model. Model 2a shows that neither E nor GABRG1 is significantly related to adult substance use (controlling for use of substances during adolescence). Model 2b indicates, however, that there is a significant effect for the interaction of E and GABRG1.

Having found the expected G × E interactions, we graphed them to determine whether they display the crossover pattern predicted by the differential susceptibility hypothesis. Figure 14.2 shows this to be the case for substance use prototype. The association between E and substance use prototype is significantly steeper for respondents who carry the putative plasticity hyplotypes (b = 2.87, p = .001) than for those

Table 14.2
Correlation matrix for the study variables

	1	2	3	4	5	6	7	8	9
1. Gender (1 = Male)	—								
2. Area (1 = South)	.043	—							
3. Family SES	.003	-.176 **	—						
4. Family structure (1 = single families)	-.047	.043	-.321 **	—					
5. Protective/risk factors (W1–W4)	.043	-.206 **	-.078 †	-.017	—				
6. Substance use prototype W5	.054	-.167 **	.175 **	-.063	.140 **	—			
7. Substance use W5	.170 **	-.199 **	.180 **	-.053	.185 **	.457 **	—		
8. Block 1 of GABRG1	-.009	-.112 *	-.009	.048	.058	-.034	.064	—	
9. Block 2 of GABRG1	-.074	-.033	-.011	.082 †	-.045	-.044	.047	.432 **	—
Mean	.430	.530	.022	.530	.000	22.861	4.876	.162	.072
SD	.495	.500	1.471	.499	1.000	6.760	4.087	.369	.258

**p ≤ .01, *p ≤ .05, †p < .10 (two-tailed tests); n = 474.

Table 14.3

Social environment and genetic diversity as predictors of substance use

	Substance use prototype				Substance use			
	Model 1a		Model 1b		Model 2a		Model 2b	
	b	β	b	β	b	β	b	β
Intercept	23.510 **		23.479 **		2.867 **		2.901 **	
	(.647)		(.645)		(.398)		(.397)	
Environment and genetic variables								
Substance use protective/risk factors (E)	.868 **	.128	.431	.064	.220	.054	.032	.008
	(.327)		(.355)		(.193)		(.200)	
GABRG1	−.796	−.045	−.979	−.056	.575	.054	.493	.047
	(.770)		(.736)		(.425)		(.403)	
Two-way interaction								
GABRG1 × (E)			2.436 **	.154			1.112 **	.116
			(.645)				(.419)	
Control variables								
Prior substance use (wave 4)					.681 **	.357	.663 **	.347
					(.087)		(.097)	
Male	.710	.052	.735	.054	1.192 **	.144	1.210 **	.147
	(.593)		(.587)		(.338)		(.337)	
South	−1.592 *	−.118	−1.522 *	−.112	−.922 *	−.113	−.896 *	−.110
	(.637)		(.627)		(.348)		(.348)	
Family socioeconomic status	.760 **	.165	.739 **	.161	.400 **	.144	.392 **	.141
	(.207)		(.205)		(.111)		(.109)	
Single families	.036	.003	−.019	−.001	.314	.038	.283	.035
	(.628)		(.619)		(.345)		(.343)	
Adjusted R²	.058		.076		.221		.231	
R² increase due to interaction			.019 **				.010 *	

Note: Unstandardized and standardized coefficients are shown with robust standard errors in parentheses; the measure of substance use protective/risk factors is standardized (mean = 0 and SD = 1). SES, socioeconomic status; W, wave.

**p ≤ .01, *p ≤ .05, †p < .10 (two-tailed tests); n = 474.

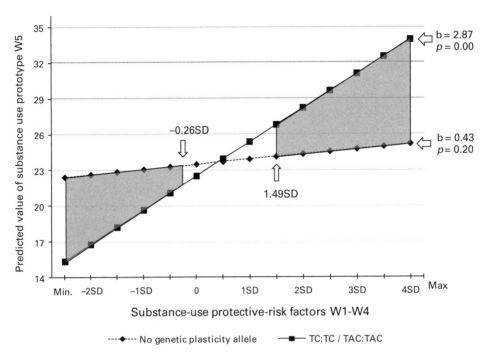

Figure 14.2
The effect of protective/risk factors on substance use prototype by *GABRG1* with the Johnson–Neyman 95% confidence bands. The gray areas are significant confidence regions. W, wave.

who do not (b = 0.43, *p* = .20). More importantly, the slopes demonstrate the expected crossing pattern so that carriers of the plasticity haplotypes demonstrate significantly more positive images of substance users than other genotypes when E is favorable to substance use but hold more negative images of substance users than other genotypes when E is adverse to substance use. Using the Johnson–Neyman [52] technique [61], genetic plasticity significantly increases an individual's adoption of a positive substance use prototype when E is greater than 1.49 standard deviations above the mean whereas it significantly decreases adoption of a positive prototype when E is less than –0.26 standard deviations below the mean. Approximately 7% of respondents scored above 1.49 standard deviations on E, and roughly 45% of respondents scored below –0.26 standard deviations on the social environmental measure.

Figure 14.3 shows a similar pattern of results when substance use is the outcome. Using the simple slope procedure [50], the slope for respondents with a plasticity hyplotype is significantly different from zero (b = 1.14, *p* < 0.01) whereas the slope for noncarriers is not (b = 0.03, *p* = 0.87). Further, the slopes show a crossing pattern with carriers of the plasticity haplotypes reporting significantly more substance use

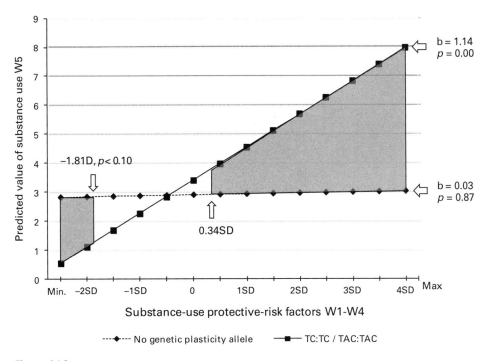

Figure 14.3
The effect of protective/risk factors on substance use by *GABRG1* with the Johnson–Neyman 95% confidence bands. The gray areas are significant confidence regions. W, wave.

than other genotypes when E encourages use but reporting less use than other genotypes when E discourages substance use. Using the Johnson–Neyman [52] technique [61], the plasticity haplotypes increase an individual's involvement in substance use when E is greater than 0.34 standard deviations above the mean ($p < 0.05$) while it marginally significantly decreases substance use when it is less than –1.81 standard deviations below the mean ($p < 0.10$). These results present weaker support for differential susceptibility than those presented in figure 14.2 for prototype.

14.3.5 The Mediating Effect of Substance Use Prototype
Next, we examined the extent to which substance use prototype mediated the effect of E and G on substance use. We began with a simple three-variable model that included E, substance use prototype, and substance use. The model showed E predicted substance use prototype, which, in turn, predicted substance use. Using bootstrap methods with 1,000 replications, the indirect effect of E on substance use through prototype was significant (indirect effect = 0.051, $p = .005$) and accounted for 65% of the total variance.

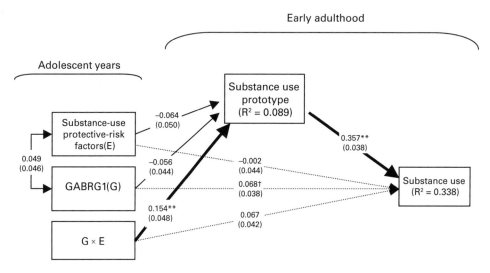

Figure 14.4

Mediated moderation model. Note: $\chi^2 = 8.701$, $df = 1$, $p = .003$; standardized root mean square residual = .016; comparative fit index = .968. The values presented are standardized parameter estimates and the standard errors are in parentheses. Using bootstrap methods with 1,000 replications, the bold lines indicate that the test of the indirect effect of interaction term is significant (indirect effect = .055 [the 45% portion of the total variance], $p < .05$). Previous status of substance use, gender, area, family socioeconomic status, and family structure are controlled in these analyses. $**p \leq .01$, $*p \leq .05$, $\dagger p < .10$ (two-tailed tests); n = 474. E, environment; G, gene.

The model was then expanded to include the main effect of *GABRG1* as well as the interaction of *GABRG1* with E (see figure 14.4). This mediated-moderation model tested the extent to which the gene–environment interaction on substance use is mediated by substance use prototype. As shown in figure 14.4, the model fit the data well by all criteria. Further, the model reveals that the relationship between the *GABRG1* gene and E is not significant, indicating no gene–environment correlation effect.

Turning to the paths in the mediated-moderator model, neither E nor *GABRG1* is significantly associated with substance use prototype. Only the interaction of *GABRG1* with E is significantly related to substance use prototype ($\gamma = 0.154$). E is not related to substance use whereas *GABRG1* shows a small association that approaches significance (p < .10). Although the interaction of E and *GABRG1* on substance use was significant in the regression model reported in table 14.3, figure 14.4 shows that this effect is no longer significant when the mediating influence of prototype is taken into account. The indirect effect of G × E on substance use through prototype is significant (indirect effect = 0.055) and accounts for 45% of the total variance. These findings support the hypothesis that substance use prototype operates as an intermediate phenotype that mediates the effect of $G \times E$ on substance use.

14.4   Discussion

The results of the present study suggest that *GABRG1* may increase the risk of substance abuse via two routes. First, we found that, after controlling for prototype, *GABRG1* minor haplotypes showed a small association with substance use that approached statistical significance. This finding supports the idea that this gene may operate as a diathesis that increases the attractiveness of substances and risk for addiction [24, 32]. Stronger confirmation, however, was found for the second avenue whereby *GABRG1* influences substance use. We found that the gene interacts with the social environment to influence learning of prototype images of substance users that, in turn, impact the use of substances. The pattern of this interaction was consistent with the differential sensitivity to context hypothesis in that carriers of the *GABRG1* minor haplotypes demonstrated significantly more positive images of substance users than other genotypes when the environment was favorable to substance use but more negative images of substance users than other genotypes when the environment was adverse to substance use. This finding indicates that, in addition to perhaps being a diathesis for substance use, *GABRG1* minor haplotypes operate as plasticity alleles that increase susceptibility to environmental context. The drug use prototype that results from this enhanced sensitivity can be positive or negative, depending on the types of information communicated by the environment.

Our results showed that *GABRG1* also interacts with environmental messages to influence substance use. The pattern of this interaction was also consistent with the differential-susceptibility-to-context hypothesis as carriers of the minor haplotypes reported significantly more substance use than other genotypes when the environment encouraged substance use but reported less use than other genotypes when the environment discouraged substance use. Importantly, however, this interaction effect was no longer significant once the impact of prototype was taken into account. This pattern of mediated moderation indicates that substance use prototype operates as an intermediate phenotype that mediates the effect of $G \times E$ on substance use.

Recent studies indicate that genes such as *5HTT*, *DRD4*, *DRD2*, and *MAOA* are plasticity genes in that they increase, for better or worse, sensitivity to environmental influence [2]. Our results suggest that *GABRG1* may also be a plasticity gene. To the extent to that this is the case, carriers of minor haplotypes are likely vulnerable to a wide variety of problem behaviors, and not simply substance abuse, in response to adverse environments. On the other hand, they would be expected to flourish compared to other genotypes when environmental conditions are favorable.

This increased responsiveness to a favorable environment suggests a more optimistic view regarding potential for change. Whereas the diathesis–stress perspective paints such persons as difficult to change given their genetic tendency to be hyperresponsive to adversity, the differential susceptibility model argues that that their environmental sensitivity makes them good candidates for intervention. They are more likely than those with differing genotypes to learn the lessons being taught by a new, more positive environment.

This idea is supported by recent intervention studies. Bakermans-Kranenburg et al. [62] found, for example, that children with the l-allele *DRD4* showed the largest decline in conduct problems in response to parent training. Brody et al. [63] recently reported that a family based-intervention with African American teens was most effective in reducing risky behavior for those with s-allele *5HTTLPR*, and Beach, Brody, Lei, and Philibert [39] reported similar findings for l-allele *DRD4* and substance use. Together, these intervention studies provide strong support for the contention that those genotypes most likely to develop problem behaviors in response to adversity are also the ones most likely to benefit from intervention.

These studies focused upon teens, and it is not clear that genetic plasticity extends into adulthood [3]. However, consistent with the differential susceptibility perspective, Beaver, Wright, DeLisi, and Vaughn [64] found that men with *DRD2*, *DRD4*, and *MAOA* "risk alleles" showed greater desistance from crime following marriage than other genotypes. This finding indicates that differential susceptibility persists into adulthood. Hopefully, future studies will examine the extent to which various putative plasticity genes, including *GABRG1*, enhance the learning of attitudes, values, and beliefs regarding substance use, including the messages conveyed in adult treatment programs.

Acknowledgments

This research was supported by the National Institute of Mental Health (MH48165, MH62669), the Center for Disease Control (U01CD001645), the National Institute on Drug Abuse (DA021898, 1P30DA027827), the National Institute on Alcohol Abuse and Alcoholism (2R01AA012768, 3R01AA012768–09S1), and both the Center for Contextual Genetics and Prevention Science and the Center for Gene–Social Environment Transaction at the University of Georgia.

References

1. Dick, D. M. (2011). Gene–environment interaction in psychological traits and disorders. *Annual Review of Clinical Psychology, 7*, 383–409.

2. Belsky, J., & Pluess, M. (2009). Beyond diathesis stress: differential susceptibility to environmental influences. *Psychological Bulletin, 135,* 885–908.

3. Ellis, B. J., Boyce, W. T., Belsky, J., Bakermans-Kranenburg, M. J., & van Ijzendoorn, M. H. (2011). Differential susceptibility to the environment: an evolutionary-neurodevelopmental theory. *Development and Psychopathology, 23,* 7–28.

4. Freese, J. (2008). Genetics and the social science explanation of individual outcomes. *AJS, 114*(Suppl), S1–S35.

5. Shanahan, C. A., Gaffney, B. L., Jones, R. A., & Strobel, S. A. (2011). Differential analogue binding by two classes of c-di-GMP riboswitches. *Journal of the American Chemical Society, 133,* 15578–15592.

6. Shanahan, M. J., & Hofer, S. M. (2005). Social context in gene–environment interactions: retrospect and prospect. *Journals of Gerontology. Series B, Psychological Sciences and Social Sciences, 60*(Spec No 1), 65–76.

7. Belsky, J., Bakermans-Kranenburg, M. J., & van IJzendoorn, M. H. (2007). For better and for worse: differential susceptibility to environmental influences. *Current Directions in Psychological Science, 16,* 300–304.

8. Carver, C. S., Johnson, S. L., & Joormann, J. (2008). Serotonergic function, two-mode models of self-regulation, and vulnerability to depression: what depression has in common with impulsive aggression. *Psychological Bulletin, 134,* 912–943.

9. Frank, M. J., D'Lauro, C., & Curran, T. (2007). Cross-task individual differences in error processing: neural, electrophysiological, and genetic components. *Cognitive, Affective & Behavioral Neuroscience, 7,* 297–308.

10. Simons, R. L., Lei, M. K., Beach, S. R. H., Brody, G. H., Philibert, R. A., & Gibbons, F. X. (2012). Social adversity, genetic variation, street code, and aggression: a genetically informed model of violent behavior. *Youth Violence and Juvenile Justice,* 10, 3–24.

11. Simons, R. L., Lei, M. K., Beach, S. R., et al. (2011). Social environmental variation, plasticity genes, and aggression: evidence for the differential susceptibility hypothesis. *American Sociological Review, 76,* 833–912.

12. Shanahan, M. J., & Macmillan, R. (2008). *Biography and the Sociological Imagination: Contexts and Contingencies.* New York: Norton.

13. Gibbons, F. X., & Gerrard, M. (1997). Health images and their effects on health behavior. In B. P. Buunk & F. X. Gibbons (Eds.), *Health, Coping, and Well-Being: Perspectives from Social Comparison Theory* (pp. 63–94). Mahwah, NJ: Erlbaum.

14. Gibbons, F. X., Gerrard, M., & Lane, D. J. (2003). A social reaction model of adolescent health risk. In J. M. Suls & K. A. Wallston (Eds.), *Social Psychological Foundations of Health and Illness* (pp. 107–136). Oxford, UK: Blackwell.

15. Denscombe, M. (2001). Uncertain identities and health-risking behaviour: the case of young people and smoking in late modernity. *British Journal of Sociology*, *52*, 157–177.

16. McFadyen, M. P., Kusek, G., Bolivar, V. J., & Flaherty, L. (2003). Differences among eight inbred strains of mice in motor ability and motor learning on a rotorod. *Genes, Brain, and Behavior*, *2*, 214–219.

17. Blanton, H., Gibbons, F. X., Gerrard, M., Conger, K. J., & Smith, G. E. (1997). Role of family and peers in the development of prototypes associated with substance use. *Journal of Family Psychology*, *11*, 271–288.

18. Gerrard, M., Gibbons, F. X., Lune, L. S., Pexa, N. A., & Gano, M. L. (2002). Adolescents' substance-related risk perceptions: antecedents, mediators and consequences. *Risk Decision and Policy*, *7*, 175–191.

19. Gerrard, M., Gibbons, F. X., Stock, M. L., Lune, L. S., & Cleveland, M. J. (2005). Images of smokers and willingness to smoke among African American pre-adolescents: an application of the prototype/willingness model of adolescent health risk behavior to smoking initiation. *Journal of Pediatric Psychology*, *30*, 305–318.

20. Spijkeman, R., Larsen, H., Gibbons, F. X., & Engels, R. (2010). Students' drinker prototypes and alcohol use in a naturalistic setting. *Alcoholism, Clinical and Experimental Research*, *34*, 64–71.

21. Long, J. C., Knowler, W. C., Hanson, R. L., et al. (1998). Evidence for genetic linkage to alcohol dependence on chromosomes 4 and 11 from an autosome-wide scan in an American Indian population. *American Journal of Medical Genetics*, *81*, 216–221.

22. Reich, T., Edenberg, H. J., Goate, A., et al. (1998). Genome-wide search for genes affecting the risk for alcohol dependence. *American Journal of Medical Genetics*, *81*, 207–215.

23. Zinn-Justin, A., & Abel, L. (1999). Genome search for alcohol dependence using the weighted pairwise correlation linkage method: interesting findings on chromosome 4. *Genetic Epidemiology*, *17*(Suppl 1), S421–S426.

24. Enoch, M. A., Hodgkinson, C. A., Yuan, Q., et al. (2009). GABRG1 and GABRA2 as independent predictors for alcoholism in two populations. *Neuropsychopharmacology*, *34*, 1245–1254.

25. Krystal, J. H., Staley, J., Mason, G., et al. (2006). Gamma-aminobutyric acid type A receptors and alcoholism: intoxication, dependence, vulnerability, and treatment. *Archives of General Psychiatry*, *63*, 957–968.

26. Agrawal, A., Edenberg, H. J., Foroud, T., et al. (2006). Association of GABRA2 with drug dependence in the collaborative study of the genetics of alcoholism sample. *Behavior Genetics*, *36*, 640–650.

27. Covault, J., Gelernter, J., Hesselbrock, V., Nellissery, M., & Kranzler, H. R. (2004). Allelic and haplotypic association of GABRA2 with alcohol dependence. *American Journal of Medical Genetics. Part B, Neuropsychiatric Genetics*, *129B*, 104–109.

28. Edenberg, H. J., Dick, D. M., Xuei, X., et al. (2004). Variations in GABRA2, encoding the alpha 2 subunit of the GABA(A) receptor, are associated with alcohol dependence and with brain oscillations. *American Journal of Human Genetics*, *74*, 705–714.

29. Fehr, C., Sander, T., Tadic, A., et al. (2006). Confirmation of association of the GABRA2 gene with alcohol dependence by subtype-specific analysis. *Psychiatric Genetics*, *16*, 9–17.

30. Lappalainen, J., Krupitsky, E., Remizov, M., et al. (2005). Association between alcoholism and gamma-amino butyric acid alpha2 receptor subtype in a Russian population. *Alcoholism, Clinical and Experimental Research*, *29*, 493–498.

31. Soyka, M., Preuss, U. W., Hesselbrock, V., et al. (2008). GABA-A2 receptor subunit gene (GABRA2) polymorphisms and risk for alcohol dependence. *Journal of Psychiatric Research*, *42*, 184–191.

32. Covault, J., Gelernter, J., Jensen, K., Anton, R., & Kranzler, H. R. (2008). Markers in the 5'-region of GABRG1 associate to alcohol dependence and are in linkage disequilibrium with markers in the adjacent GABRA2 gene. *Neuropsychopharmacology*, *33*, 837–848.

33. Drgon, T., D'Addario, C., & Uhl, G. R. (2006). Linkage disequilibrium, haplotype and association studies of a chromosome 4 GABA receptor gene cluster: candidate gene variants for addictions. *American Journal of Medical Genetics. Part B, Neuropsychiatric Genetics*, *141B*, 854–860.

34. Matthews, A. G., Hoffman, E. K., Zezza, N., Stiffler, S., & Hill, S. Y. (2007). The role of the GABRA2 polymorphism in multiplex alcohol dependence families with minimal comorbidity: within-family association and linkage analyses. *Journal of Studies on Alcohol and Drugs*, *68*, 625–633.

35. Pirker, S., Schwarzer, C., Wieselthaler, A., Sieghart, W., & Sperk, G. (2000). GABA(A) receptors: immunocytochemical distribution of 13 subunits in the adult rat brain. *Neuroscience*, *101*, 815–850.

36. Schwarzer, C., Berresheim, U., Pirker, S., et al. (2001). Distribution of the major gamma-aminobutyric acid(A) receptor subunits in the basal ganglia and associated limbic brain areas of the adult rat. *Journal of Comparative Neurology*, *433*, 526–549.

37. Heckman, J. J. (1979). Sample selection bias as a specification error. *Econometrica*, *47*, 153–161.

38. Gibbons, F. X., Reimer, R. A., Gerrard, M., Yeh, H. C., Houlihan, A. E., Cutrona, C., et al. (2007). Rural–urban differences in substance use among African-American adolescents. *Journal of Rural Health*, *23*, 22–28.

39. Beach, S. R., Brody, G. H., Lei, M. K., & Philibert, R. A. (2010). Differential susceptibility to parenting among African American youths: testing the DRD4 hypothesis. *Journal of Family Psychology*, *24*, 513–521.

40. Stewart, E. A., Schreck, C. J., & Simons, R. L. (2006). "I ain't gonna let no one disrespect me": does the code of the street reduce or increase violent victimization among African American adolescents? *Journal of Research in Crime and Delinquency*, *43*, 427–458.

41. Stewart, E. A., & Simons, R. L. (2010). Race, code of the street, and violent delinquency: a multilevel investigation of neighborhood street culture and individual norms of violence. *Criminology*, *48*, 569–605.

42. Brody, G. H., Chen, Y. F., Murry, V. M., et al. (2006). Perceived discrimination and the adjustment of African American youths: a five-year longitudinal analysis with contextual moderation effects. *Child Development*, *77*, 1170–1189.

43. Sampson, R. J. V. (2006). Collective efficacy theory: lessons learned and directions for future inquiry. In F. T. Cullen, J. P. Wright, & K. R. Blevins (Eds.), *Taking Stock: The Status of Criminological Theory. Advances in Criminological Theory* (pp. 149–167). New Brunswick, NJ: Transaction.

44. Nunnally, J. C. (1978). *Psychometric Theory*. New York: McGraw-Hill.

45. Barrett, J. C., Fry, B., Maller, J., & Daly, M. J. (2005). Haploview: analysis and visualization of LD and haplotype maps. *Bioinformatics (Oxford, England)*, *21*, 263–265.

46. Stephens, M., Smith, N. J., & Donnelly, P. (2001). A new statistical method for haplotype reconstruction from population data. *American Journal of Human Genetics*, *68*, 978–989.

47. Philibert, R. A., Beach, S. R., Gunter, T. D., et al. (2011). The relationship of deiodinase 1 genotype and thyroid function to lifetime history of major depression in three independent populations. *American Journal of Medical Genetics. Part B, Neuropsychiatric Genetics*, *156B*, 593–599.

48. Muthén, L. K., & Muthén, B. O. (2010). *Mplus 6.0 User's Guide*. Los Angeles, CA: Authors.

49. Efron, B., & Tibshirani, R. (1991). Statistical data analysis in the computer age. *Science*, *253*, 390–395.

50. Dawson, J. F., & Richter, A. W. (2006). Probing three-way interactions in moderated multiple regression: development and application of a slope difference test. *Journal of Applied Psychology*, *91*, 917–926.

51. Hayes, A. F., & Matthes, J. (2009). Computational procedures for probing interactions in OLS and logistic regression: SPSS and SAS implementations. *Behavior Research Methods*, *41*, 924–936.

52. Johnson, P. O., & Neyman, J. (1936). Tests of certain linear hypotheses and their applications to some educational problems. *Statistical Research Memoirs*, *1*, 57–93.

53. Preacher, K. J., Rucker, D. D., & Hayes, A. F. (2007). Addressing moderated mediation hypotheses: theory, methods, and prescriptions. *Multivariate Behavioral Research*, *42*, 185–227.

54. Muller, D., Judd, C. M., & Yzerbyt, V. Y. (2005). When moderation is mediated and mediation is moderated. *Journal of Personality and Social Psychology*, *89*, 852–863.

55. Baron, R. M., & Kenny, D. A. (1986). The moderator–mediator variable distinction in social psychological research: conceptual, strategic, and statistical considerations. *Journal of Personality and Social Psychology*, *51*, 1173–1182.

56. Mallinckrodt, B., Abraham, W. T., Wei, M., & Russell, D. W. (2006). Advances in testing the statistical significance of mediation effects. *Journal of Counseling Psychology, 53,* 372–378.

57. Browne, M. W., & Cudeck, R. (1992). Alternative ways of assessing model fit. *Sociological Methods & Research, 21,* 230–258.

58. Bentler, P. M. (1990). Comparative fit indexes in structural models. *Psychological Bulletin, 107,* 238–246.

59. Caspi, A., Sugden, K., Moffitt, T. E., et al. (2003). Influence of life stress on depression: moderation by a polymorphism in the 5-HTT gene. *Science, 301,* 386–389.

60. Moffitt, T. E. (2005). The new look of behavioral genetics in developmental psychopathology: gene–environment interplay in antisocial behaviors. *Psychological Bulletin, 131,* 533–554.

61. Preacher, K. J., Curran, P. J., & Bauer, D. J. (2006). Computational tools for probing interaction effects in multiple linear regression, multilevel modeling, and latent curve analysis. *Journal of Educational and Behavioral Statistics, 31,* 437–448.

62. Bakermans-Kranenburg, M. J., Van, I. M. H., Pijlman, F. T., Mesman, J., & Juffer, F. (2008). Experimental evidence for differential susceptibility: dopamine D4 receptor polymorphism (DRD4 VNTR) moderates intervention effects on toddlers' externalizing behavior in a randomized controlled trial. *Developmental Psychology, 44,* 293–300.

63. Brody, G. H., Beach, S. R., Philibert, R. A., Chen, Y. F., & Murry, V. M. (2009). Prevention effects moderate the association of 5-HTTLPR and youth risk behavior initiation: gene × environment hypotheses tested via a randomized prevention design. *Child Development, 80,* 645–661.

64. Beaver, K. M., Wright, J. P., DeLisi, M., & Vaughn, M. G. (2008). Desistance from delinquency: the marriage effect revisited and extended. *Social Science Research, 37,* 736–752.

15 From Genes to Behavior Change: Treatment Response as an Intermediate Phenotype

Courtney J. Stevens, Hollis C. Karoly, Renee E. Magnan, and Angela D. Bryan

Substance abuse is a major public health concern due in part to the substantial burden it imposes on the health care system. According to the National Institute on Drug Abuse, approximately $600 billion per year is spent on factors directly related to substance use, abuse, and dependence, including productivity-, health- and crime-related costs [1]. At around $235 billion per year, alcohol use disorders are the most costly of this group [2]. In fact, one estimate suggests that 30% of Americans will experience an alcohol use disorder at some point during their lifetimes [3]. This statistic is particularly troubling because alcohol dependence is associated with multiple life threatening illnesses including cirrhosis of the liver and several types cancers [4, 5]. According to the Centers for Disease Control [6] and a report by the United States Department of Health and Human Services (USDHHS; [7]), cigarette smoking and excessive alcohol consumption are the first and third most common behaviorally based causes of death among Americans, respectively.

Substance abuse is also highly resistant to change, at least in the long term [8]. For example, despite major policy changes concerning the degree to which cigarette smoking and tobacco use are advertised, taxed, or permitted [9], around 25% of Americans continue to smoke [10]. Further, according to the 2010 National Survey on Drug Use and Health published by the USDHHS, every day nearly 3,800 adolescents smoke their first cigarette (USDHHS, 2010). For these reasons, it is extremely important to develop effective and targeted prevention, intervention, and treatment programs for reducing and preventing the negative influence of health-compromising behaviors on individuals as well as on society as a whole.

Developing effective prevention, intervention, and treatment programs to reduce the negative impacts associated with substance abuse has proven to be a tremendously difficult and complicated task. Part of the problem may be the lack of tailored treatment options available. For instance, the majority of treatment options for addictive behaviors only provide a general framework for recovery, without taking into account the role of individual variations. In the context of addictive behaviors, *individual variations* might refer to the personal-level reasons, desires, social circumstances, abilities, or mental health comorbidities influencing an individual's use of a substance. Although there are a number of empirically supported treatments for addiction [11], few of these treatments make use of personalized information to guide the course of therapy.

Current intervention strategies from the behavioral sciences face similar criticism. A recent review revealed that most behavioral interventions are only modestly effective for changing behavior (as opposed to changing intentions for behavior) [12]. These small effects may be attributable to the "one size fits all" nature of intervention content [13] that are often assumed to produce similar results for all individuals.

To address this issue, researchers conducting theory-based interventions to decrease risk behavior and promote health behavior have begun to investigate the role of potential moderators that may determine for whom behavioral interventions are most likely to be successful [14, 15]. For example, individuals who possess a biological predisposition (e.g., genetic, neurocognitive) for impulsivity may be more likely to engage in risky behaviors such as heavy alcohol consumption or drug experimentation. In this way, behavioral interventions could become more appropriately tailored to the specific needs of individuals who score highly on measures of impulsivity as they attempt to change their behaviors. Similarly, certain psychotherapy treatments for addictive behaviors have demonstrated success when individual client motives are considered central to the treatment plan [16]. Motivational Enhancement Therapy (MET; [17]), for example, is a psychotherapy commonly used to treat behaviors of addiction using a client-centered approach that focuses on each client's personal motivations for change (e.g., reasons, needs, abilities, desires) and uses these to help each client resolve ambivalence about behavior change (e.g., drinking, using cocaine) [18]. However, incorporating personalized behavioral interventions tailored to fit the specific needs of individuals into treatment plans currently lacks sufficient empirical support. One reason for this lack of knowledge is that the factors that will determine the forms of treatment that will work for different individuals are not fully understood. It is expected that an individual's unique treatment response will be influenced by biological, psychological, and social factors, as well as some degree of interaction between these complex factors.

Personalized medicine, from a pharmacological standpoint, is becoming standard practice in some domains [19, 20] and is far ahead in terms of seeking and finding genetic loci that are associated with response to treatment. In terms of behavioral interventions, the link between genetic variation and differential treatment response is less clear and is indeed an emerging field of research.

Researchers are increasingly looking for ways to integrate behavioral genetic information in substance use treatment in order to improve the likelihood of successful initiation and sustained behavior change. This approach is supported by genetically informative studies using twin and/or family designs which have consistently demonstrated that a substantial proportion of the variance in health and risk behaviors can be attributed to heritable factors. For example, genetic factors have been shown to account for approximately 50% of the variance in smoking [21] and drinking behaviors [22]. It is therefore reasonable to suggest that a more sophisticated understanding of genetic factors underlying health behaviors will give rise to more effective and personally tailored behavioral treatments [23]. However, understanding that some proportion of the variance in behavior is linked to genetic factors is only a starting point. This information is not inherently useful if the question is how best to *change* behavior.

Addressing the issue of behavior change requires development of intermediate phenotypes that help to inform which genetic markers might be associated with factors that ultimately relate not only to the initiation and maintenance of substance use but, importantly, response to behavioral interventions (i.e., MET, twelve-step programs, etc.) to prevent and treat substance use. A central goal of this chapter will be to discuss the state of the science regarding the current understanding of intermediate mechanisms by which genetic factors influence response to treatments for substance abuse and to propose ideas for the incorporation of genetic factors into treatment protocols with the goal of behavior change.

15.1 Terminology

For clarity, we first provide definitions of frequently used terms in this chapter. The term *genotype* refers to an individual's DNA sequence, encompassing all of the genes he or she possesses, even if the genes are not all expressed. In contrast, a *phenotype* refers to an individual's expression (behavioral, physical, and psychological) of his or her genotype. Previously we introduced the notion that certain mechanisms intervene between the genotype and the phenotype and we will refer to this mechanism as an *intermediate phenotype*. Historically, the intervening mechanism between genotype and phenotype was referred to as an *endophenotype* and was used to describe the physiological, behavioral, or cognitive processes that are more proximal to genes than are diagnostic phenotypes. Endophenotypes are defined as biological/physiological traits that can be found not only in individuals affected by a particular phenotype, but also in their unaffected family members [24]. Intermediate phenotypes are a broader class of variables that include biological, physiological, and psychological traits that are presumably more proximal to basic mechanisms, more genetically informative, more narrowly defined, and more amenable to empirical validation than are complex diagnostic phenotypes [24]. Further, intermediate phenotypes can relate to a broader range of phenotypes including specific classes of behavior (e.g., risky sexual behavior in the context of alcohol use) that are not necessarily a form of diagnosed psychopathology. Thus, intermediate phenotypes provide a possible solution to the problem of discovering genetic factors that are responsible for complex behavioral phenotypes because the genotype explains more of the variance in the intermediate phenotype than it does in the expression of the complex behavior [25]. Accordingly, we utilize the term *intermediate phenotype* rather than *endophenotype* in this chapter.

Single nucleotide polymorphisms (SNPs), are single base pair positions in genomic DNA at which variation in different sequence alternatives (A, C, T, or G), also known as alleles, exist in the population. In order for a polymorphism to be considered a SNP, the least frequent allele must have a frequency of 1% or greater [26]. For instance, if 95% of individuals possess the T nucleotide while the remaining 5% possess the C nucleotide at a particular genetic locus, then that locus will be considered a SNP. SNPs can occur at both coding and noncoding regions of the genome and arise about once in every 1,000 base pairs across the human genome.

15.1.1 Genetic Study Designs

Genome-wide association studies (GWASs) refer to those studies in which hundreds of thousands of genetic variants across the entire genome are analyzed in one study with the goal of uncovering genetic variation associated with susceptibility to a particular disease or with disease-related quantitative traits. To perform analyses in GWASs, researchers collect genetic material (e.g., saliva) and look for associations between SNPs and disease status (diagnosed alcohol use disorder versus no alcohol use disorder) or look for associations with a quantitative trait (number of symptoms of alcohol use disorder). Genetic targets in these analyses are common SNPs that vary in at least 5% of the population. The idea behind this strategy is that many common SNPs contribute to the expression of a particular disease state or quantitative trait, and by testing a large number of SNPs at a time, these disease-associated common SNPs can be identified [27].

Since the first GWAS was performed in 2005, numerous studies have successfully identified common SNPs that are significantly associated with various medical and psychiatric diseases, such as schizophrenia [28, 29], type-2 diabetes [30], and prostate cancer [31]. However, a weakness of GWASs is that they generally fail to reveal rare variants that may contribute to the expression of particular phenotypes [32]. Further, the analytic techniques assume that each test is independent of the next, leading to massive overcorrection of significance values and a misspecification of what is likely a polygenic effect of multiple SNPs acting in concert. In addition, because the identification of common SNPs is unlikely to explain all or even most of the genetic risk underlying common disorders, the addition of rare variants (those that vary in less than 5% of the population) to the genotyping platforms used in GWAS research would likely improve identification of causal SNPs [33] although we acknowledge there is certainly active debate about this point [34].

In general, GWASs have yielded much less impressive results than was initially hoped by experts in the field [35, 36]. The use of GWASs in lines of research aimed at explaining genetic factors associated with treatment response is even more limited, as the sample sizes for treatment and intervention studies with well-characterized behavioral phenotypes are typically much smaller than what is recommended for GWASs. Given the limitations of GWASs, particularly in intervention studies, candidate gene approaches may be more fruitful in this domain.

A *candidate gene* is one that has received support by previous genetic linkage or association studies for a relationship with a particular intermediate phenotype or broader phenotype. Thus, candidate gene studies are hypothesis-driven investigations (in contrast to the non–hypothesis driven GWAS method) and focus attention on particular genes of interest rather than on the entire genome. In order to identify candidate genetic systems that are likely to be involved in behavioral processes, it will be important to understand the underlying biological systems involved with both the development of health-compromising versus health-promoting behaviors and with treatment response. In the domain of substance abuse research, aspects of executive function that are associated with problematic alcohol use, in particular inhibitory control and cue-elicited craving for alcohol, have been studied extensively for their association with candidate genes [37–39].

At a neurobiological level, cue-elicited craving has been linked to the mesolimbic dopamine pathway in the brain, a substrate thought to be an important mechanism in the etiology of alcohol and drug dependence [40–42]. Thus, the dopamanergic system has become a prime target for investigations of candidate genes. For instance, a recent candidate gene study by Dahlgren et al. [43] found that a variant in the dopamine receptor D2 (*DRD2*) gene predicts increased risk of relapse in alcoholic-dependent men and women undergoing a twelve-step alcohol treatment program.

15.2 The Relationship of Genetic Factors to Treatment Response

Genetic factors may moderate response to pharmacotherapy and behavior intervention treatments for addiction. Several studies have explored underlying genetic variants hypothesized to predict response to opioid receptor agonists used for the treatment of alcohol dependence. One by Oslin et al. [44] found that individuals possessing the AG or GG allele of the *OPRM1* gene responded more favorably to naltrexone (an opioid antagonist drug) than those possessing the AA allelic variant. More specifically, AG or GG individuals stayed alcohol abstinent longer and were less likely to relapse than were AA individuals.

Olanzapine is another opioid antagonist drug that, similar to naltrexone, works by disrupting reward centers in the dopamanergic pathways of the brain. Hutchison and colleagues [45] showed that olanzapine is differentially effective at reducing cue-elicited craving based on individual differences in the dopamine receptor D4 (*DRD4*) variable number of tandem repeats (VNTR) polymorphism. While olanzapine reduced craving for alcohol among individuals with both the long (determined by the presence of seven or more repeat alleles) and the short allelic variants of the *DRD4* gene (*DRD4* L versus *DRD4* S, respectively), subjective levels of craving following exposure to alcohol relevant cues was attenuated only among individuals possessing the long allelic variant [45].

Hutchison and colleagues [46] followed up these findings with a longitudinal, twelve-week-long randomized clinical trial (RCT) testing the effectiveness of olanzapine among alcohol-dependent individuals. Consistent with their prior findings, *DRD4* L individuals in the olanzapine condition reported less subjective craving for alcohol during a cue-elicited craving task, as well as greater reductions in alcohol consumption over the course of the study compared to *DRD4* S individuals [46].

Tailoring of pharmacological interventions based on individual genotypes is a common strategy for treating various health disorders such as acquired immunodeficiency syndrome (AIDS) and certain cancers [47–49]. AIDS patients are often urged to be tested for HLA-B*5701, an allele variation in the human leukocyte antigen complex of genes that predicts toxic side effects of the medication abacavir [47]. Among breast cancer patients, a test of genetic variation in the *HER* gene is used clinically to predict response to the commonly used drug Herceptin [49], and patients with chronic myeloid leukemia are tested for variation in the *BCL-ABR* gene, which has been found to predict response to the drug Glivec [48].

Although tailoring pharmaceutical interventions by genotype is relatively common, work investigating the moderating role of genetics in behaviorally based treatments has thus far been limited. In fact, we were able to find only one example in the extant literature. Problem drinkers who received MET [17] took greater steps toward reducing their drinking following MET if they possessed the short allelic variant in the dopamine D4 receptor gene (*DRD4* S) [50].

A review of neurobiological and behaviorally based phenotypes related to substance use and addiction, specifically in a SNP in the *DRD4* gene, found converging evidence across several different methodologies (i.e., cellular assays, neuroimaging, and behavioral studies) supporting the existence of a robust relationship between this polymorphism and the urge to use addictive substances [51]. Based on these results, McGeary [51] argues for a genetically based, urge-related phenotype, which may increase the risk for developing addictive behaviors among certain individuals. This phenotype would likely influence response to behavioral treatments focusing on craving reduction related to substance use.

Additional evidence for addressing the role of genetic moderators in addiction comes from the smoking literature. Nicotine withdrawal symptoms are often associated with attentional deficits [52]. A recent study found that current smokers who were asked to abstain from cigarettes overnight demonstrated reduced attentional abilities compared to nonabstaining smokers [53]. Importantly, this effect was moderated by whether an individual carried the A1 allele at the *DRD2 Taq1A* polymorphism site. Specifically, abstaining smokers with the A1 allele demonstrated worse attentional processing compared to abstaining smokers without the A1 allele. Although this study does not address genetic moderation in a treatment context, it does emphasize the role of genetic variability in substance use withdrawal symptoms.

Taken together these investigations suggest that the pursuit of genetic variables as moderators of treatment response is a worthwhile and promising future avenue of research. In order to increase the efficacy of treatments for substance dependence, it may be extremely useful to know a priori what genetic factors moderate treatment response. With this information, researchers could begin to focus treatment content around those intermediate phenotypes in an effort to increase the likelihood of overall success for individuals possessing particular phenotypic and genotypic characteristics.

15.3 Developing Integrative Models of Genetics, Neurocognition, and Treatment Response

Although the limited work investigating genetic moderation of behaviorally based treatments is promising, much more work is needed to determine the role of genetic factors improving behaviorally based treatments. In order to increase the rate of progress in this area, it would be helpful to have some guidelines indicating which intermediate phenotypes are most relevant to treatment response. Researchers have called for the development of multivariate models which would link genes to more proximal variables acting as intermediate phenotypes. Those intermediate phenotypes could then be utilized to help characterize response to treatments that target more complex phenotypes (e.g., alcohol or cigarette addiction). For instance, intermediate phenotypes might be related to the behavioral activation and behavioral inhibition systems which influence sensation seeking and impulsivity [54, 55] and which may also more reliably predict behavior. Sensation seeking and impulsivity are two personality factors which have been shown to differentially predict engagement in substance use on a broad spectrum [56], as well as use of specific substances including alcohol [55, 57], stimulants (cocaine and amphetamines) [54], and tobacco [57]. The incorporation of intermediate phenotypes such as impulsivity and sensation seeking into a multivariate model linking genetic factors to substance use outcomes (cf. [56]) has the potential to more fully inform the content, structure, and goals of treatment and intervention [58].

One such translational model of substance dependence has been recently developed (see figure 15.1). In this model, genetic variants, consisting of individual SNPs that give rise to allelic variation, influence the strength of two key neurocognitive pathways [23]: the incentive motivation network and the control network. Specific neuronal structures within each of these networks are believed to be activated during substance use. The incentive motivation network is made up of the neural structures involved

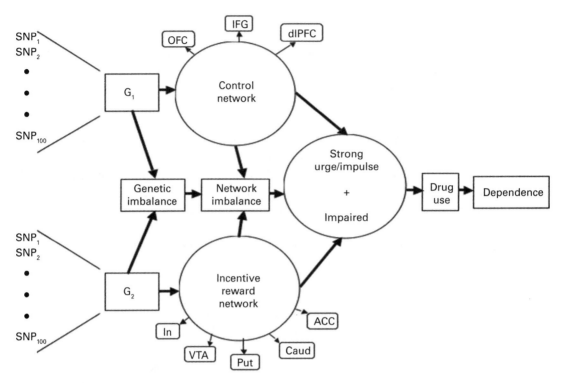

Figure 15.1

A translational model of substance abuse showing the hypothesized pathways through which genetic variation influences addiction behaviors. SNP, single nucleotide polymorphism; G, genetic variant; OFC, orbitofrontal cortex; IFG, inferior frontal gyrus; dlPFC, dorsolateral prefrontal cortex; Ins, insula; VTA, ventral tegmental area; Put, putamen; Caud, caudate nucleus; ACC, anterior cingulate cortex. Adapted from "Substance Use Disorders: Realizing the Promise of Pharmacogenomics and Personalized Medicine," by K. E. Hutchison, 2010, *The Annual Review of Clinical Psychology, 6,* 577–589.

in the drive to seek rewards, such as feeling "high" from drug use. Conversely, the neuronal control network is comprised of the structures involved in the ability to inhibit impulses and thus, abstain from substance use [23]. According to the model, when there is an imbalance such that the incentive motivation network overrides the control network, the urge/impulse to engage in addictive behaviors is made increasingly potent [23, 59, 60]. At the first level of the model, SNPs come together to form aggregate measures of genetic variation, represented as G_1 and G_2 in the model. These variations are related to differential expression within the control and incentive networks which may lead to more impulsive behavior, impaired control, and eventually, substance use.

To this end, neuroimaging paradigms have been critical for identifying which regions and pathways are activated within each system in response to substance-relevant stimuli. For example, cue-elicited craving tasks are those that show participants various images related to their substance of choice or that provide participants with a small sample of their substance of choice and thereby activate the incentive motivation network [61]. Claus and colleagues [61] demonstrated that alcohol taste cues intended to elicit alcohol craving resulted in increased activation in the dorsal striatum, insula/orbitofrontal cortex, anterior cingulate cortex, and ventral tegmental area. Additionally, alcohol use disorder severity correlated moderately with structures involved in the incentive motivation network such as the nucleus accumbens, amygdala, precuneus, insula, and dorsal striatum.

In addition to neuroimaging tasks that target specific substances, other tasks target more general constructs related to risky decision making and the ability to delay gratification. Both risk-taking and delayed-gratification paradigms tap into an individual's ability to inhibit actions that are rewarding and inherently risky and thereby activate the control network [61, 62].The Balloon Analogue Risk Task [63] is a commonly used risk-taking task that was recently employed to demonstrate the association between risk taking and activation in the dorsal anterior cingulate cortex dorsal anterior cingulate cortex and anterior insula—two regions of the brain that are involved in the control network [62]. In a separate study, Claus and colleagues [64] replicated these findings among high-risk adolescents. Greater response was found in the dorsal anterior cingulate cortex/supplementary motor area, left pre/postcentral gyrus, right precentral gyrus, superior temporal gyrus, and bilateral lateral occipital cortex during "risky" choices. In contrast, safe choices were associated with greater response in regions of the default mode network, which refers to a set of brain regions that typically deactivate during cognitive tasks [65]. The default mode regions associated with safe choices include the medial prefrontal cortex, posterior cingulate/precuneus, and angular gyrus, and also through lateral orbitofrontal cortex, bilateral middle frontal gyrus, and bilateral middle temporal gyrus.

Returning to the components of the translational model, at the most distal end, neurocognitive components influence the expression of translational phenotypes. These translational phenotypes can be useful for predicting how an individual will respond to environmental influences, including interventions to prevent and treat substance abuse. This theoretical conceptualization is nascent in its development, and, certainly, tightly controlled investigations spanning various populations and substances of abuse are needed to inform the validity of this approach and its utility for guiding targeting and tailoring of behavioral interventions.

As an example of how genetic and neurocognitive factors can be integrated into substance use treatment, we recently completed an RCT intervention funded by the National Institute on Alcohol Abuse and Alcoholism (NIAAA) among juvenile-justice-

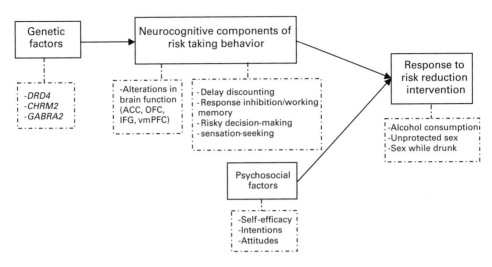

Figure 15.2

Integrative model of alcohol-related risky sexual behavior. ACC, anterior cingulate cortex; OFC, orbitofrontal cortex; IFG, inferior frontal gyrus; vmPFC, ventromedial prefrontal cortex (R01 AA017390; PI: Bryan).

involved youth (ages 14–18), comparing a sexual risk reduction intervention (SRRI) plus alcohol risk reduction intervention (SRRI + EtOH) with an information-only contact control (INFO). We refer to this trial as Project SHARP (Sexual Health and Adolescent Risk Prevention). As with other work in substance use treatment, interventions focused on risk behaviors in adolescence are only modestly effective [66, 67], and the effects tend to decay over time [68]. Project SHARP was based on an integrative framework (figure 15.2) that posits linkages between genetic factors and neurocognitive components of risk taking, between neurocognitive components of risk taking and response to the SRRI + EtOH intervention, and a combined effect of neurocognitive components with psychosocial factors resulting in risky behaviors.

Participants (N = 269; 27.9% female) completed a baseline session including a functional magnetic resonance imaging (fMRI) scanning protocol, providing a saliva sample for DNA, and a questionnaire assessment of individual differences (e.g., impulsivity) and past and current alcohol use and sexual behavior. Participants were then randomized to receive one of the two interventions. Follow-up assessments were conducted every three months for one year (three, six, nine, and twelve months post-intervention). Participants randomized to SRRI + EtOH engaged in a single MET group session that explored motivation to change alcohol use and sexual behavior, positive and negative effects of drinking and risky sex, and individualized goals for changing alcohol consumption and risky sexual behaviors. The intervention also included a previously tested theory-based sexual risk prevention protocol [68, 69]. Participants randomized to INFO also engaged in a single-session group intervention; however, the content focused on safe sex practices and information regarding different sexually transmitted illnesses.

As an example of the test of the integrative framework, we provide preliminary findings using a modified version of the go/no-go task (Go1; [70]) which assesses response inhibition. While in the fMRI scanner, participants were instructed to respond by pressing a button to the target stimulus "X" (go events = 206) and to inhibit the response (i.e., not press the button) to the stimulus "K" (no-go events = 37). A "hit" refers to correctly pressing the button during X trials (proponent response), and a "correct reject" refers to correctly inhibiting the button-push response.

Regarding the link between genetic and neurocognitive factors, we found preliminary evidence that SNPs in the *GABRA2* (rs9291282 and rs279871) and *CHRM2* (rs1455858) genes were significantly associated with neurocognitive response to the Go1 task. Higher genetic risk was associated with lower neurocognitive activation. Regarding the link between neurocognitive components and response to the SRRI + EtOH, we found inferior frontal gyrus activation during the Go1 task moderated condition on alcohol consumption and frequency reported at the three-month follow-up. Specifically, individuals with lower inferior frontal gyrus activation (and thus with greater genetic risk) responded worse to the SRRI + EtOH intervention than individuals with higher inferior frontal gyrus activation. Thus, our preliminary findings provide evidence that focusing on genetic and neurocogntivie moderators of *behavioral interventions* (and not just pharmacological interventions) may be important to better understand for whom substance treatment interventions work best.

15.4   Challenges Facing Behavioral Genetic Integration

The integration of genetics into the science of behavior change still presents several important challenges. For instance, research aimed at identifying the genetic correlates of substance abuse and moderators of interventions designed to change behavior currently suffers from some of the same issues that plague genetic research in general.

One significant problem present in this line of research is that there are likely many genetic variables at work, each of which contributes only a small fraction of the variance in a complex phenotype [34, 71]. Current statistical approaches are not presently able to capture the aggregate effects of these variations. This issue may be remedied with the use of aggregate genetic risk scores. Even without a complete understanding of the causal mechanism of particular genetic variants, combinations of risk alleles can be used to form aggregate genetic risk scores that effectively predict individual disease risk. This polygenic framework, in which qualitative disorders can be interpreted as being the extremes of quantitative dimensions, has been applied to a wide range of diseases [72]. For example, Derringer et al. [73] used a genetic risk score comprised of SNPs from four dopamine genes to predict sensation-seeking behavior in humans. In the field of health behavior, aggregate genetic risk scores might be useful for predicting whether individuals are likely to engage in risky behaviors such as drinking alcohol or smoking.

15.5   Summary

Addictive behaviors are a prime target of behavioral interventions because of the enormous costs they impose for society, not only financially but also in terms of the emotional and physical consequences of poor health, impaired social and cognitive functioning, and premature mortality. Very few treatments available for addictive behaviors utilize a personalized approach, likely because of the difficulties associated with determining which interventions work best for whom. Although there is strong evidence to suggest that substance use and abuse are influenced by genetic factors, this information is not inherently useful when the goal of treatment is to change behavior. Treatment response to behavioral interventions likely involves a complex interplay between biological and psychosocial variables, as well as their interactive effects, making efforts to personalize treatments and behavioral interventions less clear and more difficult in comparison to pharmacological treatments.

Although much more work is needed in the realm of behavioral interventions, there is some evidence to suggest that treatment response to behavioral interventions may be moderated by underlying genetic factors. A multivariate model linking genetic factors to substance use outcomes demonstrated the means through which intermediate phenotypes such as impulsivity and sensation seeking have the potential to better inform treatment design and implementation. Preliminary findings from a NIAAA-funded RCT among juvenile-justice-involved youth provided an example for how such an integrative model incorporating genetic and neurocognitive factors can be utilized.

15.6   Conclusions

The research discussed in this chapter supports the notion that genetic variants and genetically influenced traits may moderate the likelihood of an individual's responding favorably to one mode of treatment versus another (e.g., pharmacotherapy versus psychotherapy or psychosocial behavioral intervention). Collecting genetic information and replicating findings that support the predictive utility of particular genetic variants is important, but as the many failures to replicate from candidate gene studies highlight, it is crucial not to focus solely on the contribution of one or even several genes. Rather, multidimensional models that attempt to incorporate factors in between genes and behavior stand to more completely elucidate causal mechanisms. These intermediate phenotypes include everything from personality and neurocognitive factors to individual responses to different types of treatment for psychological disorders such as substance abuse. Ultimately, utilizing intermediate phenotypes to inform programs of research as well as treatment decisions will bring the field closer to the eventual goal of genetically based, personalized treatment programs.

References

1. National Institute on Drug Abuse. (2011). NIDA InfoFacts: understanding drug abuse and addiction [online]. Bethesda, MD: National Institute on Drug Abuse.

2. Rehm, J., Mathers, C., Popova, S., et al. (2009). Global burden of disease and injury and economic cost attributable to alcohol use and alcohol-use disorders. *Lancet, 373*, 2223–2233.

3. Hasin, D. S., Stinson, F. S., Ogburn, E., & Grant, B. F. (2007). Prevalence, correlates, disability, and comorbidity of DSM–IV alcohol abuse and dependence in the United States: results from the National Epidemiologic Survey on Alcohol and Related Conditions. *Archives of General Psychiatry, 64*, 830–842.

4. Gutjahr, E., Gmel, G., & Rehm, J. (2001). Relation between average alcohol consumption and disease: an overview. *European Addiction Research, 7*, 117–127.

5. O'Keefe, J. H., Bybee, K. A., & Lavie, C. J. (2007). Alcohol and cardiovascular health: the razor-sharp double-edged sword. *Journal of the American College of Cardiology, 50*, 1009–1014.

6. Center for Disease Control. (2011). Excessive alcohol use: addressing a leading risk for death, chronic disease, and injury [online]. Atlanta, GA: Centers for Disease Control.

7. SAMHSA. (2012). Preventing tobacco use among youth and young adults: a report from the Surgeon General. Washington, DC: Department of Health and Human Services.

8. Miller, W. R., Walters, S. T., & Bennett, M. E. (2001). How effective is alcoholism treatment in the United States? *Journal of Studies on Alcohol, 62*, 211–220.

9. Jemal, A., Thun, M. J., Ries, L. A., Howe, H. L., Weir, H. K., Center, M. M., et al. (2008). Annual report to the nation on the status of cancer, 1975–2005, featuring trends in lung cancer, tobacco use, and tobacco control. *Journal of the National Cancer Institute, 100*, 1672–1694.

10. Fiore, M. C., & Jaen, C. R. (2008). A clinical blueprint to accelerate the elimination of tobacco use. *Journal of the American Medical Association, 299*, 2083–2085.

11. Brown, J. M. (2004). The effectiveness of treatment. In N. Heather & T. Stockwell (Eds.), *The Essential Handbook of Treatment and Prevention of Alcohol Problems* (pp. 9–20). West Sussex, UK: Wiley.

12. Michie, S., Abraham, C., Whittington, C., McAteer, J., & Gupta, S. (2009). Effective techniques in healthy eating and physical activity interventions: a meta-regression. *Health Psychology, 28*, 690–701.

13. Napolitano, M. A., & Marcus, B. H. (2002). Targeting and tailoring physical activity information using print and information technologies. *Exercise and Sport Sciences Reviews*, *30*, 122–128.

14. Bryan, A. D., & Hutchison, K. E. (2012). The role of genomics in health behavior change: challenges and opportunities. *Public Health Genomics*, *15*, 139–145.

15. Bryan, A. D., Nilsson, R., Tompkins, S. A., et al. (2011). The big picture of individual differences in physical activity behavior change: a transdisciplinary approach. *Psychology of Sport and Exercise*, *12*, 20–26.

16. Miller, W. R. (1983). Motivational interviewing with problem drinkers. *Behavioural and Cognitive Psychotherapy*, *11*, 147–172.

17. Miller, W. R., Zweben, A., DiClemente, C. C., & Rychtarik, R. G. (1992). *Motivational Enhancement Therapy Manual: A Clinical Research Guide for Therapists Treating Individuals with Alcohol Abuse and Dependence*. Rockville, MD: National Institute on Alcohol Abuse and Alcoholism.

18. Rollnick, S., & Allison, J. (2004). Motivational interviewing. In N. Heather & T. Stockwell (Eds.), *The Essential Handbook of Treatment and Prevention of Alcohol Problems* (pp. 105–117). West Sussex, UK: Wiley.

19. Li, C. (2011). Personalized medicine—the promised land: are we there yet? *Clinical Genetics*, *79*, 403–412.

20. Offit, K. (2011). Personalized medicine: new genomics, old lessons. *Human Genetics*, *130*, 3–14.

21. Hoft, N. R., Stitzel, J. A., Hutchison, K. E., & Ehringer, M. A. (2011). CHRNB2 promoter region: association with subjective effects to nicotine and gene expression differences. *Genes, Brain, and Behavior*, *10*, 176–185.

22. Gelernter, J., & Kranzler, H. R. (2009). Genetics of alcohol dependence. *Human Genetics*, *126*, 91–99.

23. Hutchison, K. E. (2010). Substance use disorders: realizing the promise of pharmacogenomics and personalized medicine. *Annual Review of Clinical Psychology*, *6*, 577–589.

24. Cannon, T. D., & Keller, M. C. (2006). Endophenotypes in the genetic analyses of mental disorders. *Annual Review of Clinical Psychology*, *2*, 267–290.

25. Szatmari, P., Maziade, M., Zwaigenbaum, L., et al. (2007). Informative phenotypes for genetic studies of psychiatric disorders. *American Journal of Medical Genetics. Part B, Neuropsychiatric Genetics*, *144B*, 581–588.

26. Brookes, A. J. (1999). The essence of SNPs. *Gene*, *234*, 177–186.

27. Psychiatric GWAS Consortium Coordinating Committee, Cichon, S., Craddock, N., et al. (2009). Genomewide association studies: history, rationale, and prospects for psychiatric disorders. *American Journal of Psychiatry*, *166*, 540–556.

28. International Schizophrenia Consortium, Purcell, S. M., Wray, N. R., et al. (2009). Common polygenic variation contributes to risk of schizophrenia and bipolar disorder. *Nature, 460*, 748–752.

29. Stefansson, H., Ophoff, R. A., Steinberg, S., et al. (2009). Common variants conferring risk of schizophrenia. *Nature, 460*, 744–747.

30. Diabetes Genetics Initiative of Broad Institute of Harvard and MIT, Lund University, Novartis Institutes of BioMedical Research, et al. (2007). Genome-wide association analysis identifies loci for type 2 diabetes and triglyceride levels. *Science, 316*, 1331–1336.

31. Thomas, G., Jacobs, K. B., Yeager, M., et al. (2008). Multiple loci identified in a genome-wide association study of prostate cancer. *Nature Genetics, 40*, 310–315.

32. Barrett, J. C., & Cardon, L. R. (2006). Evaluating coverage of genome-wide association studies. *Nature Genetics, 38*, 659–662.

33. Gorlov, I. P., Gorlova, O. Y., Sunyaev, S. R., Spitz, M. R., & Amos, C. I. (2008). Shifting paradigm of association studies: value of rare single-nucleotide polymorphisms. *American Journal of Human Genetics, 82*, 100–112.

34. Wray, N. R., Purcell, S. M., & Visscher, P. M. (2011). Synthetic associations created by rare variants do not explain most GWAS results. *PLoS Biology, 9*, e1000579.

35. Hardy, J., & Singleton, A. (2009). Genomewide association studies and human disease. *New England Journal of Medicine, 360*, 1759–1768.

36. Kraft, P., & Hunter, D. J. (2009). Genetic risk prediction—are we there yet? *New England Journal of Medicine, 360*, 1701–1703.

37. Fillmore, M. T., & Vogel-Sprott, M. (1999). An alcohol model of impaired inhibitory control and its treatment in humans. *Experimental and Clinical Psychopharmacology, 7*, 49–55.

38. Heinz, A., Siessmeier, T., Wrase, J., et al. (2004). Correlation between dopamine D(2) receptors in the ventral striatum and central processing of alcohol cues and craving. *American Journal of Psychiatry, 161*, 1783–1789.

39. Weafer, J., & Fillmore, M. T. (2008). Individual differences in acute alcohol impairment of inhibitory control predict ad libitum alcohol consumption. *Psychopharmacology, 201*, 315–324.

40. Berridge, K. C., & Robinson, T. E. (1998). What is the role of dopamine in reward: hedonic impact, reward learning, or incentive salience? *Brain Research. Brain Research Reviews, 28*, 309–369.

41. Robinson, T. E., & Berridge, K. C. (1993). The neural basis of drug craving: an incentive-sensitization theory of addiction. *Brain Research. Brain Research Reviews, 18*, 247–291.

42. Wise, R. A. (1998). Drug-activation of brain reward pathways. *Drug and Alcohol Dependence, 51*, 13–22.

43. Dahlgren, A., Wargelius, H. L., Berglund, K. J., et al. (2011). Do alcohol-dependent individuals with DRD2 A1 allele have an increased risk of relapse? A pilot study. *Alcohol and Alcoholism*, *46*, 509–513.

44. Oslin, D. W., Berrettini, W., Kranzler, H. R., et al. (2003). A functional polymorphism of the mu-opioid receptor gene is associated with naltrexone response in alcohol-dependent patients. *Neuropsychopharmacology*, *28*, 1546–1552.

45. Hutchison, K. E., Wooden, A., Swift, R. M., et al. (2003). Olanzapine reduces craving for alcohol: a DRD4 VNTR polymorphism by pharmacotherapy interaction. *Neuropsychopharmacology*, *28*, 1882–1888.

46. Hutchison, K. E., Ray, L., Sandman, E., et al. (2006). The effect of olanzapine on craving and alcohol consumption. *Neuropsychopharmacology*, *31*, 1310–1317.

47. Hughes, A. R., Brothers, C. H., Mosteller, M., Spreen, W. R., & Burns, D. K. (2009). Genetic association studies to detect adverse drug reactions: abacavir hypersensitivity as an example. *Pharmacogenomics*, *10*, 225–233.

48. La Rosee, P., Corbin, A. S., Stoffregen, E. P., Deininger, M. W., & Druker, B. J. (2002). Activity of the Bcr-Abl kinase inhibitor PD180970 against clinically relevant Bcr-Abl isoforms that cause resistance to imatinib mesylate (Gleevec, STI571). *Cancer Research*, *62*, 7149–7153.

49. Piccart, M., Lohrisch, C., Di Leo, A., & Larsimont, D. (2001). The predictive value of HER2 in breast cancer. *Oncology*, *61*(Suppl 2), 73–82.

50. Feldstein Ewing, S. W., LaChance, H. A., Bryan, A., & Hutchison, K. E. (2009). Do genetic and individual risk factors moderate the efficacy of motivational enhancement therapy? Drinking outcomes with an emerging adult sample. *Addiction Biology*, *14*, 356–365.

51. McGeary, J. (2009). The DRD4 exon 3 VNTR polymorphism and addiction-related phenotypes: a review. *Pharmacology, Biochemistry, and Behavior*, *93*, 222–229.

52. Heishman, S. J., Taylor, R. C., & Henningfield, J. E. (1994). Nicotine and smoking: a review of effects on human performance. *Experimental and Clinical Psychopharmacology*, *2*, 345–395.

53. Evans, D. E., Park, J. Y., Maxfield, N., & Drobes, D. J. (2009). Neurocognitive variation in smoking behavior and withdrawal: genetic and affective moderators. *Genes, Brain, and Behavior*, *8*, 86–96.

54. Ersche, K. D., Turton, A. J., Pradhan, S., Bullmore, E. T., & Robbins, T. W. (2010). Drug addiction endophenotypes: impulsive versus sensation-seeking personality traits. *Biological Psychiatry*, *68*, 770–773.

55. Ray, L. A., Bryan, A., Mackillop, J., et al. (2009). The dopamine D receptor (DRD4) gene exon III polymorphism, problematic alcohol use and novelty seeking: direct and mediated genetic effects. *Addiction Biology*, *14*, 238–244.

56. Hendershot, C. S., Bryan, A. D., Ewing, S. W., Claus, E. D., & Hutchison, K. E. (2011). Preliminary evidence for associations of CHRM2 with substance use and disinhibition in adolescence. *Journal of Abnormal Child Psychology, 39*, 671–681.

57. Stoltenberg, S. F., Lehmann, M. K., Christ, C. C., Hersrud, S. L., & Davies, G. E. (2011). Associations among types of impulsivity, substance use problems and neurexin-3 polymorphisms. *Drug and Alcohol Dependence, 119*, e31–e38.

58. McBride, C. M., Koehly, L. M., Sanderson, S. C., & Kaphingst, K. A. (2010). The behavioral response to personalized genetic information: will genetic risk profiles motivate individuals and families to choose more healthful behaviors? *Annual Review of Public Health, 31*, 89–103.

59. Baler, R. D., & Volkow, N. D. (2006). Drug addiction: the neurobiology of disrupted self-control. *Trends in Molecular Medicine, 12*, 559–566.

60. Bickel, W. K., Miller, M. L., Yi, R., et al. (2007). Behavioral and neuroeconomics of drug addiction: competing neural systems and temporal discounting processes. *Drug and Alcohol Dependence, 90*(Suppl 1), S85–S91.

61. Claus, E. D., Ewing, S. W., Filbey, F. M., Sabbineni, A., & Hutchison, K. E. (2011). Identifying neurobiological phenotypes associated with alcohol use disorder severity. *Neuropsychopharmacology, 36*, 2086–2096.

62. Claus, E. D., Kiehl, K. A., & Hutchison, K. E. (2011). Neural and behavioral mechanisms of impulsive choice in alcohol use disorder. *Alcoholism, Clinical and Experimental Research, 35*, 1209–1219.

63. Lejuez, C. W., Read, J. P., Kahler, C. W., et al. (2002). Evaluation of a behavioral measure of risk taking: the Balloon Analogue Risk Task (BART). *Journal of Experimental Psychology. Applied, 8*, 75–84.

64. Claus, E. D., & Hutchison, K. E. (2012). Neural mechanisms of risk taking and relationships with hazardous drinking. *Alcoholism, Clinical and Experimental Research, 36*, 932–940.

65. Raichle, M. E., MacLeod, A. M., Snyder, A. Z., et al. (2001). A default mode of brain function. *Proceedings of the National Academy of Sciences of the United States of America, 98*, 676–682.

66. Rosengard, C., Stein, L. A., Barnett, N. P., et al. (2008). Randomized clinical trial of motivational enhancement of substance use treatment among incarcerated adolescents: post-release condom non-use. *Journal of HIV/AIDS Prevention in Children & Youth, 8*, 45–64.

67. Werch, C. E., Bian, H., Carlson, J. M., et al. (2011). Brief integrative multiple behavior intervention effects and mediators for adolescents. *Journal of Behavioral Medicine, 34*, 3–12.

68. Bryan, A. D., Schmiege, S. J., & Broaddus, M. R. (2009). HIV risk reduction among detained adolescents: a randomized, controlled trial. *Pediatrics, 124*, e1180–e1188.

69. Schmiege, S. J., Broaddus, M. R., Levin, M., & Bryan, A. D. (2009). Randomized trial of group interventions to reduce HIV/STD risk and change theoretical mediators among detained adolescents. *Journal of Consulting and Clinical Psychology, 77*, 38–50.

70. Kiehl, K. A., Liddle, P. F., & Hopfinger, J. B. (2000). Error processing and the rostral anterior cingulate: an event-related fMRI study. *Psychophysiology, 37,* 216–223.

71. Marian, A. J. (2012). Molecular genetic studies of complex phenotypes. *Translational Research; the Journal of Laboratory and Clinical Medicine, 159,* 64–79.

72. Plomin, R., Haworth, C. M., & Davis, O. S. (2009). Common disorders are quantitative traits. *Nature Reviews. Genetics, 10,* 872–878.

73. Derringer, J., Krueger, R. F., Dick, D. M., et al. (2010). Predicting sensation seeking from dopamine genes: A candidate-system approach. *Psychological Science, 21,* 1282–1290.

16 Using Intermediate Phenotypes to Bridge the Gap between Human and Mouse Genetics

Clarissa C. Parker and Abraham A. Palmer

This book focuses on the use of intermediate phenotypes for the study of drug abuse. Intermediate phenotypes provide a promising approach to the genetic analysis of substance abuse disorders because they may reflect underlying biological processes that are fundamental to drug abuse disorders. In this chapter, we will specifically explore how intermediate phenotypes can be used as a stepping-stone to connect results obtained from studies using mouse models to human disease (see figure 16.1). There is substantial genetic, neuroanatomical, physiological, and behavioral conservation between humans and mice [1, 2]. Therefore, mouse models have the potential to provide insight into human disease [3, 4]. Because drug abuse is a uniquely human disorder, mouse models typically focus on individual subcomponents of drug abuse. We have defined several of these subcomponents in table 16.1 and will consider how both intermediate phenotypes and mouse models fit into this framework. The integration of intermediate phenotypes and mouse models can be bidirectional: intermediate phenotypes in mouse models may lead to the identification of risk genes that are present in humans, or human studies may identify genetic associations underlying intermediate phenotypes and subsequent experiments in mice may be employed to examine the underlying mechanisms. We will explore this dynamic interplay throughout this chapter.

16.1 Advantages of Mouse Models

Using mice as a model genetic system has many advantages for elucidating the biological processes that underlie intermediate phenotypes. First, researchers can cross strains with measurable phenotypic differences and efficiently generate large litters of offspring from a limited number of founder genotypes [5]. Second, mouse models allow for environmental factors to be held constant or systematically varied in order to explore interactions between genotype and environment. In addition, the mouse genome has been thoroughly sequenced and annotated [6–8], and ~99% of mouse genes have a human homologue [6], thus allowing rapid translation between mice

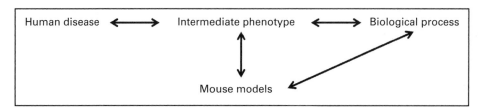

Figure 16.1
How intermediate phenotypes can connect human disease with mouse models and underlying biology.

and humans. Another advantage is that brain tissue can be obtained under idealized laboratory conditions, potentially following dangerous or invasive procedures, none of which is possible in humans [9]. Obtaining analogous brain tissue from humans is especially challenging because individuals with substance use disorders have vastly different life histories and have typically been chronically exposed to a variety of drugs; this confounds the analysis of gene expression and makes obtaining matched control tissue virtually impossible. Lastly, pharmacological manipulations can easily be applied in mouse models as additional evidence for the validity of a specific intermediate phenotype and can help distinguish between different aspects of a disorder [10, 11].

16.2   Genetic Approaches in Mouse Models of Substance Abuse Disorders

Genetic techniques used for addiction research in mouse models include both "genotype-to-phenotype" and "phenotype-to-genotype" approaches. Genotype-to-phenotype, or reverse genetics, is a method to discovering the function of a gene by examining the phenotypic effects that result from a targeted mutation. Phenotype-to-genotype, or forward genetics, begins with the measurement of the trait of interest in order to uncover the underlying genetic architecture in a population. Both techniques are useful for integrating results from human and mouse genetics studies [12].

16.2.1   Reverse Genetics
The genotype-to-phenotype approach includes gene ablation (knockouts), changes or replacement of genes (knockins), addition of extra copies of genes or addition of non-native sequences (transgenics), or the use of viral vectors or short interfering RNA segments (knockdowns) [13]. Additional possibilities include various conditional and inducible systems that allow for tissue or temporally specific changes in gene expres-

Table 16.1
Stages of addiction that may constitute independent genetic constructs

Stage	Human Intermediate Phenotype	Mouse Intermediate Phenotype
Likelihood of initial exposure	anxiety/harm avoidance	EPM/EZM/OFA/light-dark test/defensive burying
	impulsivity	impulsive choice/delay discounting/impulsive aggression/impaired response inhibition
	novelty seeking	preference for novel stimuli
Initial response to drugs	subjective descriptors of euphoria/pleasure/stimulation, brain imaging	drug-induced locomotor activation
Chronic use/ intoxication	drug /alcohol consumption	drug/alcohol self-administration/two-bottle choice
	conditioned place preference	conditioned place preference
	attentional, emotional, and behavioral conditioned responses to stimuli associated with drugs	conditioned reinforcement
Withdrawal/ negative affect Relapse	withdrawal-induced HPA axis dysregulation drug-, cue-, and stress-induced craving	withdrawal-induced HPA axis dysregulation drug-, cue-, and stress-induced reinstatement

sion or even control of by exogenous triggers (tetracycline, tamoxifen, or even light via various optogenetic systems; see [14–16] for reviews on these methods). Reverse genetic techniques are most useful for studying the influence of a single gene on a phenotype and can directly test statistical associations identified in human genetic studies. However, dramatic changes (like ablation of a gene) may exaggerate the allelic differences found in natural populations and thereby produce misleading conclusions. Hundreds of knockout and transgenic mice have been identified that show changes in traits related to drug abuse. Reverse genetics studies are particularly appropriate for examining the underlying mechanism by which genes influence intermediate phenotypes in humans.

16.2.2 Forward Genetics

Forward genetic strategies seek to identify the genes and alleles that give rise to variability in a trait of interest. Thus, forward genetics is an unbiased approach that is useful for hypothesis generation. Forward genetic approaches include selective breeding, mutagenesis, and quantitative trait locus (QTL) mapping.

Selective breeding is useful for determining the heritability of a trait and for creating lines that are highly divergent for the selection phenotype. The resulting progeny can be used to determine whether the selected alleles also influence other hypothetically related phenotypes. Selective breeding for a wide range of drug- and ethanol-related behaviors including sensitivity [17], hypothermia [18], consumption [19], preference [20], tolerance [21], and withdrawal severity [22] has been performed in mice. While this approach has been used for QTL mapping in the past, it has not proven to be of very useful for fine-scale mapping, which is required for gene identification.

Mutagens such as ethylnitrosurea (ENU) can be used to randomly create mutations (new alleles) in a population, which can then be screened to mutant individuals by screening putatively mutant mice for individuals with extreme phenotypes. Mutagenesis is analogous to studies of highly penetrant but rare familial disorders and de novo mutations in humans. ENU can produce either loss-of-function or gain-of-function mutations [23]. Many neurobehavioral mutants have been identified among large-scale ENU-mutagenized mouse populations, in some cases leading to gene identification [24]. ENU mutant mice display abnormalities across a broad range of behavioral phenotypes, including locomotion, fear learning, antidepressant activity, anxiety-related behaviors, inhibitory control, nociception, and learning/memory [25–27]. Causal mutations are relatively easy to find because they occur on an otherwise isogenic background; the advent of next-generation sequencing has vastly simplified the task of identifying such mutations [28].

QTL mapping is another forward genetic approach for identifying loci that contribute to natural variation in behavioral or physiological phenotypes. Traditionally, QTL studies have used F_2 crosses between two inbred strains, recombinant inbred (RI) lines, or similar populations. Associations between the genetic markers and phenotypes are analyzed to determine the location of the QTLs [29]. Due to limited recombination, these populations are not well suited to fine-mapping the identified loci, which is a necessary prerequisite to identifying the underlying causative gene(s). More recently, QTL mapping is being performed using highly recombinant populations [30] such as advanced intercross lines (AILs; [31, 32]), heterogeneous stocks (HS; [33]), and outbred mice [34]. Unlike F_2s and RIs, more highly recombinant populations such as AILs, HS, and outbred stocks show a rapid breakdown of linkage disequilibrium that allows for increasingly high-resolution mapping (see [30] for a review of these approaches). With advances in genotyping technology, and the development of software to address issue of relatedness [35], it has become inexpensive and relatively straightforward to perform genome-wide association studies in mice. This approach promises to streamline what has been a very slow and expensive process: definitive identification of the genes that underlie QTLs.

In summary, forward genetics provides a collection of techniques for identifying gene variants that contribute to phenotypic differences among mice. Genes identified using forward genetic approaches can then be studied in humans to determine whether they contribute to variability in corresponding intermediate phenotypes.

16.3 Mouse Models of Drug-Related Intermediate Phenotypes

In this section we will discuss phenotypes that can be examined in both mice and humans. As Stephens et al. [36] note, procedures in the laboratory that attempt to examine homologous processes may look dissimilar at a glance. However, "face validity" is not the best or only criterion for assessing the importance of intermediate phenotypes in mouse or human models of drug- and alcohol-related behaviors (see [37] for a review on validity issues in animal models). We will discuss intermediate phenotypes with well-defined neurobiology that may represent distinct stages of substance abuse (see table 16.1) and provide opportunities for translational research. In many cases, use of these intermediate phenotypes has led to identification of genes, elucidation of the causative biological mechanisms, or development of putative treatment strategies.

16.3.1 Sensitivity

One process thought to be important in the development of drug and alcohol dependence is the initial sensitivity to the effects of drugs. Many drugs of abuse, including alcohol [38], psychostimulants [39], and opioids [40] increase locomotor activity in mice at low doses. This behavior is partially mediated by dopamine release in the nucleus accumbens [41, 42]. A highly influential paper by Wise and Bozarth [43] proposes that differences in drug-induced locomotor activity in animals are related to the rewarding effects of drugs in humans (but see [44]).

This approach has been useful for human/mouse translational genetics studies; however it is not without limitations. One notable concern is that drugs of abuse do not increase locomotor activity at recreational doses in humans. Rather, initial sensitivity to drugs is commonly assessed by subjective descriptors of reward (e.g., pleasure, stimulation, euphoria). Brain-imaging studies in humans have demonstrated that drug-induced increases in dopamine in the nucleus accumbens are well correlated with these self-reported subjective responses [45, 46]. Thus, while it does not have strong face validity, locomotor activity is used to approximate the degree of dopaminergic activation in the nucleus accumbens, which is intended to reflect the same process that makes drugs rewarding in humans.

QTL mapping has identified genomic regions associated with differential sensitivity to cocaine [47, 48], ethanol [49], opiates [50], and methamphetamine [31, 32, 51]. Using a combination of forward and reverse genetic approaches, our lab has identified casein kinase 1 epsilon (*Csnk1e*) as a candidate gene associated with sensitivity to both opioids and methamphetamine in mice [17, 51, 52]. We have previously reported that a polymorphism in human *CSNK1E* was associated with differences in the rewarding effects of drugs [53]. Levran et al. [54] found that a different single nucleotide polymorphism (SNP) in *CSNK1E* was associated with heroin addiction in a population of Caucasian ancestry; however Kotaka et al. [55] did not observe a similar association with either methamphetamine-induced dependence or psychosis in a small Japanese population. Thus, while animal studies have provided good evidence that *Csnk1e* is involved in methamphetamine-induced locomotor behavior, evidence that the same gene is associated with either subjective response or risk for addiction is less compelling. This may be due to underpowered studies or a lack of common polymorphisms that alter gene function; or it may reflect a dissociation between locomotor sensitivity in mice and sensitivity to subjective effects in humans.

16.3.2 Self-Administration

Laboratory animal and human models of drug self-administration have been extensively used to measure the rewarding qualities of drugs, predict the abuse liability of novel drugs, and evaluate treatments for abuse and dependence. Self-administration paradigms seek to exploit the reinforcing effects of drugs of abuse to measure motivation for drug reward and reinforcement. The nucleus accumbens, central amygdala, and ventral pallidum are known to play a critical role for the acute reinforcing effects of drugs in mice and humans [56]. In mice, self-administration is commonly modeled using two procedures that putatively assess drug and alcohol reward (see [57–59] for reviews on self-administration models in mice and humans). The first method is called "two-bottle choice" and consists of offering mice a choice between a bottle that contains a drug solution and one that contains water. This method is especially popular for alcohol research. The second method involves training mice to perform a specific behavior (e.g., lever press, nose poke, etc.) in order to obtain a drug [59].

   Self-administration paradigms have both strengths and weaknesses. Drugs that are self-administered by rodents are also self-administered by humans, which reflect their potential for abuse [60]. Nevertheless, the utility of these models for translational research between mouse and human has been questioned for a number of reasons [61]. Despite the apparent face validity between drug self-administration in humans and mice, there is evidence to suggest that the procedures may be significantly different. For example, certain drugs that are self-administered by humans (e.g., hallucinogens) are not self-administered by mice. Additionally, the two-bottle-choice paradigm is sometimes confounded by differences in preference/aversion to the taste of the solution [62] rather than the pharmacological effects of the drug. Finally, operant procedures are extremely time- and labor-intensive and therefore may not be appropriate for forward genetic approaches.

Recently, nicotine self-administration was examined in genetically modified mice to elucidate the role of a cluster of nicotinic receptor subunits on the number of cigarettes smoked per day in humans. Multiple candidate gene and genome-wide association studies have consistently reported a significant relationship between a cluster of the nicotinic receptor subunits α3, α5, β4 (*CHRNA3–CHRNA5–CHRNB4*) and number of cigarettes smoked per day, nicotine dependence, lung cancer, and peripheral arterial disease [63–70]. However, *CHRNA3*, *CHRNA5*, and *CHRNB4* are members of a single haplotype block, making it difficult to discern which SNP(s) caused the association and whether the association was due to coding or expression differences. To explore the underlying mechanism, Fowler et al. [71] measured nicotine self-administration in mice (thought to be comparable to cigarettes smoked per day in humans) with a null mutation of *Chrna5* that did not express α5 nicotinic acetylcholine receptors. Mice with the null mutation responded far more vigorously than wild-type mice for nicotine infusions, especially when higher unit doses were available [71]. Importantly, this effect was rescued in the knockouts by reexpressing α5 subunits in the medial habenula (MHb), and the results were replicated through α5 subunit knockdown in the MHb; suggesting that *Chrna5* expression in the MHb is necessary for nicotine self-administration. Thus, a mouse model of human nicotine self-administration behavior helped to clarify the role of the *Chrna5* gene in the haplotype by manipulating *Chrna5* expression in the MHb, which in turn provided mechanistic insight about the importance of this nicotinic subunit within this particular brain region, a result that would have been difficult to obtain using human subjects.

16.3.3 Conditioned Preference

Conditional associations can be formed between drugs and cues in both mice and humans. In conditioned preference tests, drug administration is repeatedly paired with a previously neutral cue (either an environmental stimulus or a discrete cue). Through repeated pairings, the stimulus acquires the ability to act as a conditioned stimulus (CS) and is able to elicit approach or avoidance behaviors. Numerous studies have implicated the amygdala, ventral striatum, and orbitofrontal cortex in processing conditioned reinforcers in both animals [72–74] and humans [75], thus making it a promising phenotype for translational work between mice and humans.

Conditioned place preference (CPP) is a well-validated and commonly used paradigm. In this procedure, drug administration is repeatedly paired with one environment while administration of placebo is paired with a contrasting environment. Following multiple pairing sessions, mice are allowed access to both environments and the time spent in the drug-paired environment is taken as a measure of preference. Childs and de Wit [76] developed a procedure similar to CPP in humans. Individuals were exposed to amphetamine in one of two environments and were subsequently asked to express a preference for one of the environments and to rate their liking for each of the environments using Likert scales. Subjects reported greater liking for the amphetamine-paired room and also showed preference for it over the non-drug-paired room [76]. Their findings represent a rare example in which a behavioral paradigm developed in rodents has been explicitly tested and validated in humans. Despite these promising results, additional research is required to determine the extent to which the physiological and psychological processes underlying CPP in humans is homologous to those that produce CPP in mice.

Another way to measure conditioned preference is to use a discrete cue, rather than an environment as the CS. As a consequence of Pavlovian learning, a cue that is predictive of reward (CS) not only acquires predictive properties that signal the availability of the reward but may also acquire incentive properties that enable it to attract or directly reinforce appetitive behaviors [77, 78]. In mice, cues associated with drug exposure develop the ability to trigger and maintain drug seeking [79, 80]. In both mice and humans, attention to drug-paired cues is thought to rely on the same underlying neurological mechanisms and may offer a useful intermediate phenotype in which to explore the emotional, motivational, and cognitive consequences of conditioned drug cues [36]. For example, in human volunteers, previously neutral cues associated with an alcoholic drink acquire salience and produce attentional, emotional, and behavioral conditioned responses [81]. Others have demonstrated through eye tracking techniques and cognitive interference tasks that addicts allocate far greater attention to stimuli associated with drugs [82–84]. Furthermore, greater attentional bias to drug cues is associated with poorer treatment outcomes for alcoholics [85] as well as heroin [86] and cocaine [87] addicts.

Conditioned preference tests have several advantages and limitations. Notable advantages are that it is relatively easy to perform, and drugs that produce conditioned responses in mice are subjectively pleasurable in humans [57]. However, limitations include the behavioral and neurochemical effects of drugs differ depending on whether drug administration is controlled by the subject, and the failure to demonstrate a conditioned response is difficult to interpret. For example, lack of a conditioned response may be due to a lack of the rewarding effects of drugs, learning and memory impairments, differences in Pavlovian approach and cue saliency, or the anxiolytic effects of the drug [57, 88, 89]. For some of the reasons just mentioned, certain drugs are very sensitive to the exact parameters of the procedure [90].

Conditioned preference paradigms such as CPP and cue-induced craving/drug seeking have emphasized the importance of the dopamine transporter (DAT; *Slc6a3*) in mediating conditioned preference and reward in mice and humans. To clarify the extent to which the DAT was involved in cocaine reinforcement, Chen et al. [91] generated a knockin mouse line that carried a functional DAT that was insensitive to cocaine. In DAT knockin mice, cocaine did not elevate dopamine in the nucleus accumbens and did not produce reward as measured by CPP [91]. These results support the hypothesis of a critical role of DAT in the reinforcing effects of cocaine. Similarly, polymorphisms in *SLC6A3*, which codes for DAT, contribute to variability in limbic and behavioral responses to smoking cues in humans [92].

16.3.4 Withdrawal-Induced Hypothalamic–Pituitary–Adrenal (HPA) Axis Dysregulation

It is common for humans to report increased feelings of anxiety during withdrawal and protracted abstinence from numerous drugs of abuse [93–97]. Increased anxiety-like behavior is also observed in mice undergoing withdrawal [98–100].

Feelings of anxiety and other negative mood states that occur during drug withdrawal are thought to be due in part to altered levels of corticotropin-releasing hormone (CRH; also known as corticotropin releasing factor; [101]). Much of this evidence comes from experiments performed in mouse models. For example, withdrawal from either alcohol or cocaine in rodents elevates extracellular levels of CRH in the central nucleus of the amygdala, a brain region implicated in anxiety responses [102, 103]. Furthermore, administration of CRH receptor antagonists decreases anxiety-like behavior associated with withdrawal from multiple drugs of abuse [104–106]. Mice lacking the *Crhr1* gene exhibit reduced anxiety-like behavior under both basal conditions and following alcohol withdrawal [107]. Others have reported that genetic disruption of *Crhr1* pathways in mice eliminates the negative affective states of opioid withdrawal [108]. In early abstinent alcoholics, HPA axis dysregulation and increased anxiety predicted subsequent alcohol relapse and treatment outcomes [109]. Together, findings like these have prompted investigations of the effect of small molecule CRHR1 antagonists on anxiety, craving, and relapse in early abstinent addicts (reviewed in [110]). This demonstrates how coordinated human and translational mouse studies have been instrumental in developing a novel therapeutic target.

Reinstatement

The high rate of relapse to drug and alcohol use following periods of abstinence is a significant challenge in treating abuse and dependence in humans [111, 112]. In an effort to model this critical process in mice, de Wit and Stewart [113, 114] developed the reinstatement paradigm, in which animals are trained to perform an operant procedure to obtain drug infusions. Once a stable level of responding has been achieved, the drug infusions are stopped, which leads to extinction of the operant behavior. Reinstatement of responding can be inducted by certain triggers such as drug injections, drug-related cues, and stress exposure [115]. These triggers display only partial neurobiological overlap [116] and sometimes show additive effects when used together [117]. A variant of this procedure uses CPP rather than drug administration, so that drug-seeking behavior is measured as time spent in the previously drug-paired environment. The reinstatement paradigm involves the same systems that mediate the acute reinforcing effects of drugs as well as dysregulation of the HPA axis and other neurotransmitter/neuromodulator systems that include dynorphin, norepinephrine, vasopressin, neuropeptide Y, endocannabinoids, nociceptin, and substance P [56, 118, 119].

The reinstatement model has some important weaknesses. For example, reinstatement tests in mice are commonly performed under drug-free conditions. This is in contrast to relapse in the human addict, which is defined as enhanced drug/alcohol consumption following abstinence (and thus cannot happen under drug-free conditions). However, human laboratory studies of drug-, cue-, and stress-induced craving allow for direct translational comparisons by assessing drug use motivation as measured by drug craving rather than consumption. These studies have established a clear causal relationship between these triggers and motivation to use drugs and provide some evidence for the effects of trigger exposure and relapse to drug use [120]. For example, Back et al. [121] and others [122, 123] examined trigger-induced drug craving in the laboratory with subsequent assessment of relapse in the drug user's environment and demonstrated that responses in the laboratory predicted drug relapse and intake. While these studies have been instrumental in demonstrating the link between trigger-induced craving and relapse, many of the underlying neurobiological mechanisms remained largely unknown [109].

Recently, research in mice has improved our understanding of the ways in which stress contributes to relapse in abstinent humans. Substance P, along with its preferred neurokinin 1 receptor (NK1R) is highly expressed in brain areas involved in both the stress response and drug reward, including the hypothalamus, amygdala, and nucleus accumbens. Mice that are lacking the *Nk1r* gene fail to display CPP to opiates and fail to self-administer opiates, despite normal analgesic responses to morphine [124]. In addition, *Nk1r* knockout mice, as well as mice with artificial microRNA-based *Nk1r* gene silencing, displayed decreased voluntary ethanol consumption as compared to controls [125, 126], a result that was mimicked in wild-type mice given an NK1R antagonist [127]. These findings in mice, coupled with the previously known role of the substance P/NK1 system in stress responses [128] led to the investigation of the effects of NK1R antagonists on alcohol craving after exposure to alcohol-related cues and the Trier Social Stress task in detoxified anxious alcohol-dependent subjects [125]. Administration of the NK1R antagonist over a three-week period suppressed spontaneous alcohol cravings as well as the subjective craving response to the combined cue + stress challenge. Furthermore, imaging studies demonstrated that subjects treated with the NK1R antagonist displayed less activation of the insula (a brain region implicated in drug craving and relapse) in response to negative images as compared to placebo [125]. This demonstrates not only that mouse models could prove fruitful in elaborating the neurobiological substrates underlying the relationship between stress-induced craving and relapse but that the results obtained from mouse studies can be applied to the development of successful drug therapies assisting in the prevention of drug craving and relapse in humans.

16.4 Mouse Models of Personality/Temperamental Predictors of Drug Abuse

Addiction vulnerability arises from both internalizing (emotional) and externalizing (dyscontrol) behavioral dimensions [129]. Aspects of certain personality traits such as anxiousness/harm avoidance, novelty seeking, and impulsivity can be modeled in mice, share overlapping neurocircuitry with elevated drug intake, and have translational value in drug and alcohol research in humans.

16.4.1 Anxiety/Harm Avoidance

Anxious individuals who consume drugs or alcohol for the anxiolytic effects are predisposed to developing substance use disorders [130]. In mice, tests of anxiety-related behaviors are usually based on the conflict inherent in approach–avoidance situations, and most paradigms are analogous to acute anxiety episodes in humans rather than chronic anxiety. Common anxiety tests include the elevated plus maze (and its variant, the elevated zero maze), the light–dark box, the open field, and the defensive burying test (reviewed in [131, 132]).

Despite the wide array of tests measuring anxiety-like behavior in mice, they share many of the same limitations. First, tests of anxiety-like behavior may be difficult to interpret because they may reflect altered motivation to explore a novel apparatus (approach) rather than differences in avoidance. In addition, these tests can be confounded by differences in locomotor activity. Most studies that have observed increased anxiety-like behavior during withdrawal have also reported concurrent decreases in locomotor behavior and distance traveled [133].

Genes associated with anxiety-like behavior in mice have been identified that may also influence substance use disorders in humans. For example, Oliveira-Dos-Santos et al. [134] generated regulator of G protein signaling family member 2 (*Rgs2*) deficient mice that displayed increased anxiety-like behavior in the light–dark preference test as well as increased response to acoustic startle. Four years later, another group [135] used commercially available outbred mice to perform high-resolution mapping to dissect a small QTL on chromosome 1 that had been previously implicated in anxiety-like behavior as measured by open field activity and defecation [136]. They subdivided the locus into three regions and demonstrated through quantitative complementation that *Rgs2* was the quantitative trait gene [135]. Others have reported increases in *Rgs2* expression following 14 days of abstinence in rats trained to self-administer heroin. The animals displayed increased drug-seeking during the abstinence period, suggesting that increases in *Rgs2* expression may play a role in drug-seeking behavior and be important for relapse [137]. More recently, variants in *RGS2* have been associated with anxiety-related temperament [138], generalized anxiety disorder [139], posttraumatic stress symptoms [140], and panic disorder [141, 142] in human populations and may provide a promising avenue of research for exploring its role in drug-related behaviors in humans.

16.4.2 Impulsivity

Another focus of drug abuse researchers has been on the role of decision-making processes. There is an extensive literature on the relationship between drug abuse and impulsivity (see [143, 144, 145, 146] for reviews). A handful of mouse studies have focused on impulsive aggression within the context of social interactions [147] or on operant tasks measuring impulsive choice/intolerance to delays in reward (delay discounting) and impaired response inhibition/impulsive action [148].

A strength of mouse models of impulsivity is their ability to help elucidate the chain of causality for the observed correlation between impulsivity and drug abuse. Human impulsivity may precede and contribute to an enhanced risk for addiction, it may reflect a manifestation of the same susceptibility that contributes to drug abuse/dependence, or it may be a consequence of drug use itself. Appropriate mouse models of impulsivity can provide evidence supporting a causal link and, as a result, may model predictive intermediate phenotypes that can be targeted to prevent subsequent drug use in humans.

There are some important limitations to the use of mouse models of impulsive behavior. For example, the term impulsivity is often used to describe a range of behaviors including rash acts, boldness, boredom susceptibility, and risk taking [149]. These terms are commonly derived from personality inventories, may be difficult to operationalize, may reflect distinct underlying processes, and creating homologous mouse models has proven challenging [36].

Despite the small number of studies examining impulsivity in mice, promising results have been obtained with the use of selected lines, inbred, and outbred mice that suggest a causal, genetically mediated link between impulsivity and drug-related behaviors [150, 151, 152]. Others have used a reverse genetics approach to examine the influence of specific genes on measures of impulsivity and drug-related behaviors in mice. For example, transgenic male mice with a deletion in the gene encoding monoamine oxidase A (*Maoa*) displayed altered levels of brain serotonin and norepinephrine, enhanced aggression [153], and a higher resistance to acute ethanol exposure [154] as well as decreased sensitivity to the sedative–hypnotic effects of ethanol [155]. In humans, a variant in the *MAOA* gene has been linked to impulsive behavior [156], as well as to antisocial behavior, aggressiveness, and violence [157]. In healthy volunteers, the same variant predicted decreased prefrontocortical responses and increased amygdalar responses to emotional stimuli, suggesting impaired ability to control emotional responses during arousal [158]. Polymorphisms in the *MAOA* promoter have also been associated with heroin dependence [159], alcohol consumption in abused adolescents [160], alcoholism [161, 162], early onset substance use disorders [163], and methamphetamine psychosis [164] in human populations. Mouse studies have not only the potential to identify associations between polymorphisms in *Maoa* with impulsivity and drug abuse but the ability to explore the molecular events that drive these observed correlations.

16.4.3 Novelty Seeking

Novelty seeking is a heritable tendency toward exploration and excitement in response to novel stimuli [165]. Increased novelty seeking is associated with increased sensitivity to the reinforcing effects of psychostimulants and nicotine in rodents [166, 167] as well as increased vulnerability to drug abuse and dependence in humans [168, 169, 170]. In mice, novelty seeking is often measured by the degree to which a mouse explores a novel stimulus or demonstrates a preference for novel versus familiar stimuli. These stimuli may include novel environments, novel objects, novel food/drinking solutions, or access to social interaction with unknown conspecifics.

Modeling novelty seeking in mice has several limitations. First, little evidence exists to demonstrate that these partial models of novelty seeking reflect the influence of common genes or that they are capable of measuring novelty seeking in its entirety [171]. Another drawback is the potential influence of confounding factors such as anxiety, stress, and activity on novelty-seeking behavior in mice. To avoid these pitfalls, tasks that depend on a preference for novel over familiar stimuli should be considered a more direct reflection of novelty seeking than tasks involving exposure to an open field (commonly considered an anxiety-provoking environment that is also confounded with locomotor activity), and multiple tasks may be needed to capture the full genetic range of the behavioral trait.

Novelty-seeking behavior in mice has been linked to gene regions and specific loci that have also been linked to human intermediate phenotypes. In particular, the dopamine receptor D4 gene (*DRD4*) has been implicated in novelty-seeking and drug-related behaviors. For example, *Drd4* knockout mice exhibit reduced exploration of novel stimuli [172], and numerous mouse studies have implicated the importance of DRD4 receptors for the stimulant effects of ethanol [173], nicotine discrimination [174], and opiate-induced withdrawal [175]. Under laboratory conditions, polymorphisms in the *DRD4* gene are associated with subjective responses to alcohol and urge to drink [176, 177], reactivity and arousal to smoking cues [178], as well as subjective craving in opiate abusers [179]. Polymorphisms in *DRD4* have also been associated with novelty seeking [180, 181] as well as with alcoholism [182], heavy smoking [183, 184], and heroin dependence [159, 185]. While a number of reports failed to identify associations between *DRD4* and novelty seeking, a recent meta-analysis of the literature has offered qualified support for these associations [186]. In the case of *DRD4*, mouse models have provided additional support for the somewhat tentative associations reported for *DRD4* and novelty seeking in humans and have been used to directly ascertain the effect of DRD4 receptor antagonists on drug-related behaviors.

16.5 Conclusions

There is no question that combining human and mouse genetics provides a powerful platform for connecting substance abuse disorders with specific genes and polymorphisms. In the paradigms and examples that we have reviewed, a few common themes emerge. First, translation is greatly facilitated when an intermediate phenotype in humans can be paired with a similar phenotype in mice. In addition, examination of the effect of polymorphisms in specific genes, which might be naturally occurring (humans, mice) or induced (mice), provide the ability to integrate studies across these two species. Humans offer the ability to identify correlations between disease processes and intermediate phenotypes. Mice are most useful for examining causal relationships and for deducing their underlying mechanisms. Thus, when used with care, intermediate phenotypes offer a stepping-stone between studies in mice and studies of diseases such as drug abuse, which are uniquely human.

Acknowledgments

This work was supported by National Institutes of Health grants R01GM097737, R01DA021336, R01MH079103, and T32DA07255.

References

1. Tecott, L. H. (2003). The genes and brains of mice and men. *American Journal of Psychiatry, 160*, 646–656.

2. Arguello, P. A., & Gogos, J. A. (2006). Modeling madness in mice: one piece at a time. *Neuron, 52*, 179–196.

3. Geyer, M. A., Markou, A. (2002). The role of preclinical models in the development of psychotropic drugs. In Davis, K. L., Charney, D., Coyle, J. T., & Nemeroff, C. (Eds.). *Neuropsychopharmacology: The Fifth Generation of Progress* (pp. 445–455). Philadelphia, PA: Lippincott Williams and Wilkins.

4. Jacobson, L. H., & Cryan, J. F. (2010). Genetic approaches to modeling anxiety in animals. *Current Topics in Behavioral Neurosciences, 2*, 161–201.

5. Lawson, H. A., & Cheverud, J. M. (2010). Metabolic syndrome components in murine models. *Endocrine, Metabolic & Immune Disorders Drug Targets, 10*, 25–40.

6. Waterston, R. H., Lindblad-Toh, K., Birney, E., Rogers, J., Abril, J. F., Agarwal, P., et al. (2002). Initial sequencing and comparative analysis of the mouse genome. *Nature, 420*, 520–562.

7. Keane, T. M., Goodstadt, L., Danecek, P., White, M. A., Wong, K., Yalcin, B., et al. (2011). Mouse genomic variation and its effect on phenotypes and gene regulation. *Nature, 477,* 289–294.

8. Yalcin, B., Wong, K., Agam, A., Goodson, M., Keane, T. M., Gan, X., et al. (2011). Sequence-based characterization of structural variation in the mouse genome. *Nature, 477,* 326–329.

9. Huang, G. J., Shifman, S., Valdar, W., Johannesson, M., Yalcin, B., Taylor, M. S., et al. (2009). High resolution mapping of expression QTLs in heterogeneous stock mice in multiple tissues. *Genome Research, 19,* 1133–1140.

10. Seong, E., Seasholtz, A. F., & Burmeister, M. (2002). Mouse models for psychiatric disorders. *Trends in Genetics, 18,* 643–650.

11. Sarkis, E. H. (2000). "Model" behavior. *Science, 287,* 2160–2162.

12. Palmer, A. A., & de Wit, H. (2012). Translational genetic approaches to substance use disorders: bridging the gap between mice and humans. *Human Genetics, 131,* 931–939.

13. Crawley, J. N. (2007). *What's Wrong with My Mouse?: Behavioral Phenotyping of Transgenic and Knockout Mice.* Hoboken, NJ: Wiley.

14. Mallo, M. (2006). Controlled gene activation and inactivation in the mouse. *Frontiers in Bioscience, 11,* 313–327.

15. Stieger, K., Belbellaa, B., Le Guiner, C., Moullier, P., & Rolling, F. (2009). In vivo gene regulation using tetracycline-regulatable systems. *Advanced Drug Delivery Reviews, 61,* 527–541.

16. Fenno, L., Yizhar, O., & Deisseroth, K. (2011). The development and application of optogenetics. *Annual Review of Neuroscience, 34,* 389–412.

17. Palmer, A. A., Verbitsky, M., Suresh, R., Kamens, H. M., Reed, C. L., Li, N., et al. (2005). Gene expression differences in mice divergently selected for methamphetamine sensitivity. *Mammalian Genome, 16,* 291–305.

18. Browman, K. E., Rustay, N. R., Nikolaidis, N., Crawshaw, L., & Crabbe, J. C. (2000). Sensitivity and tolerance to ethanol in mouse lines selected for ethanol-induced hypothermia. *Pharmacology, Biochemistry, and Behavior, 67,* 821–829.

19. McClearn, G. E., Tarantino, L. M., Rodriguez, L. A., Jones, B. C., Blizard, D. A., & Plomin, R. (1997). Genotypic selection provides experimental confirmation for an alcohol consumption quantitative trait locus in mouse. *Molecular Psychiatry, 2,* 486–489.

20. Belknap, J. K., Richards, S. P., O'Toole, L. A., Helms, M. L., & Phillips, T. J. (1997). Short-term selective breeding as a tool for QTL mapping: ethanol preference drinking in mice. *Behavior Genetics, 27,* 55–66.

21. Erwin, V. G., & Deitrich, R. A. (1996). Genetic selection and characterization of mouse lines for acute functional tolerance to ethanol. *Journal of Pharmacology and Experimental Therapeutics, 279,* 1310–1317.

22. Hitzemann, R., Edmunds, S., Wu, W., Malmanger, B., Walter, N., Belknap, J., et al. (2009). Detection of reciprocal quantitative trait loci for acute ethanol withdrawal and ethanol consumption in heterogeneous stock mice. *Psychopharmacology*, *203*, 713–722.

23. Keays, D. A., & Nolan, P. M. (2003). N-ethyl-N-nitrosourea mouse mutants in the dissection of behavioural and psychiatric disorders. *European Journal of Pharmacology*, *480*, 205–217.

24. Vitaterna, M. H., King, D. P., Chang, A. M., Kornhauser, J. M., Lowrey, P. L., McDonald, J. D., et al. (1994). Mutagenesis and mapping of a mouse gene, Clock, essential for circadian behavior. *Science*, *264*, 719–725.

25. Reijmers, L. G., Coats, J. K., Pletcher, M. T., Wiltshire, T., Tarantino, L. M., & Mayford, M. (2006). A mutant mouse with a highly specific contextual fear-conditioning deficit found in an N-ethyl-N-nitrosourea (ENU) mutagenesis screen. *Learning & Memory*, *13*, 143–149.

26. Cook, M. N., Dunning, J. P., Wiley, R. G., Chesler, E. J., Johnson, D. K., Miller, D. R., et al. (2007). Neurobehavioral mutants identified in an ENU-mutagenesis project. *Mammalian Genome*, *18*, 559–572.

27. Furuse, T., Wada, Y., Hattori, K., Yamada, I., Kushida, T., Shibukawa, Y., et al. (2010). Phenotypic characterization of a new Grin1 mutant mouse generated by ENU mutagenesis. *European Journal of Neuroscience*, *31*, 1281–1291.

28. Takahashi, J. S., Shimomura, K., & Kumar, V. (2008). Searching for genes underlying behavior: lessons from circadian rhythms. *Science*, *322*, 909–912.

29. Peters, L. L., Robledo, R. F., Bult, C. J., Churchill, G. A., Paigen, B. J., & Svenson, K. L. (2007). The mouse as a model for human biology: a resource guide for complex trait analysis. *Nature Reviews. Genetics*, *8*, 58–69.

30. Parker, C. C., & Palmer, A. A. (2011). Dark matter: are mice the solution to missing heritability? *Frontiers in Genetics*, *2*.

31. Cheng, R., Lim, J. E., Samocha, K. E., Sokoloff, G., Abney, M., Skol, A. D., et al. (2010). Genome-wide association studies and the problem of relatedness among advanced intercross lines and other highly recombinant populations. *Genetics*, *185*, 1033–1044.

32. Parker, C. C., Cheng, R., Sokoloff, G., & Palmer, A. A. (2012). Genome-wide association for methamphetamine sensitivity in an advanced intercross mouse line. *Genes, Brain, and Behavior*, *11*, 52–61.

33. Svenson, K. L., Gatti, D. M., Valdar, W., Welsh, C. E., Cheng, R., Chesler, E. J., et al. (2012). High-resolution genetic mapping using the mouse diversity outbred population. *Genetics*, *190*, 437–447.

34. Ghazalpour, A., Doss, S., Kang, H., Farber, C., Wen, P. Z., Brozell, A., et al. (2008). High-resolution mapping of gene expression using association in an outbred mouse stock. *PLOS Genetics*, *4*, e1000149.

35. Cheng, R., Abney, M., Palmer, A. A., & Skol, A. D. (2011). QTLRel: an R package for genome-wide association studies in which relatedness is a concern. *BMC Genetics*, *12*, 66.

36. Stephens, D. N., Crombag, H. S., & Duka, T. (2013). The challenge of studying parallel behaviors in humans and animal models. *Current Topics in Behavavioral Neurosciences*, *13*, 611–645.

37. Hitzemann, R. (2000). Animal models of psychiatric disorders and their relevance to alcoholism. *Alcohol Research & Health*, *24*, 149–158.

38. Erwin, V. G., Radcliffe, R. A., Gehle, V. M., & Jones, B. C. (1997). Common quantitative trait loci for alcohol-related behaviors and central nervous system neurotensin measures: locomotor activation. *Journal of Pharmacology and Experimental Therapeutics*, *280*, 919–926.

39. Phillips, T. J., Huson, M. G., & McKinnon, C. S. (1998). Localization of genes mediating acute and sensitized locomotor responses to cocaine in BXD/Ty recombinant inbred mice. *Journal of Neuroscience*, *18*, 3023–3034.

40. Belknap, J. K., & Crabbe, J. C. (1992). Chromosome mapping of gene loci affecting morphine and amphetamine responses in BXD recombinant inbred mice. *Annals of the New York Academy of Sciences*, *654*, 311–323.

41. Di Chiara, G., & Imperato, A. (1988). Drugs abused by humans preferentially increase synaptic dopamine concentrations in the mesolimbic system of freely moving rats. *Proceedings of the National Academy of Sciences of the United States of America*, *85*, 5274–5278.

42. Koshikawa, N., Mori, E., Oka, K., Nomura, H., Yatsushige, N., & Maruyama, Y. (1989). Effects of SCH23390 injection into the dorsal striatum and nucleus accumbens on methamphetamine-induced gnawing and hyperlocomotion in rats. *Journal of Nihon University School of Dentistry*, *31*, 451–457.

43. Wise, R. A., & Bozarth, M. A. (1987). A psychomotor stimulant theory of addiction. *Psychological Review*, *94*, 469–492.

44. Phillips, T. J., Kamens, H. M., & Wheeler, J. M. (2008). Behavioral genetic contributions to the study of addiction-related amphetamine effects. *Neuroscience and Biobehavioral Reviews*, *32*, 707–759.

45. Volkow, N. D., Wang, G. J., Fowler, J. S., Gatley, S. J., Ding, Y. S., Logan, J., et al. (1996). Relationship between psychostimulant-induced "high" and dopamine transporter occupancy. *Proceedings of the National Academy of Sciences of the United States of America*, *93*, 10388–10392.

46. Martinez, D., Slifstein, M., Broft, A., Mawlawi, O., Hwang, D. R., Huang, Y., et al. (2003). Imaging human mesolimbic dopamine transmission with positron emission tomography: II. amphetamine-induced dopamine release in the functional subdivisions of the striatum. *Journal of Cerebral Blood Flow and Metabolism*, *23*, 285–300.

47. Jones, B. C., Tarantino, L. M., Rodriguez, L. A., Reed, C. L., McClearn, G. E., Plomin, R., et al. (1999). Quantitative-trait loci analysis of cocaine-related behaviours and neurochemistry. *Pharmacogenetics*, *9*, 607–617.

48. Boyle, A. E., & Gill, K. (2001). Sensitivity of AXB/BXA recombinant inbred lines of mice to the locomotor activating effects of cocaine: a quantitative trait loci analysis. *Pharmacogenetics, 11,* 255–264.

49. Downing, C., Carosone-Link, P., Bennett, B., & Johnson, T. (2006). QTL mapping for low-dose ethanol activation in the LXS recombinant inbred strains. *Alcoholism, Clinical and Experimental Research, 30,* 1111–1120.

50. Bryant, C. D., Chang, H. P., Zhang, J., Wiltshire, T., Tarantino, L. M., & Palmer, A. A. (2009). A major QTL on chromosome 11 influences psychostimulant and opiod sensitivity in mice. *Genes, Brain, and Behavior, 8,* 795–805.

51. Bryant, C. D., Graham, M. E., Distler, M. G., Munoz, M. B., Li, D., Vezina, P., et al. (2009). A role for casein kinase 1 epsilon in the locomotor stimulant response to methamphetamine. *Psychopharmacology, 203,* 703–711.

52. Bryant, C. D., Parker, C. C., Zhou, L., Olker, C., Chandrasekaran, R. Y., Wager, T. T., et al. (2012). Csnk1e is a genetic regulator of sensitivity to psychostimulants and opioids. *Neuropsychopharmacology, 37,* 1026–1035.

53. Veenstra-VanderWeele, J., Qaadir, A., Palmer, A. A., Cook, E. H. J., & de Wit, H. (2006). Association between the casein kinase 1 epsilon gene region and subjective response to D-amphetamine. *Neuropsychopharmacology, 31,* 1056–1063.

54. Levran, O., Londono, D., O'Hara, K., Nielsen, D. A., Peles, E., Rotrosen, J., et al. (2008). Genetic susceptibility to heroin addiction: a candidate gene association study. *Genes, Brain, and Behavior, 7,* 720–729.

55. Kotaka, T., Ujike, H., Morita, Y., Kishimoto, M., Okahisa, Y., Inada, T., et al. (2008). Association study between casein kinase 1 epsilon gene and methamphetamine dependence. *Annals of the New York Academy of Sciences, 1139,* 43–48.

56. Koob, G. F., & Volkow, N. D. (2010). Neurocircuitry of addiction. *Neuropsychopharmacology, 35,* 217–238.

57. Shippenberg, T. S., & Koob, G. F. (2002). Recent advances in animal models of addiction. In K. L. Davis, D. Charney, J. T. Coyle, & C. Nemeroff (Eds.), *Neuropsychopharmacology: The Fifth Generation of Progress* (pp. 1381–1397). Philadelphia, PA: Lippincott, Williams and Wilkins.

58. Haney, M. (2009). Self-administration of cocaine, cannabis and heroin in the human laboratory: benefits and pitfalls. *Addiction Biology, 14,* 9–21.

59. Crabbe, J. C., & Phillips, T. J. (2004). Pharmacogenetic studies of alcohol self-administration and withdrawal. *Psychopharmacology, 174,* 539–560.

60. Collins, R. J., Weeks, J. R., Cooper, M. M., Good, P. I., & Russell, R. R. (1984). Prediction of abuse liability of drugs using IV self-administration by rats. *Psychopharmacology, 82,* 6–13.

61. Haney, M., & Spealman, R. (2008). Controversies in translational research: drug self-administration. *Psychopharmacology, 199,* 403–419.

62. Stephens, D. N., Duka, T., Crombag, H. S., Cunningham, C. L., Heilig, M., & Crabbe, J. C. (2010). Reward sensitivity: issues of measurement, and achieving consilience between human and animal phenotypes. *Addiction Biology*, *15*, 145–168.

63. Saccone, S. F., Hinrichs, A. L., Saccone, N. L., Chase, G. A., Konvicka, K., Madden, P. A., et al. (2007). Cholinergic nicotinic receptor genes implicated in a nicotine dependence association study targeting 348 candidate genes with 3713 SNPs. *Human Molecular Genetics*, *16*, 36–49.

64. Thorgeirsson, T. E., Geller, F., Sulem, P., Rafnar, T., Wiste, A., Magnusson, K. P., et al. (2008). A variant associated with nicotine dependence, lung cancer and peripheral arterial disease. *Nature*, *452*, 638–642.

65. Thorgeirsson, T. E., Gudbjartsson, D. F., Surakka, I., Vink, J. M., Amin, N., Geller, F., et al. (2010). Sequence variants at CHRNB3-CHRNA6 and CYP2A6 affect smoking behavior. *Nature Genetics*, *42*, 448–453.

66. Hung, R. J., McKay, J. D., Gaborieau, V., Boffetta, P., Hashibe, M., Zaridze, D., et al. (2008). A susceptibility locus for lung cancer maps to nicotinic acetylcholine receptor subunit genes on 15q25. *Nature*, *452*, 633–637.

67. Amos, C. I., Wu, X., Broderick, P., Gorlov, I. P., Gu, J., Eisen, T., et al. (2008). Genome-wide association scan of tag SNPs identifies a susceptibility locus for lung cancer at 15q25.1. *Nature Genetics*, *40*, 616–622.

68. Liu, J. Z., Tozzi, F., Waterworth, D. M., Pillai, S. G., Muglia, P., Middleton, L., et al. (2010). Meta-analysis and imputation refines the association of 15q25 with smoking quantity. *Nature Genetics*, *42*, 436–440.

69. Truong, T., Hung, R. J., Amos, C. I., Wu, X., Bickeboller, H., Rosenberger, A., et al. (2010). Replication of lung cancer susceptibility loci at chromosomes 15q25, 5p15, and 6p21: a pooled analysis from the International Lung Cancer Consortium. *Journal of the National Cancer Institute*, *102*, 959–971.

70. Tobacco and Genetics Consortium. (2010). Genome-wide meta-analyses identify multiple loci associated with smoking behavior. *Nature Genetics*, *42*, 441–447.

71. Fowler, C. D., Lu, Q., Johnson, P. M., Marks, M. J., & Kenny, P. J. (2011). Habenular alpha5 nicotinic receptor subunit signalling controls nicotine intake. *Nature*, *471*, 597–601.

72. Cardinal, R. N., Parkinson, J. A., Hall, J., & Everitt, B. J. (2002a). Emotion and motivation: the role of the amygdala, ventral striatum, and prefrontal cortex. *Neuroscience and Biobehavioral Reviews*, *26*, 321–352.

73. Cardinal, R. N., Parkinson, J. A., Lachenal, G., Halkerston, K. M., Rudarakanchana, N., Hall, J., et al. (2002b). Effects of selective excitotoxic lesions of the nucleus accumbens core, anterior cingulate cortex, and central nucleus of the amygdala on autoshaping performance in rats. *Behavioral Neuroscience*, *116*, 553–567.

74. Pears, A., Parkinson, J. A., Hopewell, L., Everitt, B. J., & Roberts, A. C. (2003). Lesions of the orbitofrontal but not medial prefrontal cortex disrupt conditioned reinforcement in primates. *Journal of Neuroscience*, *23*, 11189–11201.

75. Cox, S. M., Andrade, A., & Johnsrude, I. S. (2005). Learning to like: a role for human orbito-frontal cortex in conditioned reward. *Journal of Neuroscience, 25,* 2733–2740.

76. Childs, E., & de Wit, H. (2009). Amphetamine-induced place preference in humans. *Biological Psychiatry, 65,* 900–904.

77. Brown, P. L., & Jenkins, H. M. (1968). Auto-shaping of the pigeon's key-peck. *Journal of the Experimental Analysis of Behavior, 11,* 1–8.

78. Robinson, T. E., & Flagel, S. B. (2009). Dissociating the predictive and incentive motivational properties of reward-related cues through the study of individual differences. *Biological Psychiatry, 65,* 869–873.

79. O'Connor, E. C., Crombag, H. S., Mead, A. N., & Stephens, D. N. (2010a). The mGluR5 antagonist MTEP dissociates the acquisition of predictive and incentive motivational properties of reward-paired stimuli in mice. *Neuropsychopharmacology, 35,* 1807–1817.

80. O'Connor, E. C., Stephens, D. N., & Crombag, H. S. (2010b). Modeling appetitive Pavlovian-instrumental interactions in mice. *Current Protocols in Neuroscience,* chapter 8: unit 8.25.

81. Hogarth, L., & Duka, T. (2006). Human nicotine conditioning requires explicit contingency knowledge: is addictive behaviour cognitively mediated? *Psychopharmacology, 184,* 553–566.

82. Rosse, R. B., Johri, S., Kendrick, K., Hess, A. L., Alim, T. N., Miller, M., et al. (1997). Preattentive and attentive eye movements during visual scanning of a cocaine cue: correlation with intensity of cocaine cravings. *Journal of Neuropsychiatry and Clinical Neurosciences, 9,* 91–93.

83. Bauer, D., & Cox, W. M. (1998). Alcohol-related words are distracting to both alcohol abusers and non-abusers in the Stroop colour-naming task. *Addiction (Abingdon, England), 93,* 1539–1542.

84. Lubman, D. I., Peters, L. A., Mogg, K., Bradley, B. P., & Deakin, J. F. (2000). Attentional bias for drug cues in opiate dependence. *Psychological Medicine, 30,* 169–175.

85. Cox, W. M., Hogan, L. M., Kristian, M. R., & Race, J. H. (2002). Alcohol attentional bias as a predictor of alcohol abusers' treatment outcome. *Drug and Alcohol Dependence, 68,* 237–243.

86. Marissen, M. A., Franken, I. H., Waters, A. J., Blanken, P., van den Brink, W., & Hendriks, V. M. (2006). Attentional bias predicts heroin relapse following treatment. *Addiction (Abingdon, England), 101,* 1306–1312.

87. Carpenter, K. M., Schreiber, E., Church, S., & McDowell, D. (2006). Drug Stroop performance: relationships with primary substance of use and treatment outcome in a drug-dependent outpatient sample. *Addictive Behaviors, 31,* 174–181.

88. Cunningham, C. L., & Patel, P. (2007). Rapid induction of Pavlovian approach to an ethanol-paired visual cue in mice. *Psychopharmacology, 192,* 231–241.

89. Mead, A. N., Brown, G., Le Merrer, J., & Stephens, D. N. (2005). Effects of deletion of gria1 or gria2 genes encoding glutamatergic AMPA-receptor subunits on place preference conditioning in mice. *Psychopharmacology, 179,* 164–171.

90. Cunningham, C. L., Niehus, J. S., & Noble, D. (1993). Species difference in sensitivity to ethanol's hedonic effects. *Alcohol (Fayetteville, N.Y.)*, *10*, 97–102.

91. Chen, R., Tilley, M. R., Wei, H., Zhou, F., Zhou, F. M., Ching, S., et al. (2006). Abolished cocaine reward in mice with a cocaine-insensitive dopamine transporter. *Proceedings of the National Academy of Sciences of the United States of America*, *103*, 9333–9338.

92. Franklin, T. R., Lohoff, F. W., Wang, Z., Sciortino, N., Harper, D., Li, Y., et al. (2009). DAT genotype modulates brain and behavioral responses elicited by cigarette cues. *Neuropsychopharmacology*, *34*, 717–728.

93. Gawin, F. H. (1991). Cocaine addiction: psychology and neurophysiology. *Science*, *251*, 1580–1586.

94. Gawin, F. H., & Kleber, H. D. (1986). Abstinence symptomatology and psychiatric diagnosis in cocaine abusers: clinical observations. *Archives of General Psychiatry*, *43*, 107–113.

95. Resnick, R. B., & Resnick, E. B. (1984). Cocaine abuse and its treatment. *Psychiatric Clinics of North America*, *7*, 713–728.

96. Schuckit, M. A., & Hesselbrock, V. (1994). Alcohol dependence and anxiety disorders: what is the relationship? *American Journal of Psychiatry*, *151*, 1723–1734.

97. Koob, G. F. (2009). Neurobiological substrates for the dark side of compulsivity in addiction. *Neuropharmacology*, *56*(Suppl 1), 18–31.

98. Wilson, J., Watson, W. P., & Little, H. J. (1998). CCK(B) antagonists protect against anxiety-related behaviour produced by ethanol withdrawal, measured using the elevated plus maze. *Psychopharmacology*, *137*, 120–131.

99. Jung, M. E., Wallis, C. J., Gatch, M. B., & Lal, H. (2000). Abecarnil and alprazolam reverse anxiety-like behaviors induced by ethanol withdrawal. *Alcohol (Fayetteville, N.Y.)*, *21*, 161–168.

100. Kliethermes, C. L., Cronise, K., & Crabbe, J. C. (2004). Anxiety-like behavior in mice in two apparatuses during withdrawal from chronic ethanol vapor inhalation. *Alcoholism, Clinical and Experimental Research*, *28*, 1012–1019.

101. Koob, G. F. (1999). Stress, corticotropin-releasing factor, and drug addiction. *Annals of the New York Academy of Sciences*, *897*, 27–45.

102. Merlo Pich, E., Lorang, M., Yeganeh, M., Rodriguez de Fonseca, F., Raber, J., Koob, G. F., et al. (1995). Increase of extracellular corticotropin-releasing factor-like immunoreactivity levels in the amygdala of awake rats during restraint stress and ethanol withdrawal as measured by microdialysis. *Journal of Neuroscience*, *15*, 5439–5447.

103. Richter, R. M., & Weiss, F. (1999). In vivo CRF release in rat amygdala is increased during cocaine withdrawal in self-administering rats. *Synapse (New York, N.Y.)*, *32*, 254–261.

104. Heinrichs, S. C., Menzaghi, F., Pich, E. M., Baldwin, H. A., Rassnick, S., Britton, K. T., et al. (1994). Anti-stress action of a corticotropin-releasing factor antagonist on behavioral reactivity to stressors of varying type and intensity. *Neuropsychopharmacology*, *11*, 179–186.

105. Lu, L., Liu, D., & Ceng, X. (2001). Corticotropin-releasing factor receptor type 1 mediates stress-induced relapse to cocaine-conditioned place preference in rats. *European Journal of Pharmacology*, *415*, 203–208.

106. Overstreet, D. H., Knapp, D. J., & Breese, G. R. (2004). Modulation of multiple ethanol withdrawal-induced anxiety-like behavior by CRF and CRF1 receptors. *Pharmacology, Biochemistry, and Behavior*, *77*, 405–413.

107. Timpl, P., Spanagel, R., Sillaber, I., Kresse, A., Reul, J. M., Stalla, G. K., et al. (1998). Impaired stress response and reduced anxiety in mice lacking a functional corticotropin-releasing hormone receptor 1. *Nature Genetics*, *19*, 162–166.

108. Contarino, A., & Papaleo, F. (2005). The corticotropin-releasing factor receptor-1 pathway mediates the negative affective states of opiate withdrawal. *Proceedings of the National Academy of Sciences of the United States of America*, *102*, 18649–18654.

109. Sinha R., Shaham Y., Heilig M. (2011). Translational and reverse translational research on the role of stress in drug craving and relapse. *Psychopharmacology*, *218*, 69–82.

110. Zorrilla, E. P., & Koob, G. F. (2010). Progress in corticotropin-releasing factor-1 antagonist development. *Drug Discovery Today*, *15*, 371–383.

111. Hunt, W. A., Barnett, L. W., & Branch, L. G. (1971). Relapse rates in addiction programs. *Journal of Clinical Psychology*, *27*, 455–456.

112. O'Brien, C. P., & Gardner, E. L. (2005). Critical assessment of how to study addiction and its treatment: human and non-human animal models. *Pharmacology & Therapeutics*, *108*, 18–58.

113. de Wit, H., & Stewart, J. (1981). Reinstatement of cocaine-reinforced responding in the rat. *Psychopharmacology*, *75*, 134–143.

114. de Wit, H., & Stewart, J. (1983). Drug reinstatement of heroin-reinforced responding in the rat. *Psychopharmacology*, *79*, 29–31.

115. Shaham, Y., Erb, S., & Stewart, J. (2000). Stress-induced relapse to heroin and cocaine seeking in rats: a review. *Brain Research. Brain Research Reviews*, *33*, 13–33.

116. Sutton, M. A., Schmidt, E. F., Choi, K. H., Schad, C. A., Whisler, K., Simmons, D., et al. (2003). Extinction-induced upregulation in AMPA receptors reduces cocaine-seeking behaviour. *Nature*, *421*, 70–75.

117. Liu, X., & Weiss, F. (2002). Additive effect of stress and drug cues on reinstatement of ethanol seeking: exacerbation by history of dependence and role of concurrent activation of corticotropin-releasing factor and opioid mechanisms. *Journal of Neuroscience*, *22*, 7856–7861.

118. Koob, G. F., & Le Moal, M. (2001). Drug addiction, dysregulation of reward, and allostasis. *Neuropsychopharmacology*, *24*, 97–129.

119. Koob, G. F. (2008). A role for brain stress systems in addiction. *Neuron*, *59*, 11–34.

120. McKee, S. A., Sinha, R., Weinberger, A. H., Sofuoglu, M., Harrison, E. L., Lavery, M., et al. (2011). Stress decreases the ability to resist smoking and potentiates smoking intensity and reward. *Journal of Psychopharmacology (Oxford, England)*, *25*, 490–502.

121. Back, S. E., Hartwell, K., DeSantis, S. M., Saladin, M., McRae-Clark, A. L., Price, K. L., et al. (2010). Reactivity to laboratory stress provocation predicts relapse to cocaine. *Drug and Alcohol Dependence*, *106*, 21–27.

122. Lubman, D. I., Yucel, M., Kettle, J. W., Scaffidi, A., Mackenzie, T., Simmons, J. G., et al. (2009). Responsiveness to drug cues and natural rewards in opiate addiction: associations with later heroin use. *Archives of General Psychiatry*, *66*, 205–212.

123. Sinha, R., Garcia, M., Paliwal, P., Kreek, M. J., & Rounsaville, B. J. (2006). Stress-induced cocaine craving and hypothalamic–pituitary–adrenal responses are predictive of cocaine relapse outcomes. *Archives of General Psychiatry*, *63*, 324–331.

124. Ripley, T. L., Gadd, C. A., De Felipe, C., Hunt, S. P., & Stephens, D. N. (2002). Lack of self-administration and behavioural sensitisation to morphine, but not cocaine, in mice lacking NK1 receptors. *Neuropharmacology*, *43*, 1258–1268.

125. George, D. T., Gilman, J., Hersh, J., Thorsell, A., Herion, D., Geyer, C., et al. (2008). Neurokinin 1 receptor antagonism as a possible therapy for alcoholism. *Science*, *319*, 1536–1539.

126. Baek, M. N., Jung, K. H., Halder, D., Choi, M. R., Lee, B. H., Lee, B. C., et al. (2010). Artificial microRNA-based neurokinin-1 receptor gene silencing reduces alcohol consumption in mice. *Neuroscience Letters*, *475*, 124–128.

127. Thorsell, A., Schank, J. R., Singley, E., Hunt, S. P., & Heilig, M. (2010). Neurokinin-1 receptors (NK1R:s), alcohol consumption, and alcohol reward in mice. *Psychopharmacology*, *209*, 103–111.

128. Furmark, T., Appel, L., Michelgard, A., Wahlstedt, K., Ahs, F., Zancan, S., et al. (2005). Cerebral blood flow changes after treatment of social phobia with the neurokinin-1 antagonist GR205171, citalopram, or placebo. *Biological Psychiatry*, *58*, 132–142.

129. Kendler, K. S., Schmitt, E., Aggen, S. H., & Prescott, C. A. (2008). Genetic and environmental influences on alcohol, caffeine, cannabis, and nicotine use from early adolescence to middle adulthood. *Archives of General Psychiatry*, *65*, 674–682.

130. Grant, B. F., Stinson, F. S., Dawson, D. A., Chou, S. P., Dufour, M. C., Compton, W., et al. (2004). Prevalence and co-occurrence of substance use disorders and independent mood and anxiety disorders: results from the National Epidemiologic Survey on Alcohol and Related Conditions. *Archives of General Psychiatry*, *61*, 807–816.

131. Crawley, J. N., Belknap, J. K., Collins, A., Crabbe, J. C., Frankel, W., Henderson, N., et al. (1997). Behavioral phenotypes of inbred mouse strains: implications and recommendations for molecular studies. *Psychopharmacology*, *132*, 107–124.

132. Crawley, J. N. (2008). Behavioral phenotyping strategies for mutant mice. *Neuron*, *57*, 809–818.

133. Kliethermes, C. L. (2005). Anxiety-like behaviors following chronic ethanol exposure. *Neuroscience and Biobehavioral Reviews, 28*, 837–850.

134. Oliveira-Dos-Santos, A. J., Matsumoto, G., Snow, B. E., Bai, D., Houston, F. P., Whishaw, I. Q., et al. (2000). Regulation of T cell activation, anxiety, and male aggression by RGS2. *Proceedings of the National Academy of Sciences of the United States of America, 97*, 12272–12277.

135. Yalcin, B., Willis-Owen, S. A., Fullerton, J., Meesaq, A., Deacon, R. M., Rawlins, J. N., et al. (2004). Genetic dissection of a behavioral quantitative trait locus shows that Rgs2 modulates anxiety in mice. *Nature Genetics, 36*, 1197–1202.

136. Mott, R., Talbot, C. J., Turri, M. G., Collins, A. C., & Flint, J. (2000). A method for fine mapping quantitative trait loci in outbred animal stocks. *Proceedings of the National Academy of Sciences of the United States of America, 97*, 12649–12654.

137. Kuntz-Melcavage, K. L., Brucklacher, R. M., Grigson, P. S., Freeman, W. M., & Vrana, K. E. (2009). Gene expression changes following extinction testing in a heroin behavioral incubation model. *BMC Neuroscience, 10*, 95.

138. Smoller, J. W., Paulus, M. P., Fagerness, J. A., Purcell, S., Yamaki, L. H., Hirshfeld-Becker, D., et al. (2008). Influence of RGS2 on anxiety-related temperament, personality, and brain function. *Archives of General Psychiatry, 65*, 298–308.

139. Koenen, K. C., Amstadter, A. B., Ruggiero, K. J., Acierno, R., Galea, S., Kilpatrick, D. G., et al. (2009). RGS2 and generalized anxiety disorder in an epidemiologic sample of hurricane-exposed adults. *Depression and Anxiety, 26*, 309–315.

140. Amstadter, A. B., Koenen, K. C., Ruggiero, K. J., Acierno, R., Galea, S., Kilpatrick, D. G., et al. (2009). Variant in RGS2 moderates posttraumatic stress symptoms following potentially traumatic event exposure. *Journal of Anxiety Disorders, 23*, 369–373.

141. Leygraf, A., Hohoff, C., Freitag, C., Willis-Owen, S. A., Krakowitzky, P., Fritze, J., et al. (2006). Rgs 2 gene polymorphisms as modulators of anxiety in humans? *Journal of Neural Transmission, 113*, 1921–1925.

142. Otowa, T., Shimada, T., Kawamura, Y., Sugaya, N., Yoshida, E., Inoue, K., et al. (2011). Association of RGS2 variants with panic disorder in a Japanese population. *American Journal of Medical Genetics. Part B, Neuropsychiatric Genetics, 156B*, 430–434.

143. Dick, D. M., Smith, G., Olausson, P., Mitchell, S. H., Leeman, R. F., O'Malley, S. S., et al. (2010). Understanding the construct of impulsivity and its relationship to alcohol use disorders. *Addiction Biology, 15*, 217–226.

144. de Wit, H. (2009). Impulsivity as a determinant and consequence of drug use: a review of underlying processes. *Addiction Biology, 14*, 22–31.

145. de Wit, H., & Richards, J. B. (2004). Dual determinants of drug use in humans: reward and impulsivity. *Nebraska Symposium on Motivation, 50*, 19–55.

146. Iacono, W. G., Malone, S. M., & McGue, M. (2008). Behavioral disinhibition and the development of early-onset addiction: common and specific influences. *Annual Review of Clinical Psychology, 4,* 325–348.

147. Bouwknecht, J. A., Hijzen, T. H., van der Gugten, J., Maes, R. A., Hen, R., & Olivier, B. (2001). Absence of 5-HT(1B) receptors is associated with impaired impulse control in male 5-HT(1B) knockout mice. *Biological Psychiatry, 49,* 557–568.

148. Isles, A. R., Humby, T., & Wilkinson, L. S. (2003). Measuring impulsivity in mice using a novel operant delayed reinforcement task: effects of behavioural manipulations and d-amphetamine. *Psychopharmacology, 170,* 376–382.

149. Depue R. A., & Collins P. F. (1999). Neurobiology of the structure of personality: dopamine, facilitation of incentive motivation, and extraversion. *Behavioral and Brain Sciences, 22,* 491–517; discussion 518–569.

150. Logue, S. F., Swartz, R. J., & Wehner, J. M. (1998). Genetic correlation between performance on an appetitive-signaled nosepoke task and voluntary ethanol consumption. *Alcoholism, Clinical and Experimental Research, 22,* 1912–1920.

151. Mitchell, S. H., Reeves, J. M., Li, N., & Phillips, T. J. (2006). Delay discounting predicts behavioral sensitization to ethanol in outbred WSC mice. *Alcoholism, Clinical and Experimental Research, 30,* 429–437.

152. Oberlin, B. G., & Grahame, N. J. (2009). High-alcohol preferring mice are more impulsive than low-alcohol preferring mice as measured in the delay discounting task. *Alcoholism, Clinical and Experimental Research, 33,* 1294–1303.

153. Cases, O., Seif, I., Grimsby, J., Gaspar, P., Chen, K., Pournin, S., et al. (1995). Aggressive behavior and altered amounts of brain serotonin and norepinephrine in mice lacking MAOA. *Science, 268,* 1763–1766.

154. Ivanova, E. A., & Popova, N. K. (2002). Effect of monoamine oxidase A knockout on resistance to long-term exposure to ethanol. *Bulletin of Experimental Biology and Medicine, 133,* 603–605.

155. Popova, N. K., Vishnivetskaya, G. B., Ivanova, E. A., Skrinskaya, J. A., & Seif, I. (2000). Altered behavior and alcohol tolerance in transgenic mice lacking MAO A: a comparison with effects of MAO A inhibitor clorgyline. *Pharmacology, Biochemistry, and Behavior, 67,* 719–727.

156. Brunner, H. G., Nelen, M., Breakefield, X. O., Ropers, H. H., & van Oost, B. A. (1993). Abnormal behavior associated with a point mutation in the structural gene for monoamine oxidase A. *Science, 262,* 578–580.

157. Caspi, A., McClay, J., Moffitt, T. E., Mill, J., Martin, J., Craig, I. W., et al. (2002). Role of genotype in the cycle of violence in maltreated children. *Science, 297,* 851–854.

158. Meyer-Lindenberg A., Buckholtz J. W., Kolachana B., Hariri, A. R., Pezawas, L., Blasi, G., et al. (2006). Neural mechanisms of genetic risk for impulsivity and violence in humans. *Proceedings of the National Academy of Sciences of the United States of America, 103,* 6269–6274.

159. Chien, C. C., Lin, C. H., Chang, Y. Y., & Lung, F. W. (2010). Association of VNTR polymorphisms in the MAOA promoter and DRD4 exon 3 with heroin dependence in male Chinese addicts. *World Journal of Biological Psychiatry*, *11*, 409–416.

160. Nilsson, K. W., Comasco, E., Aslund, C., Nordquist, N., Leppert, J., & Oreland, L. (2011). MAOA genotype, family relations and sexual abuse in relation to adolescent alcohol consumption. *Addiction Biology*, *16*, 347–355.

161. Gokturk, C., Schultze, S., Nilsson, K. W., von Knorring, L., Oreland, L., & Hallman, J. (2008). Serotonin transporter (5-HTTLPR) and monoamine oxidase (MAOA) promoter polymorphisms in women with severe alcoholism. *Archives of Women's Mental Health*, *11*, 347–355.

162. Saito, T., Lachman, H. M., Diaz, L., Hallikainen, T., Kauhanen, J., Salonen, J. T., et al. (2002). Analysis of monoamine oxidase A (MAOA) promoter polymorphism in Finnish male alcoholics. *Psychiatry Research*, *109*, 113–119.

163. Vanyukov, M. M., Maher, B. S., Devlin, B., Kirillova, G. P., Kirisci, L., Yu, L. M., et al. (2007). The MAOA promoter polymorphism, disruptive behavior disorders, and early onset substance use disorder: gene–environment interaction. *Psychiatric Genetics*, *17*, 323–332.

164. Nakamura, K., Sekine, Y., Takei, N., Iwata, Y., Suzuki, K., Anitha, A., et al. (2009). An association study of monoamine oxidase A (MAOA) gene polymorphism in methamphetamine psychosis. *Neuroscience Letters*, *455*, 120–123.

165. Cloninger, C. R., Svrakic, D. M., & Przybeck, T. R. (1993). A psychobiological model of temperament and character. *Archives of General Psychiatry*, *50*, 975–990.

166. Piazza, P. V., Deminiere, J. M., Le Moal, M., & Simon, H. (1989). Factors that predict individual vulnerability to amphetamine self-administration. *Science*, *245*, 1511–1513.

167. Redolat, R., Perez-Martinez, A., Carrasco, M. C., & Mesa, P. (2009). Individual differences in novelty-seeking and behavioral responses to nicotine: a review of animal studies. *Current Drug Abuse Reviews*, *2*, 230–242.

168. Grucza, R. A., Robert Cloninger, C., Bucholz, K. K., Constantino, J. N., Schuckit, M. I., Dick, D. M., et al. (2006). Novelty seeking as a moderator of familial risk for alcohol dependence. *Alcoholism, Clinical and Experimental Research*, *30*, 1176–1183.

169. Kelly, T. H., Robbins, G., Martin, C. A., Fillmore, M. T., Lane, S. D., Harrington, N. G., et al. (2006). Individual differences in drug abuse vulnerability: d-amphetamine and sensation-seeking status. *Psychopharmacology*, *189*, 17–25.

170. Lange, L. A., Kampov-Polevoy, A. B., & Garbutt, J. C. (2010). Sweet liking and high novelty seeking: independent phenotypes associated with alcohol-related problems. *Alcohol and Alcoholism (Oxford, Oxfordshire)*, *45*, 431–436.

171. Kliethermes, C. L., & Crabbe, J. C. (2006). Genetic independence of mouse measures of some aspects of novelty seeking. *Proceedings of the National Academy of Sciences of the United States of America*, *103*, 5018–5023.

172. Dulawa, S. C., Grandy, D. K., Low, M. J., Paulus, M. P., & Geyer, M. A. (1999). Dopamine D4 receptor-knock-out mice exhibit reduced exploration of novel stimuli. *Journal of Neuroscience, 19,* 9550–9556.

173. Thrasher, M. J., Freeman, P. A., & Risinger, F. O. (1999). Clozapine's effects on ethanol's motivational properties. *Alcoholism, Clinical and Experimental Research, 23,* 1377–1385.

174. Brioni, J. D., Kim, D. J., O'Neill, A. B., Williams, J. E., & Decker, M. W. (1994). Clozapine attenuates the discriminative stimulus properties of (–)-nicotine. *Brain Research, 643,* 1–9.

175. Mamiya, T., Matsumura, T., & Ukai, M. (2004). Effects of L-745,870, a dopamine D4 receptor antagonist, on naloxone-induced morphine dependence in mice. *Annals of the New York Academy of Sciences, 1025,* 424–429.

176. Hutchison, K. E., McGeary, J., Smolen, A., Bryan, A., & Swift, R. M. (2002). The DRD4 VNTR polymorphism moderates craving after alcohol consumption. *Health Psychology, 21,* 139–146.

177. Ray, L. A., Miranda, R., Jr., Tidey, J. W., McGeary, J. E., MacKillop, J., Gwaltney, C. J., et al. (2010). Polymorphisms of the mu-opioid receptor and dopamine D4 receptor genes and subjective responses to alcohol in the natural environment. *Journal of Abnormal Psychology, 119,* 115–125.

178. Hutchison, K. E., LaChance, H., Niaura, R., Bryan, A., & Smolen, A. (2002). The DRD4 VNTR polymorphism influences reactivity to smoking cues. *Journal of Abnormal Psychology, 111,* 134–143.

179. Shao, C., Li, Y., Jiang, K., Zhang, D., Xu, Y., Lin, L., et al. (2006). Dopamine D4 receptor polymorphism modulates cue-elicited heroin craving in Chinese. *Psychopharmacology, 186,* 185–190.

180. Becker, K., Laucht, M., El-Faddagh, M., & Schmidt, M. H. (2005). The dopamine D4 receptor gene exon III polymorphism is associated with novelty seeking in 15-year-old males from a high-risk community sample. *Journal of Neural Transmission, 112,* 847–858.

181. Ray, L. A., Bryan, A., Mackillop, J., McGeary, J., Hesterberg, K., & Hutchison, K. E. (2009). The dopamine D Receptor (DRD4) gene exon III polymorphism, problematic alcohol use and novelty seeking: direct and mediated genetic effects. *Addiction Biology, 14,* 238–244.

182. Sander, T., Harms, H., Dufeu, P., Kuhn, S., Rommelspacher, H., & Schmidt, L. G. (1997). Dopamine D4 receptor exon III alleles and variation of novelty seeking in alcoholics. *American Journal of Medical Genetics, 74,* 483–487.

183. David, S. P., Munafo, M. R., Murphy, M. F., Proctor, M., Walton, R. T., & Johnstone, E. C. (2008). Genetic variation in the dopamine D4 receptor (DRD4) gene and smoking cessation: follow-up of a randomised clinical trial of transdermal nicotine patch. *Pharmacogenomics Journal, 8,* 122–128.

184. Ton, T. G., Rossing, M. A., Bowen, D. J., Srinouanprachan, S., Wicklund, K., & Farin, F. M. (2007). Genetic polymorphisms in dopamine-related genes and smoking cessation in women: a prospective cohort study. *Behavioral and Brain Functions*, *3*, 22.

185. Lai, J. H., Zhu, Y. S., Huo, Z. H., Sun, R. F., Yu, B., Wang, Y. P., et al. (2010). Association study of polymorphisms in the promoter region of DRD4 with schizophrenia, depression, and heroin addiction. *Brain Research*, *1359*, 227–232.

186. Munafo, M. R., Yalcin, B., Willis-Owen, S. A., & Flint, J. (2008). Association of the dopamine D4 receptor (DRD4) gene and approach-related personality traits: meta-analysis and new data. *Biological Psychiatry*, *63*, 197–206.

Index

A2a receptor gene, 130
A118G SNP, 112*n*
A355G SNP, 112*n*
Abacavir, 332
Action tendencies, 212–213
Addiction. *See also* Alcoholism; Drug
   addiction; Nicotine addiction; Substance
   abuse; Substance abuse disorders
   adoption studies, 4
   behavioral economic approach, 158
   brain regions and, 190–191
   defined, 207
   delayed reward discounting and, 159–165,
      160*f*
   differentiator model (DM) for, 99
   family risk studies, 3
   implicit cognition as intermediate
      phenotype for, 207–221
   low level of response model (LLRM), 99
   psychomotor stimulant theory of, 99
   recovered individuals, 161
   "running in families," 3
   stages of, 345, 347*t*
   twin studies, 3, 4, 121
Addiction genetics. *See also* Animal models;
   Animal studies; Epigenetics; Intermediate
   phenotypes; Twin studies
   behavioral economic intermediate
      phenotypes and, 157–176
   evidence for genetic influences, 3–4
   heritability of discounting, 163
   implicit cognition and, 214, 215*f*, 216–219
   intermediate phenotype approach, 1–5, 6*t*,
      7–14

missing heritability, 4
phenotype and, 4–8
Plains Indians study, 21–25
susceptibility genes, 283
Addiction Research Center Inventory (ARCI),
   128*t*
Addictive drugs
   subjective responses to, 121–133
   twin studies, 3, 4, 121
ADH enzymes, 41
ADH genes, alcoholism and, 41, 47–51
ADH1A gene, 47, 50
ADH1B gene, 47, 50
   alcohol dependency and, 48–49, 50, 51
*ADH1B*1* allele, 48
*ADH1B*2* allele, 48, 49, 50
*ADH1B*3* allele, 50
*ADH1B*47HIS*, 48
*ADH1B*48Arg*, 48
ADH1C gene, 47, 49, 51
*ADH1C*1* allele, 49, 50
*ADH1C*2* allele, 49
*ADH1C*349Ile*, 49
ADH4 gene, 47, 50
ADH5 gene, 47
ADH6 gene, 47
ADH7 gene, 47, 50, 51
Adolescents
   alcoholism in, 144, 145
   behavioral undercontrol in, 266
   brain development in, 191, 197
   executive control processes and the brain,
      198–199
   Go/No-Go paradigms, 194–195, 337